Failure Analysis of Paints and Coatings

Revised Edition

Failure Analysis of Paints and Coatings

Revised Edition

Dwight G. Weldon
Weldon Laboratories, Inc, Imperial, PA, USA

WILEY
A John Wiley & Sons, Ltd., Publication

This edition first published 2009

Registered office
John Wiley & Sons Ltd, The Atrium, Southern Gate, Chichester, West Sussex, PO19 8SQ, United Kingdom

For details of our global editorial offices, for customer services and for information about how to apply for permission to reuse the copyright material in this book please see our website at www.wiley.com.

Library of Congress Cataloging-in-Publication Data

Weldon, Dwight G.
Failure analysis of paints and coatings / Dwight G. Weldon. – Rev. ed.
p. cm.
Includes bibliographical references and index.
ISBN 978-0-470-69753-5 (pbk.)
1. Paint–Testing. 2. Coatings–Testing. I. Title.
TP936.5.W45 2009
667′.90287 – dc22

2009005620

A catalogue record for this book is available from the British Library.

ISBN: 978-0-470-69753-5 (Cloth)

Set in 10 on 12 pt Times by Toppan Best-set Premedia Limited

Reprinted with corrections May 2011

To my son Andy, who can still make me smile.

Contents

Preface to the Revised Edition xiii

Preface to the First Edition xv

Acknowledgements xvii

1 General Principles of Coating Formulation 1
1.1 Introduction 1
1.2 Binders 2
1.3 Pigments 4
1.4 Solvents 5
1.5 Additives 5
1.6 Formulation Concepts: Pigment-to-Binder Ratio 6
1.7 Formulation Concepts: Pigment-Volume Concentration 6
1.8 Formulation Concepts: Density, Weight Solids and Volume Solids 7
References 8

2 Why Coatings Work and Why They Fail 9
2.1 Why Coatings Work 9
2.1.1 Adhesion 9
2.1.2 Wetting 10
2.1.3 Surface Preparation 14
2.1.4 Cohesive Strength 23
2.1.5 Permeability 23
2.2 Why Coatings Fail 24
2.2.1 Mechanical Stress 25
2.2.2 Internal Stress 29
2.2.3 Chemical Attack 31
2.2.4 Weathering Stress 33

2.2.5 Osmotic Blistering 34
2.2.6 Electroendosmotic Blistering 35
References 36

3 Pigments **39**
3.1 Inorganic Pigments 39
3.1.1 Inorganic Colour Pigments – White 39
3.1.2 Inorganic Colour Pigments – Yellow 40
3.1.3 Inorganic Colour Pigments – Orange 40
3.1.4 Inorganic Colour Pigments – Red 41
3.1.5 Inorganic Colour Pigments – Blue 41
3.1.6 Inorganic Colour Pigments – Green 41
3.2 Extender Pigments 41
3.2.1 Silica/Silicates 42
3.2.2 Calcium Carbonate 42
3.2.3 Barytes 43
3.3 Corrosion-Resistant Pigments 43
3.4 Organic Pigments 46
3.4.1 Organic Red Pigments 47
3.4.2 Organic Yellow Pigments 48
3.4.3 Organic Blue Pigments 49
3.4.4 Organic Green Pigments 50
References 50

4 Additives and Solvents **51**
4.1 Additives 51
4.1.1 Anti-settling Agents 51
4.1.2 Viscosity Modifiers 52
4.1.3 Surfactants and Emulsifying Agents 53
4.1.4 De-foaming and Anti-foaming Agents 54
4.1.5 Driers 54
4.1.6 Plasticizers 54
4.1.7 Ultraviolet Stabilizers 55
4.1.8 Anti-skinning Agents 55
4.1.9 Biocides 56
4.1.10 Flow-Modifying Agents 56
4.2 Solvents 57
References 64

5 Coating Types and Common Failure Modes **65**
5.1 Natural Resins and Oils 65
5.1.1 Natural Resins 66
5.1.2 Oils 68

5.2	Alkyds and Epoxy Esters		70
	5.2.1	Alkyds	70
	5.2.2	Epoxy Esters	74
5.3	Epoxies		75
	5.3.1	Amine and Amide Curing Agents for Epoxy Resins	76
	5.3.2	Epoxy Failure Modes	80
5.4	Modified Epoxies		84
	5.4.1	Acrylic Epoxies	84
	5.4.2	Coal Tar Epoxies	85
	5.4.3	Epoxy Phenolics	85
5.5	Phenolics		86
	5.5.1	Resole Phenolics	86
	5.5.2	Novolac Phenolics	87
	5.5.3	Phenolic Failure Modes	87
5.6	Amino Resins		88
5.7	Acrylics		90
	5.7.1	Solution Acrylics	91
	5.7.2	Acrylic Latex Coatings	92
	5.7.3	Thermoset Acrylics	95
5.8	Polyesters		96
	5.8.1	Saturated Polyesters	97
	5.8.2	Unsaturated Polyesters	100
5.9	Polyurethanes		101
	5.9.1	Two-Component Polyisocyanate/Polyol Coatings	103
	5.9.2	Urealkyds	106
	5.9.3	Moisture-Cured Polyurethanes	107
	5.9.4	Polyurethane Lacquers and Dispersions	108
	5.9.5	Two-Component Water-Borne Polyurethanes	108
5.10	Vinyls		109
	5.10.1	Solution Vinyls	109
	5.10.2	Plastisols and Organosols	110
	5.10.3	Vinyl Fluorides	110
	5.10.4	Poly(vinyl butyral)	111
	5.10.5	Vinyl Latexes	111
5.11	Bituminous Coatings		112
5.12	Inorganic and Silicone-Modified Coatings		114
	5.12.1	Silicone Coatings	114
	5.12.2	Silicate Coatings	116
	5.12.3	Polysiloxane Coatings	118
5.13	Polyureas		120
	5.13.1	Polyaspartic Polyurea Coatings	121
5.14	Powder Coatings		122
	References		124

6 Application-Related Problems **125**
6.1 Brush and Roller 125
6.2 Spray Applications 126
6.2.1 Air (Conventional) Spray 126
6.2.2 Airless Spray 126
6.2.3 Plural Spray 127
6.2.4 Electrostatic Spray 127
6.3 Flow Coating 127
6.4 Roll Coating 128
6.5 Powder Coating 128
6.6 Coating Failures Related to Application Problems 129
References 134

7 Field Methods **135**

8 Analytical Methods **141**
8.1 Light Microscopy 141
8.2 Infrared Spectroscopy 152
8.2.1 Theory 152
8.2.2 Instrumentation 166
8.2.3 Sample Handling 169
8.2.4 Applications 177
8.3 Gas Chromatography (GC) 202
8.3.1 Theory of GC 202
8.3.2 Instrumentation 208
8.3.3 Pyrolysis-GC 216
8.3.4 Application of GC 217
8.4 Gel Permeation Chromatography (GPC) 221
8.4.1 Theory 222
8.4.2 Instrumentation 226
8.4.3 Applications 227
8.5 Ion Chromatography 232
8.5.1 Theory 232
8.5.2 Applications 237
8.6 Scanning Electron Microscopy 238
8.6.1 Imaging Theory 239
8.6.2 Elemental Analysis by X-ray Spectroscopy 241
8.6.3 Sample Preparation 244
8.6.4 Applications of SEM-EDS 244
8.7 Differential Scanning Calorimetry (DSC) 247
8.7.1 Theory 247
8.7.2 Calibration and Sample Preparation 250
8.7.3 Applications of DSC 252
8.8 Miscellaneous Methods of Analysis 258
References 259

9 Physical Methods **261**
9.1 Adhesion 261
9.2 Flexibility and Impact Resistance 263
9.3 Solvent Resistance 265
9.4 Weathering Resistance 265
9.5 Chemical Resistance 266
9.6 Freeze–Thaw 269
9.7 Application Variables 269
References 273

10 Examples of Coating Failures **275**
10.1 Urethane Topcoat Disbonding from Epoxy 275
10.2 Blistering of an Epoxy Tank Coating 279
10.3 Cracking and Delamination of Epoxy Coating Systems on Masonry Walls 282
10.4 Discolouration of Coil-Coated Aluminium Siding 285
10.5 Discolouration and Delamination of Plastisol 288
10.6 Delamination of Floor Coating 293
10.7 Delamination of Dry-Fall Alkyd 296
10.8 Coal Tar Epoxy Failure 300
10.9 Splitting of Inorganic Zinc-Rich Primer 304
10.10 Defects on Electro-Coated Panels 309
10.11 Failure of Coated Light Poles 309
10.12 Blistering of Coating on Concrete Floor, Number 1 314
10.13 Blistering of Coating on Concrete Floor, Number 2 318
10.14 Peeling of Paint from Metal Chairs 322
10.15 Failure of Railroad Car Liner 327
10.16 Failure of Calcium Sulfonate Modified Alkyd Topcoat from a Bridge 331
10.17 Discolouration of Furniture Lacquer 333
10.18 Failure of Tank Car Lining 337
References 345

11 The Repair of Coating Failures **347**

Index **351**

Preface to the Revised Edition

It has been over 8 years since the first edition of *Failure Analysis of Paints and Coatings* was published. During this period, many coating types have remained the same, and a few new ones have been introduced, or at least have been more heavily marketed. Advances have been made in surface preparation technology, and more opportunities for training and education have been made available than ever before. However, in spite of all of this, the coatings industry still experiences its fair share of failures.

A coating failure does not just involve the cost of some paint. Repair of an existing structure can be extremely expensive, perhaps even entailing lost production time if the facility has to be shut down for repairs. If a coating lining a process vessel or a railroad tank car fails, large amounts of product may become unmarketable and may have to be disposed of. Reputations can be damaged. The cost of a paint failure is seldom cheap.

The first edition of this book introduced some basic concepts in paint formulation and chemistry. This has been updated in the second edition to include some additional types of coatings, such as polyureas, polysiloxanes and powder coatings. Some additional discussion has been included concerning surface preparation, the mechanism of blistering and the chemistry of amine blush. Analytical techniques such as gas chromatography–mass spectroscopy and differential scanning calorimetry have been discussed in more detail. However, as with the first edition, the subjects of coatings chemistry and analytical chemistry are presented at the introductory level. Although essential to conducting a failure analysis, many thorough texts already exist on these subjects.

As pointed out in the preface to the first edition, the failure analysis of paints and coatings is all about problem solving. Problem solving is a difficult subject to teach, and probably relies on curiosity and experience more than anything else. Therefore, in addition to the basics of coatings science and analytical chemistry, which form the foundation upon which curiosity and experience can build, this book contains numerous practical examples of solving real-world coating failures. Indeed, the biggest difference between the first and second editions is the much greater number of coating failures discussed in Chapter 10. Eight additional examples of coating failure investigations have been added to this chapter, almost doubling its size. I hope the reader enjoys studying them as much as the author did conducting them.

Dwight G. Weldon
Weldon Laboratories, Inc.
PO Box 40
Imperial, PA 15126, USA

Preface to the First Edition

From the author's point of view, there are few things more satisfying than solving a problem. The failure analysis of paints and coatings is all about problem solving, and is both a fascinating and a challenging endeavour. It requires the blending of polymer chemistry, analytical chemistry, coatings science and microscopy with the curiosity of a detective. All of these skills, with the possible exception of the last, can be studied in school or learned from books. It is putting them all together that is the trick, and which was the impetus for writing this present book. Perhaps the most difficult part of writing this book was deciding how to organize it. A thorough knowledge of analytical techniques, including weaknesses as well as strengths, is essential in solving a coating failure problem. However, an understanding of the chemical and physical properties of the substrate/coating system is equally important, as are the more mundane areas of samples and background information. The approach finally decided upon was to begin with a basic review of coatings technology. Without this knowledge, it is virtually impossible to decide what to look for in the laboratory.

While there are several excellent and expansive texts on the subject, it seemed impossible to write a manuscript on failure analysis without first describing the nature of what it is that we are investigating. Readers familiar with the technology may wish to skip some of these chapters, although it should be pointed out that the chapter on resins and polymers contains some specific information on failure mechanisms. Likewise, it seemed impossible to write about the various analytical methods without first discussing their theoretical background. These discussions are admittedly very brief, as entire volumes have been written about each and every technique. Indeed, as I am personally particularly fond of analytical chemistry, it was tempting not to be more expansive here. I hope that I have adequately resisted the temptation, and have written a failure analysis book rather than an analytical chemistry one. It is, however, important to understand the basics of the various techniques, in order to appreciate their limitations. Without knowing what a technique cannot do, it is possible to make serious errors in judgement.

In a perfect world, there would be no paint failures and I would have to find a different line of work. In a nearly perfect world, one would have unlimited time, excellent samples and a handsome budget with which to solve the problems that occur. Usually, none of these three conditions exist. Furthermore, although great strides have been made, there are still substantial gaps in our basic knowledge of paints and coatings. Without knowing exactly why something works, it can be difficult to say exactly why it doesn't. As a consequence of the above, there are times when it is not possible to unequivocally determine the cause of a coating failure. However, starting with a firm understanding of coatings technology, applying the many powerful analytical techniques available, and keeping an open and inquisitive mind, the majority of coating failures can be solved. It is hoped that this text will be a step forward in that direction.

Dwight G. Weldon

Acknowledgements

Words of thanks must go to Carolyn Smith for all of her help with the initial manuscript, to my mother and father who gave me my love of books, and whom I will never forget, and to the encouragement of friends.

1

General Principles of Coating Formulation

1.1 INTRODUCTION

Before discussing the chemistry or mechanics of formulating a coating, the reader needs to take a minute or two to think about all of the coated articles encountered in everyday life. The list is a very long one, including bridges, water tanks (both inside and out), cars, planes, structural steel for any number of uses, chemical process vessels, wood of all types, plastic bumpers, furniture, concrete block, cement floors, ships' hulls, electrical equipment, plaster, drywall, appliances and food containers. The list seems almost endless. It is not surprising that the first step in formulating a coating is to ask a simple question – that is, what is this coating going to be used for?

This is a critical question, since every coating is a compromise. A coating cannot be both soft and hard, or both glossy and flat. A coating that might have superb corrosion resistance on structural steel, such as an ethyl silicate-zinc-rich primer, would be a failure as a coating for a backyard deck. Once the purpose of the coating has been identified, there are certain basic principles that can be followed to produce an effective coating.

Most coatings consist of four basic ingredients, namely, a binder, pigments, solvents and additives. Some coatings may not have all of these ingredients. For example, a 100% solids plural spray epoxy contains no solvents, while an automotive clear coat contains no pigments. Nevertheless, it is important for the formulator to understand the roles of these basic ingredients and how they interact with one another.

Failure Analysis of Paints and Coatings, Revised Edition Dwight G. Weldon

The purpose of this present chapter is not to enable the reader to formulate a coating, but rather to serve as a useful starting point in understanding the mechanisms of how paints work and therefore, how they can fail.

1.2 BINDERS

In simple terms, a coating's binder is the 'glue' which holds it together, and which is principally responsible for providing adhesion to the substrate. With a few exceptions, the binder is nearly always organic, consisting of natural resins or man-made polymers or pre-polymers.

There are numerous types of binders for various applications such as alkyds, vinyls, natural resins and oils, epoxies and urethanes. Specific properties of many of these resins and polymers will be discussed later in this volume. However, some general properties of binders will be discussed here.

It is common practice to classify binders into one of two types, depending on how they cure. Those that cure solely by solvent evaporation are referred to as thermoplastic, or sometimes non-convertible. Those that cure via a chemical reaction during or after application are referred to as thermosetting.

As simple as this nomenclature seems, there are still a few areas of confusion in the industry. Two such examples are enamels and alkyds. To a chemist, the term 'enamel' is synonymous with a thermoset coating, that is, one that cures by chemical reaction. To a marketing or advertising person, however, the term is used to conjure up the vision of a hard, glossy coating, regardless of its curing mechanism. It is possible to have thermoset coatings that are soft, and even to have thermoplastic coatings that are relatively hard. Therefore, 'enamel' is more of a descriptive term than a scientific one.

Alkyds are basically polyesters modified with natural oils, such as linseed oil. They are widely used for a variety of applications, including exterior house paint, and are often referred to as thermoplastic, or non-convertible, coatings. This terminology, however, is usually incorrect.

While it is true that a linseed oil alkyd dries by solvent evaporation, this is only the first step in its curing process. The second step is not a chemical reaction with another ingredient in the paint, but rather a reaction with oxygen in the air. The reaction is not a rapid one, and may take weeks or months before it is essentially complete. The fact that the co-reactant is not part of the paint, and that the reaction is so slow, has led to some errors in classifying these types of coatings as thermoplastic, whereas they are actually thermosetting.

Another fundamental aspect of binders that needs to be understood by the formulator is viscosity. In intuitive terms, viscosity is a material's resistance to flow. Thick, viscous materials such as maple syrup do not flow out of a container nearly as easily as low-viscosity materials such as water. There are numerous ways of measuring a coating's viscosity, and therefore numerous units to describe it. While it is beyond the scope of this current text to present a detailed discussion of viscosity and rheology, an understanding of their practical significance is important.

A coating must be low enough in viscosity to be applied by conventional equipment (brush, roller or spray) and high enough in viscosity so as not to sag or drip. It must also maintain, for at least a short length of time after application, a suitable viscosity such that the coating can flow and level. This is necessary not only for the elimination of imperfections such as brush marks, but also so that the coating can thermodynamically wet the substrate. This latter property is critical in achieving good adhesion.

The primary factor responsible for the viscosity of a binder is its molecular weight. Other things being equal, a polymer of high molecular weight will be more viscous than one of low molecular weight. This is an important fact, and basically leaves the coating formulator with two means of controlling viscosity: varying the molecular weight of the polymer or resin, or diluting it with solvent. The path taken can have significant consequences regarding the physical and chemical properties of the coating.

In simpler times, the choice was usually an easy one, that is, dilute. Higher-molecular-weight polymers generally have superior properties to low-molecular-weight ones, and solvents are relatively cheap. However, in the modern era of regulation and environmental awareness, the move is towards lower-molecular-weight, higher-solids coatings. In many ways, this is synonymous with saying the trend is towards thermoset coatings and away from thermoplastic ones.

A typical poly(vinyl acetate)/poly(vinyl chloride) (PVA/PVC) coating, which has many architectural and light industrial uses, might have a molecular weight of 200 000 or more. It would have to, in fact, because, as a thermoplastic coating, 'what you get is what you start with', and lower-molecular-weight polymers would simply not have adequate physical properties once they dry and cure by solvent evaporation. A consequence of such high-molecular-weight material is its concomitant high viscosity. Such coatings may contain 70% solvent to bring the viscosity into a workable range.

In contrast, let us consider amine-cured two-component epoxies. These coatings consist of an epoxy component (which may be clear or pigmented) and a separate can of activator' component that contains an amine. The two ingredients are mixed together shortly before application, and the amine reacts with the epoxy to produce a hard, tough, highly cross-linked coating. The molecular weight of the epoxy starting material might be only 500–1000, and that of the amine even less. Therefore, their initial viscosity is very low and such coatings might only contain 10 or 20% solvent.

The previous discussion has focused on molecular weight as it relates to viscosity. Another important variable in formulation, at least for two-component coatings, is the equivalent weight. The equivalent weight of a polymer is its molecular weight divided by the number of reactive functional groups on the polymer molecule. Staying with the example of an amine-cured epoxy, the

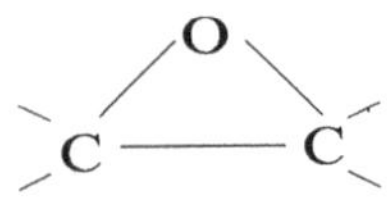

Figure 1.1 Structure of the epoxy (oxirane) group.

primary reactive functional group of an epoxy resin is the three-membered epoxy, or oxirane, group (Figure 1.1).

If a particular epoxy resin has a molecular weight of 380 and has two oxirane groups per molecule, its equivalent weight is 190. This is very important because although it is *sometimes true* that 'one molecule reacts with one molecule', it is *always true* that 'one equivalent reacts with one equivalent'.

If one were to select an amine-curing agent that has an equivalent weight of 43, then in order to completely react or cross-link the above epoxy, one would have to mix 190 g of epoxy resin with 43 g of the amine-curing agent. While in some cases it may be desirable to have a slight excess of one component versus another, it is obvious that a knowledge of equivalent weights is critical in the formulation process.

1.3 PIGMENTS

Pigments can affect a coating's corrosion resistance, physical properties and appearance. They are commonly grouped into two categories, namely, inorganic and organic.

Inorganic pigments consist of discreet particles, often crystalline in nature, which are dispersed in paints, often with the aid of special additives which improve their compatibility with various resin systems. They can contribute to all three of the pigment functions listed above, that is, corrosion resistance, physical properties and aesthetics.

Two of the most common inorganic pigments are titanium dioxide and iron oxide. Titanium dioxide is the most widely used white pigment, particularly for exterior coatings. It has a high refractive index, which means that it has excellent hiding strength, and also provides a measure of stability against the harsh ultraviolet rays of sunlight, which can degrade many coating binders. Iron oxide, of which there are many varieties, is perhaps the most common inorganic red pigment and is used in both primers and topcoats. Both synthetic and natural iron oxides exist.

Extender pigments are a subdivision of inorganic pigments. They provide little colour or corrosion resistance, but can affect such coating properties as density, flow, hardness and permeability. The three most common extender pigments are calcium carbonate (also known as whiting or chalk), various silicates such as china clay (kaolin) and talc, and barium sulfate (barytes).

There are a very large number of organic pigments on the market. These are generally fairly complicated organic molecules, some of which may be partially soluble in certain resins and solvents. They are seldom used in primers or heavy-duty industrial coatings, as their main advantage is the pallet of vibrant, rich colours that they can produce.

1.4 SOLVENTS

With very few exceptions, most coatings require solvents to dissolve the binder and to modify the viscosity so that the coating can be applied by conventional methods. Solvents evaporate after the coating has been applied and aid in the flow and levelling of the coating, as well as the wetting of the substrate. Solvents are generally thought of as organic liquids, although in a latex paint the main solvent is water.

Certain solvents can dissolve or 'cut' some resins more effectively than others. The ability of a solvent to do this is reflected in its solubility parameter, a concept that can be applied to both solvents and resins. Solubility parameters share one of organic chemistry's simplest rules, namely, the concept of 'like dissolves like'. For instance, polar solvents are more effective than non-polar ones in dissolving polar resins. Solubility parameters will be discussed in more detail later (see Section 4.2).

In addition to compatibility, the rate of evaporation of solvents is also very important. Solvents that have a high vapour pressure and evaporate rapidly are considered fast, or hot, solvents, while those that evaporate slowly are referred to as slow solvents. The rate of solvent evaporation can have a major influence on the properties of the coating, and several common defects can be traced to an inappropriate choice of solvents. If the solvent evaporates too quickly, the coating may not have a chance to flow into a smooth, continuous film. A common example of this is the 'orange peel' texture occasionally seen on certain automobile finishes. If the solvent evaporates too slowly, sagging can occur. Orange peel and sagging are just two of the more visually apparent defects related to improper choice of solvents. As we shall see in later chapters, there are several less obvious effects that can also have serious consequences.

1.5 ADDITIVES

Although additives will be dealt with more thoroughly in Section 4.1, they span such a wide range of chemistry and functions that they probably deserve their own text. Additives are various chemicals, typically added in small amounts, which can greatly affect the properties of a coating. These include surfactants, anti-settling agents, coalescing agents, anti-skinning agents, catalysts, defoamers, ultraviolet light absorbers, dispersing agents, preservatives, driers and plasticizers. Some chemists, however, would not consider a plasticizer to be an additive since sometimes they constitute a major ingredient of the formulation.

1.6 FORMULATION CONCEPTS: PIGMENT-TO-BINDER RATIO

Once the proper ingredients of a coating have been selected, they must be put together in the proper amounts. A fundamental parameter of a coating is its ratio of pigment to binder. The pigment-to-binder ratio (*P/B*) is simply the ratio of the weight of pigment to the weight of binder. A gallon of paint containing 8.0 lb of pigment and 4.0 lb of binder therefore has a *P/B* of 2.0. Topcoats typically have a *P/B* of 1.0 or less, while primers often have a *P/B* of 2.0–4.0 [1]. All other things being equal, gloss coatings will have a lower *P/B* than flat coatings.

1.7 FORMULATION CONCEPTS: PIGMENT-VOLUME CONCENTRATION

The pigment-to-binder ratio is an easily measurable and helpful concept in paint formulation. A concept with more far-reaching consequences is the pigment-volume concentration (*PVC*). The latter is the ratio of the volume of pigment (V_p) to the volume of pigment plus the volume of solid binder (V_b), as shown in the following:

$$PVC = \frac{V_p}{V_p + V_b} \tag{1.1}$$

When multiplied by 100, this parameter is referred to as the % *PVC*. From the above, it is obvious that two coatings can have an identical *P/B* value, but dramatically different *PVC* values, simply by using pigments of different densities.

A term related to the *PVC* is the critical pigment-volume concentration (*CPVC*). The latter is the *PVC* at which there is precisely the right amount of binder to provide a thin coating over the complete surface of each particle of pigment. At levels above the *CPVC*, there is insufficient binder to wet all of the pigment, while below the *CPVC*, there is an excess of binder.

Several properties of coatings can be significantly affected by varying the *PVC* of the formulation. Abrasion resistance and tensile strength, for example, are usually best below the *CPVC* and can rapidly decrease when the *CPVC* is approached. Permeability is usually low below the *CPVC* and rapidly increases as the *CPVC* is approached or exceeded. Therefore, because of the deterioration of a number of physical properties, most high-performance or exterior coatings should not be formulated above the *CPVC*.

1.8 FORMULATION CONCEPTS: DENSITY, WEIGHT SOLIDS AND VOLUME SOLIDS

Quantities such as density, weight solids and volume solids, along with % pigment and % binder, are often referred to as a coating's 'physical constants'. The formulator needs to understand these quantities, and how to measure or calculate them.

The density of a coating, or its weight-gallon, is usually expressed as pound per gallon, and is conveniently measured by using a pyncnometer. The latter is simply a device that holds a precisely known volume of liquid. In the paint industry, it is often called a weight-gallon cup. Since the various ingredients in the formula have a wide range of densities, the density of the finished product is a quick and easy way to tell if a substantial error occurred when making a particular batch of coating.

The weight solids of a coating, sometimes referred to as the % non-volatile by weight, is simply the weight of solid material divided by the coating's total weight. For example, if a 100-g sample of liquid paint contained 20 g of pigment, 30 g of resin solids, 2 g of non-volatile additives and 48 g of solvent, its weight solids is 52%. Weight solids can be calculated from the formula of the coating, or measured by determining the weight of non-volatile material remaining after a known initial weight of coating has been evaporated to dryness in an oven.

The volume solids of a coating is the volume % of non-volatile material. If one knows the formula and the density of the various ingredients, it is tempting to try to calculate the volume solids of the coating. At first glance, this would seem to be an easy task. One would simply calculate the volume of all of the ingredients by dividing their weights by their densities, add up the volumes of all the non-volatile ingredients (resins, pigments, etc.) and then divide by the total volume of all the ingredients (including, of course, any solvents). The only problem with this procedure is that it usually gives the wrong answer. Sometimes the error is small, although sometimes it can be significant.

The reason for the error is the simple fact that usually 100 ml of 'A' plus 100 ml of 'B' does not equal 200 ml of 'C'. This would only be the case in a perfect solution where there is no heat of mixing and no interaction of the various ingredients. This is seldom the case, and the final volume can be either more or less than expected.

The American Society for Testing and Materials (ASTM) publishes a procedure to determine the volume solids of a coating (ASTM D2697 – Test Method for Volume Nonvolatile Matter in Clear or Pigmented Coatings). This basically involves weighing metal discs, both before and after coating them, in both air and water. By employing the concept of buoyancy and by measuring the coating's density and weight solids, one can then calculate its volume solids.

Volume solids are an important property of a coating, especially to the person using it. Most painting specifications require that the coating be applied at a certain thickness. A gallon of paint which is 60% solid by volume will cover twice the area, at the same dry film thickness, as will a gallon of paint which is only 30% solid by volume.

The % resin and % pigment are weight percentage values and can therefore be easily calculated from the formula. However, confusion occasionally occurs over the distinction between % resin (sometimes called % vehicle) and % resin solids (sometimes called % vehicle solids). The quantity that is really meaningful is the % resin solids.

The difference between % resin and % resin solids is quite simple. Consider an alkyd coating that has a weight-gallon of $10\,\mathrm{lb\,gal^{-1}}$ and contains 6 lb of commercially available alkyd resin, which is supplied as an 80% solution in toluene, a common solvent. The coating, therefore, does not contain 6 lb of actual alkyd resin, but only 4.8 lb. Its % resin solids is therefore 48%, not 60%. Sometimes the term % resin is used synonymously with % resin solids, which only adds to the confusion.

REFERENCES

1. J. Boxall and J. A. Von Fraunhofer, *Paint Formulation, Principles and Practice*, Industrial Press, New York, 1981, p. 16.

2

Why Coatings Work and Why They Fail

Why do coatings work? In broad terms, they work because they adhere well to the substrate, have good cohesive strength or integrity, resist weathering and chemical deterioration, and impede the permeability of harmful or corrosive agents. In the current chapter, some of the fundamental factors that make coatings work and those that cause them to fail will be discussed.

2.1 WHY COATINGS WORK

2.1.1 Adhesion

Adhesion, or the lack thereof, is not the only means by which a coating can fail. However, without good adhesion, the coating will likely not get the chance to fail by some other mechanism.

While there is some disagreement as to the nature of adhesion, most workers agree that it is comprised of three components: primary chemical bonding, secondary (or polar) bonding and mechanical bonding. The primary chemical bonds that contribute to adhesion are of the same type which hold molecules together and are therefore either ionic or covalent. Ionic bonds are based on the strong electrical forces that attract negatively charged ions (such as the chloride anion) to positively charged ions (such as the sodium cation). These forces are very strong, as attested to by the very high melting points of inorganic salts such as sodium chloride (801 °C). Covalent bonds form between atoms that have less tendency to ionize, and are due to a sharing of electrons rather than an actual transfer of them. Organic compounds are held together by covalent bonds.

Failure Analysis of Paints and Coatings, Revised Edition Dwight G. Weldon

Primary chemical bonding of a coating to a substrate is highly desirable due to the strength of the bond, whether ionic or covalent. The term 'substrate' refers to the surface being coated. For a primer, the substrate might be steel, wood or aluminium. For a topcoat, the substrate is the primer, or perhaps an intermediate coat of paint.

Primary chemical bonding to a substrate such as steel is relatively rare. Phosphoric acid-based vinyl butyral wash primers and siloxirane-based coatings are two such examples. Both can form primary chemical bonds with active sites on the metal surface. In most cases, however, the largest component of adhesion is secondary, or polar, bonding.

Secondary chemical bonding relies on polar interactions such as hydrogen bonding and van der Waals forces. Such forces rely on an uneven distribution of electrical charge within the molecule, such that dipoles are produced. That end of the molecule which has a partial positive charge will be attracted to the site on a neighbouring molecule which has a partial negative charge. Coatings that have functional groups composed of atoms with substantially different electronegativities, such as hydroxyl (OH) and amine (NH) groups, have a greater tendency to form polar bonds with a substrate.

If substrates were totally inert, chemical bonding could not occur. Fortunately, such a situation is not the case. Even clean steel is not a monolithic, neutral surface; it will have extremely thin areas of oxides, which can impart polarity. If the steel is pretreated (see Section 2.1.3.1), additional functional groups can be introduced that can interact strongly with polar functional groups on the paint binder. Wood is certainly not inert, as it consists of a complicated mixture of cellulose, hemicellulose and lignin. If the substrate is another coat of paint, reactive functional groups can be intentionally engineered into it.

Mechanical forces are the third factor contributing to adhesion. It is generally acknowledged that a coating will stick better to a rough surface than to a smooth one. Some workers attribute this to the production of more chemically active sites as a consequence of the roughening operation. Others contend that the coating adheres better because of a simple mechanical interlocking as the coating flows into the pores and crevices of the rough substrate. It is likely that there is some validity to both arguments.

2.1.2 Wetting

Regardless of whether adhesion is due to primary chemical bonds, secondary bonding or mechanical effects, the coating must come into intimate physical contact with the substrate in order to achieve a good bond. Indeed, chemical bonding, whether through primary valence bonds or secondary electrostatic effects, occurs over distances measured in mere angstroms (10^{-10} m). The process of a liquid spreading over a substrate and coming into intimate contact with it is called wetting. In order to understand wetting and the factors that influence it, one must have an understanding of surface tension or surface energy.

Thermodynamics teaches us that a system has a natural tendency to seek a state of minimum free energy and a state of maximum entropy (disorder). Why then does a falling droplet of water assume a spherical shape? It certainly does not do it to maximize entropy, since a sphere (or any well-defined geometrical shape) is a relatively ordered structure. The answer, then, must be because a sphere somehow minimizes the free energy. Since a sphere encloses the maximum volume within the minimum surface area, the implication is that the surface of water has excess free energy associated with it. Therefore, surface energy is the excess free energy associated with molecules at the surface compared to molecules in the interior, or bulk, of the liquid.

Surface energy is defined as the amount of work (energy) required to create a unit area of surface, and therefore has units such as joules per square metre or ergs per square centimetre. Since 1 erg is equal to 1 dyn cm, an equivalent unit is also dyne per centimetre, which is the unit commonly used in describing surface tension. Therefore, surface energy and surface tension are numerically and dimensionally equivalent.

Polar molecules, such as water, have higher surface tensions than non-polar molecules, such as hexane. For a water molecule to be at the surface, it must overcome the relatively large hydrogen bonding forces exerted on it by other water molecules beneath it in the bulk phase. This explains why a volume of water contained in a narrow cylinder has a very distinct meniscus, while hexane has only a slight one. The surface tension of water is a relatively high 72.6 dyn cm^{-1}, versus a value for hexane of 18 dyn cm^{-1}.

The surface tension of a liquid can be measured experimentally by such techniques as the Du Nouy ring method or the Wilhelmy plate method. Solids also have surface tensions, but they are more difficult to measure. Typically, the surface tension of a solid surface is determined indirectly through contact angle measurements, by using liquids of known surface tension.

The contact angle refers to the angle of incidence between a solid surface and a small droplet of liquid placed on that surface (see Figure 2.1).

If the surface tension of the liquid is much less than the surface energy of the solid, the liquid will completely spread, or wet, the solid, and the contact angle will be zero. If the surface tension of the liquid is greater than that of the solid, it will not wet it; rather, it will bead up and the contact angle will be very high.

It is apparent that for a coating to wet a substrate and subsequently bond to it, it must have a surface tension lower than that of the substrate. There are several factors that influence the surface tension of a liquid coating. These include temperature, solvents, additives and resin types. The surface tension of the substrate is dictated largely by its chemical composition (steel, aluminium, etc.) and by its degree of cleanliness. The surface tensions of a number of common solvents are shown in Table 2.1, while those of several resin types are shown in Table 2.2 [1].

The driving force for a liquid to spread over a solid is the surface tension gradient between the two surfaces. Opposing this flow of liquid are viscoelastic

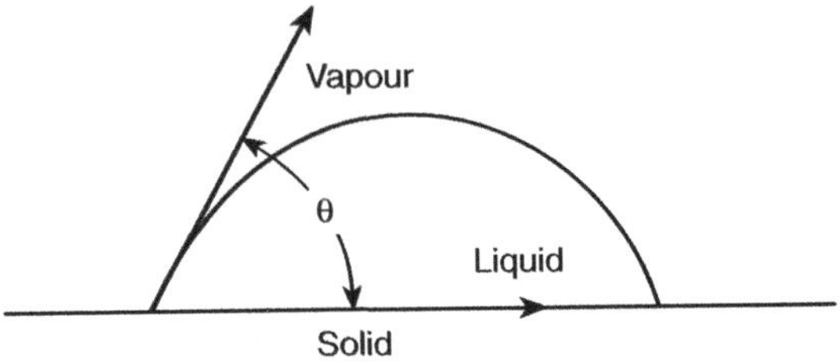

Figure 2.1 Contact angle formed between a drop of liquid and a solid substrate.

Table 2.1 Surface tensions of a number of common solvents [1]. Reproduced by permission of the Federation of Societies for Coatings Technology, from P. E. Pierce and C. K. Schoff, *Coating Film Defects*, 1988.

Solvent	**Surface tension (dyn cm^{-1})**
Water	72.7
Ethylene glycol	48.4
Propylene glycol	36.0
o-Cylene	30.0
Toluene	28.4
Dowanol PM®[a]	27.7
Dowanol PM acetate[a]	26.4
n-Butyl acetate	25.2
n-Butanol	24.6
Mineral spirits	24.0
Methyl isobutyl ketone	23.6
Methanol	22.6
VM&P naphtha[b]	22.0
Lactol spirits	19.9

[a] Dowanol PM is a registered trade name of Dow Chemical Corporation.
[b] VM&P.

forces. A highly viscous liquid will not flow, or wet, as readily as a low-viscosity liquid having the same surface tension.

Luckily for coatings formulators, most metallic surfaces have high values of surface tension. Zinc has been shown to have a surface tension of 770 dyn cm^{-1} [2]. Even relatively high surface tension resins such as epoxies, when dissolved in relatively high surface tension oxygenated solvents, can wet metallic surfaces, unless opposed by strong viscous forces.

The influence of viscous forces should not be overlooked when considering wetting and adhesion phenomena. In simpler times, when government regulations regarding solvent emissions, that is, volatile organic compounds (VOCs), were less stringent or non-existent, coatings could be formulated with high-molecular-

Table 2.2 Surface tensions of a number of different resin types [1]. Reproduced by permission of the Federation of Societies for Coatings Technology, from P. E. Pierce and C. K. Schoff, *Coating Film Defects*, 1988.

Polymer or resin	Surface tension $(dyn\,cm)^{-1}$
Melamine resin (HMMM type)	57.6
Poly(vinyl butyral)	53.6
Benzoguanamine resin	52.0
Epon 828®[a]	46.0
Urea resin	45.0
Polyester melamine film	44.9
Polystyrene	42.6
Poly(vinyl chloride)	41.9
Poly(methyl methacrylate)	41.0
65% Soya FA alkyd	38.0
Poly(vinyl acetate)	36.5
Poly(butyl methacrylate)	34.6
Modaflow®[b]	32.0
Polydimethylsiloxane	19.8

[a] Epon 828 is a registered trade name of Shell Chemical Company.
[b] Modaflow is a registered trade name of Monsanto Company.

weight resins and large quantities of solvent. Not only did the solvent usually serve to lower the surface tension, but it also ensured low-viscosity coatings that offered little resistance to wetting and flow.

Such is not the case today. Many of the modern, high-performance industrial coatings, such as epoxies and urethanes, are relatively high solids, viscous products. Not only are their initial viscosities relatively high, but they can increase rapidly upon mixing and curing. A typical two-component polyamide epoxy bridge paint may have a perfectly acceptable viscosity immediately upon mixing the epoxy component with the polyamide converter, but 2 h later when the material is nearing the end of its so-called 'pot life', the viscosity may have increased significantly. Although its surface tension may have increased only marginally as a result of cross-linking or even solvent evaporation, the viscous forces opposing flow will have increased dramatically. The end result may be an adhesion failure.

It should also be kept in mind that the substrate is not always metallic. Wood, plastic and other coatings, such as primers, will have significantly lower surface tensions than metals, and may therefore prove considerably more difficult to wet.

The above discussion has assumed that the surfaces to be painted are clean. If this is not the case, and the contaminant is of the low-surface-tension variety such as oil or grease, wetting may prove much more difficult to achieve.

It should not be assumed that an oily contaminant necessarily leads to a wetting problem with subsequent loss of adhesion. It is common in the coatings industry to hear remarks such as 'even the slightest residue of oil will cause a coating to fail'. At least one study [3] has shown that, on blast-cleaned carbon steel, quantities of oil ranging up to several hundred milligrams per square centimetre did not result in adhesion failure. This study has suggested that, if the contaminant is soluble in the resin/solvent system of the coating, it can be dissolved into the body of the coating and can essentially be lifted from the steel surface. However, the study did not address smooth substrates or inter-coat adhesion problems.

One way to simultaneously lower a coating's surface tension and its viscosity is to heat it. Certain 100% solids (highly viscous two-component coatings) are routinely applied by heated airless spray or plural component airless spray equipment. Although manufacturers and applicators of such coatings may think that heating them is required simply in order to apply them by conventional techniques, they are inadvertently and fortuitously lowering their surface tension and thus reducing the viscous forces opposing wetting.

2.1.3 Surface Preparation

Surface preparation is a critical factor in obtaining good adhesion. Such preparation can affect not only the fundamental forces of bonding, but also the surface tension, and hence wetting.

Surface preparation means different things to different people. To a house painter, it may mean scraping off loose paint with a putty knife; to a bridge painter, it may mean sandblasting off mill scale or old rust, while to an aluminium extruder, it may mean anodizing.

2.1.3.1 Surface preparation for steel

Since most industrial structures that are to be painted are fabricated of steel, we will begin by addressing methods of preparing steel surfaces for painting. Typical methods include solvent cleaning, hand and power tool cleaning, blast cleaning, pickling and conversion coating.

Common sense suggests that dirt, debris, salts and oil-like contamination should be removed from steel surfaces prior to painting. The Society for Protective Coatings (SSPC), formerly known as Steel Structures Painting Council, has published a two-volume painting manual, Volume 2 of which [4] contains their surface preparation specifications. The SSPC Surface Preparation Specification No. 1 (SSPC-SP1) describes solvent cleaning as a means of removing oil, grease, dirt, soil, drawing compounds and other similar organic compounds. The procedures vary from simple solvent wiping with a clean rag, through vapour degreasing, to emulsion cleaning. The choice of technique depends largely on the type of contamination. Solvent cleaning is specified as a prelude to the other, more aggressive, methods of surface preparation described by the SSPC.

Hand tool cleaning (SSPC-SP2) and power tool cleaning (SSPC-SP3) are often adequate for removing loose rust and previous coats of unsound paint, especially in relatively small areas and in interior or normal atmospheric environments. There are a variety of power tools described, such as needle guns and abrasive wheels and discs.

For more aggressive environments, larger areas or high-performance coatings that require cleaner and/or rougher surfaces, blast cleaning is preferred. There are four degrees of blast cleaning described by the SSPC. They are, in order of thoroughness: brush-off blast (SSPC-SP7), commercial blast (SSPC-SP6), near-white blast (SSPC-SP10) and white metal blast (SSPC-SP5). The specifications vary primarily in the amount of tight rust, tight mill scale, and amount of mill scale and rust staining that is allowed. The type of surface is usually specified by the coating supplier, and may vary for the same coating, depending on the severity of the service environment. White metal cleaning is generally reserved for very corrosive environments or immersion service. None of the SSPC blast cleaning standards specify a particular abrasive, or require a specific surface texture or profile. There are a wide range of abrasives available. These include steel shot and grit, natural products such as silica sand, olivine sand and garnet, by-product abrasives from smelter and utility generators such as nickel, coal and copper slags, agricultural by-products such as crushed corncobs and walnut shells, and manufactured abrasives such as aluminium oxide and glass beads.

These abrasives vary in such properties as size, shape, hardness and density, and therefore also vary in cleaning rate, dirt production, embedment, surface texture and profile depth. It is possible to obtain an SP-5 white metal blast which has a peened, or rounded, profile of 1.5 mil (0.0015 in.), or one which has a sharp, angular surface with a profile of 3.0 mil. In both cases, the surface would meet the requirements of SP-5. However, the performance of a coating could vary substantially over these two surfaces. It is generally agreed that coatings adhere better to a sharp, angular surface with a deep profile, compared with a surface with a rounded, shallow profile. This is especially true of very thick coatings. However, applying a thin coating to a surface with a deep profile can result in the tips of the peaks protruding above the coating surface. This can result in premature corrosion.

Not only do the SSPC surface preparation standards not specify the type of abrasive or the depth of profile, but they do not specify the surface roughness. In terms of blast-cleaned steel, most people have an intuitive understanding of roughness. Deep, angular peaks correspond to a 'rough' surface, and shallow, rounded peaks correspond to a smoother surface. However, instruments which can help to measure some aspects of surface roughness have been commercially available for some time.

ASTM D7127 (Standard Test Method for Measurement of Surface Roughness of Abrasive Blast Cleaned Metal Surfaces using a Portable Stylus Instrument) is a new standard which describes how to measure surface roughness in terms of R_{max}, R_t, P_c and R_z. The parameters R_{max}, R_t and R_z are various ways of characterizing the peak-to-valley distances of a roughened substrate, while P_c (peak count) refers to the actual number of peaks in a given linear distance. Having more peaks of the same height in a given distance or area (higher peak density) has long been assumed to be beneficial in terms of coating adhesion and hence performance.

A recent study [5] has indicated that coatings do indeed tend to perform better at higher peak count. The authors prepared panels using a set profile height of approximately 2.4 mils but with peak counts varying from low (P_c of 56) to medium (P_c of 101) to high (P_c of 126). For most of the coatings studied, adhesion to steel and resistance to undercutting after salt fog exposure showed a substantial improvement at medium and high peak counts compared to the low peak count panels. However, the study did not necessarily reflect a simple linear relationship between peak count and performance. In some of the testing, the medium peak count panels actually performed somewhat better than the high peak count panels. The authors theorized that this might be due to the inability of certain higher viscosity coatings to fully flow into and wet all the steel in the higher, narrower valleys of the high peak count panels. In other words, the coating may have partially bridged these gaps, and thus may not have been able to take full advantage of the potentially greater amount of bonding area.

In addition to conventional blast cleaning procedures, equipment and techniques are available for preparing surfaces using a mixture of abrasive and water, known as 'wet blasting', or 'wet abrasive blasting'. The technique is described in SSPC-TR 2/National Association of Corrosion Engineers (NACE) 6G198 Joint Technical Report – Wet Abrasive Blast Cleaning, which is contained in Volume 2 of SSPC Systems and Specifications. It is a method of cleaning where water is injected into the air/abrasive stream generated by conventional abrasive blast cleaning equipment. According to the SSPC/NACE report, 'Water helps to remove contaminants from the substrate, to wet the abrasive, and to substantially reduce dispersion of fine particulates (dust) … The objective is to remove contaminants and suppress the dusting effect caused by the impact of the abrasive on the substrate, while retaining the blasting characteristics of dry abrasive, including creation of anchor profile'.

In contrast to wet abrasive blasting, SSPC-SP12 describes high and ultra-high-pressure water jetting as a means of preparing or cleaning steel or other hard surfaces for re-coating. SSPC defines water jetting as the use of water at pressures above 70 MPa (10 000 psi). Pressures below 10 000 psi are considered water cleaning, not water jetting. Ultra-high-pressure water jetting is performed at 170 MPa (25 000 psi) or higher.

Water jetting differs from wet abrasive blasting in that it does not impart any surface profile to the substrate, since it does not use any abrasive in the water stream. What it can do, however, is to remove contamination such as dirt and rust, as well as old paint. If the substrate initially had a profile, this profile will be retained by the water jetting process.

Pickling is the removal of rust and mill scale by chemical rather than physical means. It is usually carried out in a workshop environment on somewhat smaller and/or thinner sections of steel, or on more complicated shapes than a simple girder or angle. Although some pickling is done by using electrolytic methods, the more usual procedure is to employ an acid. Common acids used for this purpose include sulfuric, hydrochloric and phosphoric. Hydrochloric acid removes scale faster than sulfuric acid, but is rarely used at elevated temperatures because of the noxious fumes that are produced. Inhibitors are employed when acid pickling is carried out in order to minimize attack on the base metal, which will usually develop an etch. Residues must be thoroughly rinsed from the steel prior to painting.

Conversion coatings, or pretreatments, usually refer to the formation of a thin insoluble crystalline or amorphous layer of inorganic material which is integral or bonded to the metal surface. The purpose of conversion coating is twofold, namely, to increase paint adhesion and to improve corrosion resistance. They are usually employed on relatively thin-gauge cold-rolled steel, such as that used in the automotive or appliance industries, or on aluminium or galvanized steel. There are numerous types of pretreatments available, and a thorough discussion of each of these is beyond the scope of this present volume. However, the basic principles of some of the more common ones will be discussed in the following paragraphs.

The most common types of pretreatments used on steel are phosphate conversion coatings. The steel is first cleaned of oil and grease by using a hot, alkaline cleaning solution, after which it is rinsed, treated in a bath (or via a spray process) of phosphoric acid and various additives, rinsed again, and finally sealed or passivated. The sealer, which coats any bare substrate and also removes loose or poor-quality phosphate, is often a chromic acid rinse, although there is now a move towards less toxic agents. Most processes in use today result either in a thin coating of iron phosphate or of zinc phosphate.

An iron phosphate bath employs a solution of phosphoric acid, sometimes with other additives. Oxidation-reduction reactions occur, which result in the dissolution of some of the metal as a phosphate salt from the surface of the article. Hydroxyl ions, which are one of the by-products of the electrochemical reaction, serve to raise the pH at the metal–solution interface, resulting in the formation of an insoluble metal phosphate which bonds to the metal surface.

Coatings deposited in an iron phosphate bath are amorphous, iridescent and range in colour from a faint gold to a blue or violet colour. They are very thin, with coating weights of 20–60 $mg\,ft^{-2}$ being common.

Iron phosphate coating baths are relatively easy to maintain and require less expensive equipment than other types of pretreatments. Corrosion resistance is improved, as is paint adhesion, and the pretreatments are used with organic coatings in low-to-moderate-exposure environments [6].

Zinc phosphate conversion coatings require more expensive application equipment and more rigorous control of the bath parameters than iron phosphate pretreatments, although such procedures also produce higher-quality coatings. These are extensively used in the automobile industry, where a solution of primary zinc phosphate ($Zn(H_2PO_4)_2$) and excess phosphoric acid is common.

The coating weight of zinc phosphate is typically 150–300 mg ft^{-2} [7]. Depending on the composition and parameters of the bath, two different phosphate phases can form on the steel surface, namely, hopeite ($Zn_3(PO_4)_2{\cdot}4H_2O$), which is rich in zinc, and phosphophyllite ($FeZn_2(PO_4)_2{\cdot}4H_2O$), which is rich in iron. Some pretreatments also contain other dissolved ions besides zinc, such as manganese and nickel.

Although corrosion resistance improves with increasing coating weight [8], the initial layer of phosphate, which is one crystal thickness deep and very dense, is more firmly adherent than the less dense, somewhat brittle subsequent layers. The dendritic growth of this second layer is also somewhat porous, and may extend far enough above the plane of the base metal to impart a rough appearance to subsequently applied liquid coatings [9]. It is desirable to stop at the stage of single-crystal thickness.

After the final sealing step using chromic acid, which enhances corrosion resistance, it is common practice to bake the pretreated metal for several minutes at approximately 350 °F (177 °C). This not only dries the surface, but also releases some of the water of hydration associated with the phosphate conversion coating. If this is not done, and if one subsequently applies a coating with a bake schedule near or above this temperature, the water of hydration may be released into the coating, thus potentially causing small craters in the paint.

2.1.3.2 Surface preparation for aluminium

In contrast to steel, aluminium surfaces are often prepared for painting by an electrolytic process known as anodizing. Even though the oxide layer produced in anodizing has good corrosion resistance, this property, along with paintability, are improved via chemical pretreatments.

For decades, the most common pretreatments for aluminium were the use of chromates. However, hexavalent chromium is a toxic metal. Alternatives include aluminium phosphate (approximately 20–100 mg ft^{-2}), zinc phosphate (approximately 150–250 mg ft^{-2}) and various proprietary polycrystalline coatings [7].

2.1.3.3 Surface preparation for galvanized steel

Galvanized steel is steel that has been coated with zinc metal in order to improve corrosion resistance. In hot dip galvanizing, the metal part is immersed in a bath of molten zinc, resulting in a metallurgical bond between the iron and zinc. In continuous strip galvanizing, steel strips (coils) are drawn through the bath of molten zinc, and often post-formed. In electrolytic or electroplated galvanizing, the zinc is deposited electrolytically on the steel surface. There is no alloying effect; only a single continuous phase of zinc metal is bonded to the steel surface.

Zinc can be a relatively difficult metal to paint. The surfaces are typically hard and smooth, and often have waxes or oils on them (from the mills) that must be removed. Thin layers of oxides and carbonates can form which can alternately hinder or improve paint adhesion. Some pretreatments used to prevent 'storage stain' can actually interfere with the adhesion of subsequent coatings.

ASTM D7396 (Standard Guide for Preparation of New, Continuous Zinc Coated (Galvanized) Steel Surfaces for Painting) describes several methods of surface preparation for preparing new galvanized steel for painting. These include zinc phosphate, chromate and amorphous complex oxide (mixed oxide) pretreatments, as well as abrasive blast cleaning.

Zinc phosphate pretreatments produce relatively dense, thick, crystalline deposits that provide a good base for paint adhesion and a barrier to moisture penetration into the zinc substrate. They are typically applied at the rate of 150–250 mg ft^{-2} [10]. Because of their porosity, which can result in some retention of moisture, and their water of hydration, they can sometimes release small amounts of water during the curing of baked coatings, thus resulting in blistering of the finish. Soluble reaction residues from the phosphating process can also be difficult to remove by rinsing, thus potentially resulting in blistering and lifting of the coatings in moist environments.

Chromate pretreatments for galvanized steel have the obvious disadvantage of employing toxic chemicals. There are a variety of proprietary types, with coating weights ranging from approximately 50 to 160 mg ft^{-2}. The pretreatment imparts very good corrosion resistance, although paint adhesion may be variable. The erratic adhesion results are probably caused by the relatively low tolerance of chromate treating baths to variation in composition and contaminants [10].

Complex oxide treatments for galvanized steel are widely used processes. The general method involves reacting the zinc surface in a hot, alkaline solution containing heavy metal ions for a few seconds. The zinc surface is converted to a non-metallic, amorphous complex oxide coating that provides slightly improved corrosion resistance and greatly improved paint adhesion.

Galvanized surfaces can also be prepared for subsequent painting by blast cleaning. As ASTM D7396 points out, 'care must be taken to leave the zinc layers intact'. This is done by what is known as 'sweep' or 'brush' blasting, which uses a rapid movement of the blasting nozzle, coupled with the use of a non-aggressive abrasive. Materials which are suggested include a fine grade of aluminium/magnesium silicate, soft mineral sands with a Mohs hardness of 5 or less, and organic materials such as crushed corncobs or walnut shells. Relatively low operating pressures of less than 0.40 MPa (60 psi) are also recommended.

As the title of ASTM D7396 suggests, it is primarily concerned with the preparation of new galvanized surfaces for painting. While some of the procedures listed in D7396 can be performed in the field, many cannot. Furthermore, there are additional factors which need to be considered in the painting of weathered galvanized steel.

Zinc is a very reactive metal, and the surface of a galvanized article begins to change soon upon removal from the galvanizing bath. Most of these changes involve oxidation of the zinc upon exposure to the atmosphere and moisture. The first oxidation product to form is zinc oxide, followed by zinc hydroxide and eventually, zinc carbonate. Zinc oxide and zinc hydroxide particles are not well adherent to the underlying zinc surface and can interfere with paint adhesion. The patina of zinc carbonate, on the other hand, is well adherent and provides a good surface for paint to adhere to. Because of these different oxidation processes, the American Galvanizers Association ('Duplex Systems: Painting Over Hot Dip Galvanized Steel', American Galvanizers Association, 1998) distinguishes three types of galvanized steel: new galvanized steel, partially weathered galvanized steel and fully weathered galvanized steel. New galvanized steel is that which has been hot dip galvanized within the past 48 hours. The surface is usually very smooth and may need to be slightly roughened prior to painting. This is especially true if the galvanizing was chromate quenched. Indeed, galvanized steel should not be water or chromate quenched, nor should it be oiled, if it is to be subsequently painted. The surface of new galvanizing has little to no zinc oxide or zinc hydroxide, so it does not require major cleaning. Profiling may be done by sweep blasting, as described above or by the application of a zinc phosphate pretreatment. Some paint manufacturers make low solids acid-based wash primers which can be applied in the shop, and which produce a slightly rough surface for subsequent topcoating. Of course, if there is reason to believe that there are any type of oils or waxes on the surface of the zinc, these should be removed by solvent cleaning prior to roughening of the surface. According to the American Galvanizers Association, partially weathered galvanized steel has begun to form the protective zinc carbonate patina, but has not completed the process. If the steel had been chromate quenched, this layer should be removed by sweep blasting or sanding. Depending on how it was stored, a whitish zinc corrosion by-product known as storage stain may have developed. This should also be removed prior to painting. Suggested methods include brushing with a mild ammonia solution or using a mild acidic solution such as dilute acetic acid. In both cases, a final rinse with clean, warm water should be performed. Finally, it is also recommended that partially weathered galvanized steel should be roughened slightly to improve paint adhesion. Depending on the environment to which it was exposed, if fatty or oily contaminates are present, they should be removed first by solvent cleaning or through the use of an appropriate alkaline cleaning solution.

It may take a year or more to achieve the completely formed patina of fully weathered galvanized steel. This patina, due mostly to a very thin layer of zinc carbonate, results in a very stable and finely etched surface suitable for painting, and requires only power washing to remove loose debris ('Duplex Systems:

Painting Over Hot Dip Galvanized Steel', American Galvanizers Association, 1998). Again, if fatty or oily contaminates are present, they should be removed as described above.

2.1.3.4 Surface preparation for concrete

Large quantities of coatings are sold annually for the painting of concrete. While concrete is relatively inert when compared to steel, it can still be damaged by various chemicals and, in certain northern regions, by freeze–thaw conditions in the winter. Floors in warehouses may be sealed to provide a barrier against spills, and top-coated for both aesthetic and anti-skid properties. Many other cementitious surfaces are coated for either waterproofing purposes or simply for aesthetics.

Concrete is a mixture of cement dust and aggregate that cures to a hard, dense matrix of hydrated calcium silicate when mixed with water. If small aggregates such as sand are used, mortar is obtained. Larger aggregates result in the formation of what is termed concrete. The bulk properties of concrete are determined to a large extent by the type of aggregate, the cement-to-aggregate ratio and the amount of water added. The type of trowelling can even affect the surface properties of the finished product. Steel trowelling, for instance, typically produces a very hard, dense, smooth surface which can affect paint adhesion. Since concrete requires moisture in order to cure, certain chemicals such as oils and waxes are sometimes applied to the surface to ensure moisture retention. Such membrane-curing compounds, as they are called, can affect paint adhesion if not removed by appropriate methods.

Another reason why cementitious substrates require surface preparation is laitance. Laitance is a fine, loose, dusty layer that is often present on the surface of new concrete. The particles are cement fines carried to the surface by excess water exuding from the bulk of the curing concrete. This is possible because, in order to mix and work the concrete, much more water is added than what is simply required for complete hydration.

Efflorescence is another type of material which should be removed from concrete surfaces prior to coating application. It consists of a crystalline or powdery deposit that forms when soluble salts migrate to the surface of older concrete. These salts can react with carbon dioxide in the air. Unfortunately, efflorescence can also occur underneath existing coatings, especially in areas subject to water ingress, such as at cracks or near windows.

The same fundamental forces that result in adhesion of a coating to steel, that is, primary chemical bonds, secondary (polar) bonding and mechanical effects, are also at work at the coating/concrete interface. As with steel, the surface preparation of concrete should result in a rough surface free of contamination.

For light-duty service, particularly with relatively thin coatings that develop little internal stress on curing, simply scrubbing with a detergent and a bristle brush, followed by vacuuming, will remove light deposits of laitance and result in an adequate bond. For heavy deposits of laitance, for areas where membrane-curing compounds or form-release agents have been used, or in areas subjected to heavy usage or where thicker coatings are to be applied, more aggressive methods are needed. Usually these consist of either abrasive blasting, power tool cleaning or acid etching.

Abrasive blasting can be carried out with the same conventional blast cleaning equipment that might be used on steel or by various specialized equipment designed particularly for concrete. Centrifugal blasting equipment is often used. This method is convenient and portable, and also reduces dust. The result should be a significantly roughened surface, with any bug holes or pores opened up so the coating can flow into them.

Several electric and pneumatic power tools are available for roughening concrete surfaces. These are usually one of three types: impact cleaning tools, rotary cleaning tools and rotary impact cleaning tools. Some of these have vacuum attachments for collecting debris. While many of these devices can be hand-held, large, sometimes self-propelled mechanical tools often known as scarifiers or scabblers are available. These devices can remove relatively thick layers of concrete and are especially useful on old concrete which is badly deteriorated.

Acid etching is a fairly popular method for cleaning and roughening concrete. The technique involves applying a dilute solution of muriatic (hydrochloric) acid to the concrete. Since the surface of new concrete is highly alkaline, this causes a vigorous reaction, with copious bubbling caused by the evolution of hydrogen gas as the acid reacts with the alkaline cement. After the bubbling has subsided, the acid is washed off with large amounts of fresh water. If any residue remains, the surface should be vacuumed prior to the coating application.

Acids other than muriatic can be used to etch concrete, such as citric and phosphoric acids. Citric acid in particular has the advantage of being lower in odour and also of being biodegradable and easier to discard. Unlike muriatic acid, it contains no chloride ions, which might penetrate into porous concrete and possibly damage rebar. However, while muriatic acid can etch concrete in perhaps 3 or 5 minutes, citric acid may require 20 minutes or so to work.

It should be stressed that the goal of acid etching is not to pour acid on the concrete but to obtain a clean, rough surface with a texture similar to that of a fine-to-medium grade of sandpaper. Too many times it is found that acid etching is not adequate, either because the acid was too dilute, the surface was not rinsed or cleaned well enough, or a form-release agent or membrane-curing compound was present, which impeded the action of the acid.

Published guidelines for preparing concrete for painting include ASTM D4258 (Practice of Surface Cleaning Concrete for Coating) and D4259 (Practice of Abrading Concrete), as well as the International Concrete Repair Institute (ICRI) Guideline 03732 (Selecting and Specifying Concrete Surface Preparation for

Sealers, Coatings, and Polymer Overlays). Replica coupons, representing different degrees of surface roughness, are also available from ICRI.

2.1.4 Cohesive Strength

It is fairly common to encounter coating systems that have failed because of either cracking or splitting within a single coat of paint. This is often referred to as a cohesive failure. The factors that have the greatest impact on cohesive strength include the pigment-to-binder ratio (*P*/*B*), the molecular weight of the binder and the cross-link density of the binder.

As previously pointed out (Section 1.7), the ratio of pigment to binder can affect several properties of the coating. As the critical pigment volume concentration (*CPVC*) is approached, the cohesive strength of the coating will suffer. Thus, coatings that have very high pigment loadings, such as masonry block fillers and inorganic zinc-rich primers, typically have poor cohesive strength. When these types of coatings fail, they often do so cohesively.

Higher molecular weights, at least for polymers which are linear or only lightly branched, contribute to improved cohesive strengths. This is because the longer the polymer chains, the more they can twist and intertwine with one another. Thus, cohesive strength is improved by a simple mechanical effect.

Cohesive strength also improves as the molecular weight of highly cross-linked coatings goes up. Such polymers result from the bonding of polymer chains with one another at several sites to form what is sometimes called a 'network' polymer. The bulk of the polymer essentially consists of one gigantic interconnected molecule of an extremely high molecular weight.

Because of the number of chemical links on the chain (similar to the effect of the physical links or 'loops' on linear polymers), network polymers have very good cohesive strengths, and it therefore takes a considerable force to pull them apart. Paradoxically, however, they have poor impact resistance and can shatter more easily than the more flexible linear or lightly branched polymers.

2.1.5 Permeability

All coatings are permeable to at least some extent. Often this is a critical factor in how well the coating protects the substrate. The obvious example is a coating that is protecting steel by acting as a barrier to the passage of water, oxygen and salts. The main factors that affect the permeability of a coating include the pigment-to-binder ratio, the type of pigment and the cross-link density of the coating. Of course, the coating must also be chemically resistant to the elements that it is exposed to.

Permeability is usually low below the *CPVC*, and rapidly increases as the latter is approached or exceeded. This is because there is insufficient binder to seal off the microscopic areas or voids between the pigment particles.

Although this trend is true regardless of the type of pigment used, at a given pigment loading, some pigments have better barrier properties than others. Generally, pigments that are platy or flake-like, such as mica, aluminium flake and micaceaous iron oxide, contribute to coatings of lower permeability than do spherical pigments.

Thermosetting binders of high cross-link density, such as amine-cured epoxies, are more impermeable than are linear or lightly cross-linked binders. This is simply a physical phenomenon, since greater cross-link density means less space for the various chemicals to diffuse through.

Of course, thicker coatings present more of a barrier than do thinner ones. Indeed, some coatings in harsh service environments may be applied as thick as 100 mil. However, apart from the obvious expense in applying greater amounts of coating material, in the coatings industry, 'more is not always better'. Not only will thicker coatings generate more stress than thinner ones (see Section 2.2.2), but they can retain more solvent, which can sometimes result in its own set of problems (see Section 2.2.5).

2.2 WHY COATINGS FAIL

There are a number of reasons why a particular coating may fail. It is fairly common to ascribe these failures to one of four causes, namely, mis-application of the coating, defective coating, the wrong choice of coating, or exposure to an unanticipated environmental excursion. While it is true that coating failures fall into one of these four categories, or perhaps a combination of them, there are still more fundamental forces that can cause coatings to fail.

For a coating to fail, it has to be stressed. In the context of this discussion, there are many forms of stress. The most obvious is the literal, scientific variety, that is, an applied force that tends to strain or deform a body. However, there are other factors and forces which can stress a coating, such as chemical attack, degradation by ultraviolet (UV) light and even the fundamental tendency for all closed systems to reach a state of equilibrium. In subsequent chapters, we will deal with failure modes specific to certain coating types, and the analytical and physical techniques available to investigate them. However, taking such an approach to failure analysis without understanding some of the underlying factors is somewhat akin to studying history by memorizing dates and places, with no understanding of the social and political forces which shaped the events.

Although for a coating system to fail it must be stressed, this should not be taken to mean that it is stress which should always be 'blamed' for a coating failure. One factor or another will stress coatings. If the degree of stress exceeds that which might normally be expected of the particular service in which the coating has been placed (e.g. a temperature excursion in a chemical process vessel), then it is probably correct to say that the coating has failed because of this unexpected stress. If the coating were poorly designed or misapplied, and subsequently failed under the normal stress of the environment, the stress cannot be blamed for the failure.

2.2.1 Mechanical Stress

A dried or cured coating may be considered as being a viscoelastic solid. The term 'viscoelasticity' reflects the fact that certain materials, such as coatings, have properties of both viscous liquids and elastic solids. When a force is exerted on a material, it will flow if it is a liquid, while it will deform if it is a solid. The study of the flow and deformation of materials under an applied force is called rheology. When a paint film cures, it gradually changes from a viscoelastic liquid to a viscoelastic solid. Additionally, whether a material behaves more like a liquid or a solid can depend on both the temperature and rate at which the experiment is carried out. If a piece of 'silly' putty is slowly pulled apart, it may extend quite a distance before finally breaking, thus resembling a viscous liquid. However, if it is rapidly pulled apart, it may snap in half almost immediately, thus resembling 'a not very elastic' solid.

The physical properties of materials are often studied by pulling them apart, using what are termed universal testing machines. The sample, which usually has either a dog-bone or a rectangular shape, is clamped within the jaws of the testing device. Typically, one jaw remains stationary while the other moves away at a constant rate. The instrument measures the force (often in pounds) required to elongate the sample at a constant rate. By dividing the force by the cross-sectional area of the sample, one obtains the 'stress' being applied to the sample, which is often reported in lbs/in^2. By measuring the elongation of the sample, which is usually expressed as a fractional elongation or a percentage change from the original length, one obtains the 'strain'. Often the data are expressed graphically as a stress–strain curve (see Figure 2.2).

With most materials, the strain (elongation) initially varies directly with the stress (applied force), and therefore the ratio of stress to strain is a constant within this region. When the ratio of stress to strain is constant regardless of the value of the stress, the material is said to be Newtonian in behaviour.

The ratio of stress to strain is termed the modulus, and may be viewed as the 'stiffness' of the material. This parameter can be calculated from the stress–strain curve by determining the slope of the line in the initial (Newtonian) phase of the testing.

In addition to measuring the modulus, a stress–strain experiment also allows one to determine the tensile strength of a coating and its elongation at break. Tensile strength is the value of the stress at which the sample breaks, divided by

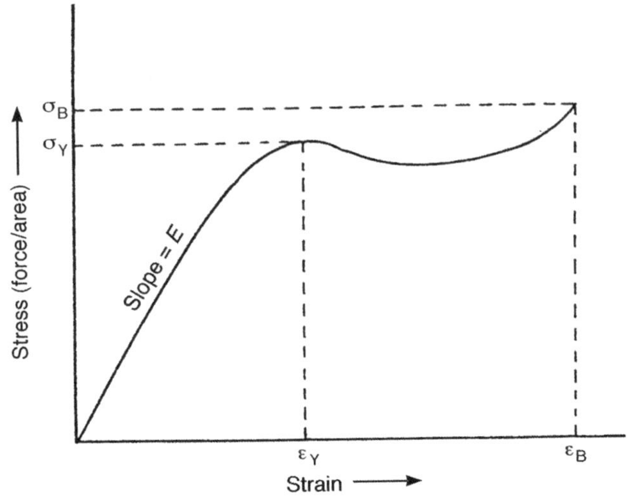

Figure 2.2 A hypothetical stress–strain curve for ductile film. The tensile properties are defined as follows: E, tensile modulus; ε_Y, elongation at yield; ε_B, elongation at break; σ_Y, yield stress; σ_B, tensile strength. From J. V. Koleske (Ed.), *The New GARDNER–SWARD Handbook (Paint and Coatings Testing Manual)*, 14th Edn, 1995. Copyright American Society for Testing and Materials (ASTM). Reprinted with permission of the ASTM.

the sample's cross-sectional area, while the elongation at break is the percentage elongation at the point where the sample breaks.

The shape of the stress–strain curve can change dramatically as a function of both the rate of application of stress (the speed of the jaw separation) and the temperature of the experiment. Materials tend to behave more like viscoelastic solids as opposed to viscoelastic liquids, when the time frame of the experiment decreases (such as in the 'silly putty' example above). If the rate of jaw separation is dramatically increased (the time frame of the experiment is decreased), the modulus will increase (the initial slope of the line in Figure 2.2 will increase) and the elongation at break will decrease. This behaviour can have very practical consequences. A common test in the paint industry is to determine a coating's elongation by bending a coated substrate about a conical mandrel in the time frame of 15 seconds. A coating may appear to have sufficient flexibility by this method, yet will crack when used to coat aluminium soffit and facia that are bent (formed) in a fraction of a second on a workman's 'brake'.

Based on everyday experience, we have come to expect that materials are more brittle at lower temperatures. A coating may appear flexible at 70 °F (21 °C), yet will crack badly when tested at 0 °F (718 8C). This difference in behaviour is a consequence of the material's glass transition temperature, T_g.

There are two nearly equivalent definitions of the glass transition temperature. One states that T_g is that temperature at which an amorphous polymer changes from a hard, glassy state to a rubbery, flexible state. The other, perhaps more fundamental, description is that the T_g is that temperature above which there is a sudden and rapid increase in the free volume of the material. The first definition is really a consequence of the second. Basically, the T_g is that temperature at which the polymer chains, which are highly intertwined, achieve enough thermal energy to begin to move and 'relax' into a more preferred orientation. Carrying out a stress–strain experiment above the T_g will result in a lower modulus than doing it below T_g.

Physical stress can be imparted to a coating in a number of ways. These include post-forming coil coated stock, thermal expansion/contraction, vibration of the substrate due to nearby machinery, impact of falling objects, flexing due to people walking on thin-gauge galvanized roof decking, expansion–contraction of coated wood due to changes in moisture content, and so forth. This stress has to be relieved, and how the coating (or coating system) does this will determine whether or not it fails, and the mode of failure. In some instances, a coating may have a stress–strain curve similar to that shown in Figure 2.2. The point on this curve where the slope becomes zero is termed the yield point. Stress–strain curves similar to that of Figure 2.2 are usually obtained with coatings that have a T_g well below the temperature of the experiment. Up to the yield point, all of the strain (elongation) is recoverable. If the stress is removed, the coating will contract back to its original shape. However, elongation beyond the yield point is not recoverable, and the coating has thus been permanently deformed.

A second type of coating having a much higher T_g may have a stress–strain curve similar to that shown in Figure 2.3a. Rather than yielding, the coating breaks (it undergoes brittle failure). The two coating types with stress–strain curves depicted in Figures 2.2 and 2.3a have much different ways of relieving stress. The two different means of stress relief will have very practical consequences in terms of their failure mechanisms.

Consider a coating of the first type (low T_g, with a yield point) applied to exterior wooden house sidings. As the wood expands and contracts due to temperature and humidity changes, so too does the coating. At stress values corresponding to the strain (elongation) prior to the yield point, the coating will expand and contract along with the wood. Since this also stresses the bond to the wood, the adhesive strength must be great enough to withstand the stress in order to avoid delamination. If this is the case, everything is fine until the expansion of the wood is large enough that the yield point of the coating is passed. The coating has now been permanently and irreversibly deformed and the strain has dissipated. When the wood subsequently shrinks, the coating will not, and a sag or wrinkle will result.

This same stress–strain behaviour may be desirable, however, if the substrate expands but does not subsequently contract. This would be the case with post-formed coil coated metal sidings. The bending of the siding may result in the coating being extended past its yield point. The coating is now permanently deformed; the molecular relaxation accompanying the yield point has dissipated

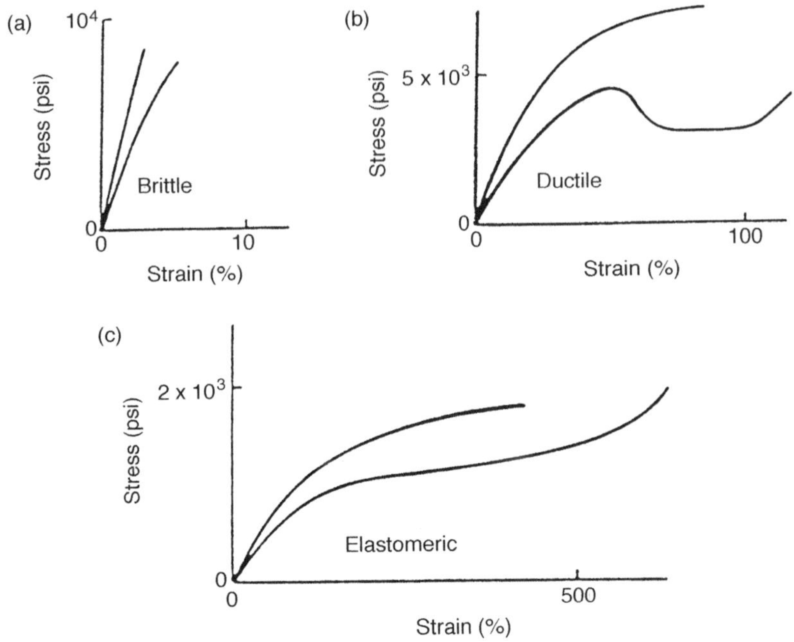

Figure 2.3 General kinds of stress–strain curves for various types of coatings: (a) brittle, (b) ductile, (c) elastomeric. The scales indicate order-of-magnitude values. From J. V. Koleske (Ed.), *The New GARDNER–SWARD Handbook (Paint and Coatings Testing Manual)*, 14th Edn, 1995. Copyright American Society for Testing and Materials (ASTM). Reprinted with permission of the ASTM.

the stress; there are no restoring forces acting at the steel–coating interface, and the coating therefore retains adhesion to the metal, with no cracking.

Now consider a coating that has a stress–strain curve depicted by that shown in Figure 2.3a (high T_g, break point), again applied to exterior wooden house sidings. If subjected to enough stress due to expansion and contraction of the wood, the stress can be relieved in one of two ways, namely, the coating will either crack, or it will delaminate. Cracking will occur if the coating has very good adhesion to the substrate. In this case, the only way that the stress can be relieved is through cracking of the coating (brittle failure). On the other hand, if the adhesion to the substrate is poor, the stress can be relieved via delamination.

What would happen in the above case if the substrate was not metal or wood but a coat of primer? Ignoring the untidy complication that the primer also has its own stress–strain behaviour, cracking of the topcoat and/or delamination of the topcoat from the primer may result, just as described above. However, if the primer has poor cohesive strength, another way for the stress to be relieved is for the primer to split cohesively. Stress is transferred throughout a multi-coat coating system and to the ultimate primer/substrate bond. If the coatings are strong enough cohesively and the adhesion at all of the interfaces are adequate, the stress may or may not be gradually dissipated through internal molecular reorientation, but in either case failure will not result. If there is a cohesive or adhesive 'weak link', that link will be the point at which stress-induced failure occurs.

A viscoeleastic coating's ability to return to its original shape once the stress is removed is a function of its elastic component. In a classic relaxation experiment, one applies an instantaneous strain (elongation) to a sample, and then measures the change in stress with time. For a material at the elastic end of the viscoelastic continuum, the stress stays constant with time – there is no relaxation. This is often the case with a highly cross-linked thermoset system. For a material at the viscous (Newtonian liquid) end of the continuum, there is no stress because the material has permanently flowed or deformed. A thermoplastic coating will tend to show this type of behaviour. For viscoelastic samples that fall somewhere within the continuum rather than at the ends, the stress is initially high and will drop to lower values (it relaxes) with time. As a consequence, when a cross-linked coating on a metal substrate is elongated through impact or forming of the metal, if the adhesion is sufficient, the coating is held in the deformed state by its adhesion to the substrate. There is a continual, never-ending stress acting to pull the film off of the substrate. With time, or with added stress (vibrations), this force may finally overcome the adhesive forces and delamination may result.

Another form of external stress is swelling of the coating film by water, solvents or other chemicals. If the swelling is severe enough, the stress due to the volume expansion can result in delamination as a form of stress dissipation.

The above discussion concentrated on the response of the coating system to external stress. However, coating systems can also react to internal stress.

2.2.2 Internal Stress

When coatings cure, they also shrink. The shrinkage is caused by solvent loss during drying and/or by cross-linking during curing. When thermoset coatings cure, the length of the covalent bonds being formed is less than the distance between the neighbouring molecules prior to their reaction. In the early stages of solvent evaporation and/or cross-linking, the coating is often above its glass transition temperature (T_g) and the polymer chains have sufficient mobility to allow for shrinkage. However, as film formation via solvent evaporation or continual cross-linking proceeds, the T_g will continue to rise; the polymer chains will have reduced mobility, and internal stress will result from the coating's inability to undergo further shrinkage.

When examined carefully, the above discussion seems to contradict the statement commonly made in the coatings industry, that 'shrinkage causes internal stress'. It is, in fact, quite the opposite: internal stress is the consequence of a coating's inability to shrink. If the coating shrinks, this is actually a form of stress relaxation, and the stress or internal energy has been dissipated (much like a crack forming in a brittle coating which dissipates the stress that has built up or has been applied to that coating).

An experiment that adds to this confusion is that of applying a two-component, thermosetting epoxy coating to thin aluminium foil. As the epoxy cures, the foil curls up, and it is said that this is a consequence of the internal stress generated when the epoxy cures and shrinks. What is actually happening is that the epoxy is relieving its internal stress by shrinking, and the forces are being transferred to the aluminium foil substrate, which has insufficient strength to resist them. If the substrate is thicker and stronger, such as structural steel, it will not deform. In this case, when the epoxy tries to shrink to relieve its own internal stress, one of several things will happen. If the adhesion is good and the coating has low cohesive strength, it will not be able to shrink, and it will then crack to relieve the stress. If it has high cohesive strength, along with good adhesion to the steel, there will be a permanent stress, both in the coating and acting on the bond to the substrate. In this case, additional external stress (flexing, freeze/thaw, etc.) may eventually result in failure. Finally, if the adhesion to the steel is poor, the stored internal stress will result in delamination.

Another possible response to internal stress occurs if the substrate is another coating, such as a primer. If the primer is hard and strong, the possible scenarios are the same as those described above. However, if the primer has poor cohesive strength (perhaps as a result of over-pigmentation), it may be unable to resist the shrinkage of the topcoat as it relieves its internal stress, and cohesive splitting of the primer may result. This type of failure is often said to be due to 'shrinkage stress' of the topcoat. This is an accurate term, but it must be kept in mind that shrinkage stress is not the same as internal stress. Shrinkage stress is a stress that is transferred to a lower coat of paint (or substrate) as a result of the topcoat relieving its own internal stress.

A special type of internal stress is known as 'physical ageing'. The term 'ageing' generally refers to some change in a coating's physical or chemical properties as it is exposed over a long period of time to the degrading effects of chemicals, light, moisture or heat. These environmental factors can alter the chemical structure of the coating, resulting in changes in such things as gloss, colour, brittleness and adhesion. However, physical ageing is a phenomenon whereby the properties of a polymer or coating can change simply as a function of storage time, without any chemical changes. There are numerous references in the literature to physical ageing [11–14], although most of these relate to polymers and plastics rather than coatings.

The glass transition temperature (T_g) is that temperature below which a polymer is in a hard, glassy state, and above which it is in a rubbery, flexible state. It is also that temperature below which there is a significant decrease in free volume, meaning that the polymer chains and side chains have greatly restricted mobility. When a polymer is rapidly cooled and subsequently stored at a temperature below T_g, the molecular chains and segments making up the polymer (or coating, in our case) have little ability to move freely. Although they would like to achieve a more preferred orientation corresponding to lower free energy, they are, in a sense, 'frozen' in place. However, while molecular motion is slow, it is not zero. Indeed, there is actually an excess of free volume in the polymer. Over time, the polymer will rearrange, causing collapse of this excess free volume, and a densification of the coating as the material approaches its true equilibrium state. This phenomenon is known as physical ageing.

While the above discussion sounds highly theoretical, the consequences of physical ageing can be very practical. The consequences are primarily observed as an increase in the brittleness of the coating. It is important to note that this brittleness is not due to any chemical change. The coating has not continued to cross-link, nor has it become embrittled due to any other type of chemical reaction or degradation. Indeed, one of the distinguishing features between physical ageing and other types of chemical ageing is that chemical changes are irreversible, whereas changes due to physical ageing are reversible. They can be erased by annealing the polymer at a temperature above its T_g. It is expected that internal stress problems related to physical ageing would be more prevalent with coatings which are baked rather than with ambient cure coatings.

2.2.3 Chemical Attack

At the beginning of this section, it was stated that for a coating to fail, it has to be stressed. That discussion focused on classical views of stress as an applied force that tends to strain or deform a body. However, anything that attacks or degrades the integrity of the coating can also be regarded as a stress. Certainly, chemical degradation falls under this heading.

For the most part, coatings are made up of organic polymers and are therefore subject to the same types of chemical reactions as small organic molecules. The reactivity of organic molecules, and the polymers made from them, depends largely on the types of functional groups that are present. In broad terms, the reactivity of an individual functional group is related to the polarity of the bond holding the two atoms together.

The covalent bonds that hold organic molecules together consist of one or more shared pairs of electrons. Often, the two atoms do not share these electrons equally, and consequently, one end of the molecule is slightly negative while the other end is slightly positive, and the bond is then said to possess polarity. The tendency of an atom to attract electrons is referred to as electronegativity; a bond formed between two atoms that differ substantially in electronegativity will be polar. For those atoms commonly encountered in organic molecules, the order of electronegativity, from highest to lowest, is as follows: F, O, Cl and N, Br, C and H.

Another factor that influences the reactivity of a covalent bond is the accessibility of the electrons. Generally, electrons that are delocalized are more available for reaction than those that are not. Therefore, carbon–carbon double bonds, with relativity delocalized clouds of electrons, are more reactive than carbon–carbon single bonds. Conjugated carbon–carbon double bonds, where carbon–carbon double and single bonds form one alternating structure, are even more reactive than a single isolated double bond as a consequence of even greater delocalization of the electron cloud. An exception to this rule is benzene and compounds with benzene-like (aromatic) structures. Although the electrons in the six-membered benzene ring are highly delocalized, there is also an extended stable arrangement due to resonance stabilization. However, the underlying theory behind resonance stabilization is beyond the scope of this present discussion.

In practical terms, coating binders consisting primarily of carbon–carbon single bonds, or ether (C–O–C) linkages, are relatively stable towards chemical attack. Binders containing alcohol, carboxylic acid, ester, amine and amide groups, and those containing conjugated carbon–carbon double bonds, will be more susceptible to aggressive chemicals such as acids, bases and oxidizing agents.

Some high-performance industrial coatings are specifically designed to be chemically resistant. Phenolics and vinyl esters are routinely used as linings in harsh chemical service. Such coatings have very good resistance to a wide range of chemicals, but even these are not indestructible, and will slowly degrade over time. Usually such degradation is expected, and quite often manufacturers of such coatings publish fairly comprehensive guides as to their suitability in various environments.

Coating degradation due to chemical attack is not restricted to the specialty market of high-performance coatings. A very common failure is that of alkyd coatings exposed to alkalinity and moisture. Alkyds are basically oil-modified polyesters. The polyester linkage is not very stable under moist alkaline conditions, and is subject to alkaline hydrolysis. This process is referred to as saponication, and results in the cleavage of the carbon–oxygen single bond. This results in the production of a carboxylic acid and an alcohol, but since the environment is alkaline, the acid is immediately converted to its salt. Since carboxylic acid salts have a certain degree of water solubility, the coating begins to become water-sensitive. It will eventually become soft and tacky, and catastrophic failure can result. It is for this reason that alkyds should not be applied to alkaline substrates such as concrete.

The organic binder is not the only ingredient of a coating which is subject to chemical attack. Some pigments are sensitive to acids or bases. A good example is aluminium flake, which is common in both decorative coatings and in relatively water-impermeable mastics. Aluminium is amphoteric. It can become quite active towards either end of the pH scale, and rapid deterioration of the coating can therefore result.

2.2.4 Weathering Stress

Apart from contaminated industrial atmospheres and acid rain, the simple elements of weather – heat, light and moisture – can be very destructive to coatings. The primary cause of weather-induced coating degradation is the energetic UV radiation component of sunlight.

Sunlight is composed of three types of radiation, that is, UV, visible and infrared. UV light has the highest frequency (or lowest wavelength) and is therefore the most energetic and potentially damaging component.

UV light itself can be further divided into three regions, namely, UVA, UVB and UVC, as follows (Table 2.3):

Table 2.3 Types of ultraviolet light.

Region	Wavelength range (nm)
UVA	320–400
UVB	280–320
UVC	<280

The UVC region is the most energetic and destructive. Fortunately, such radiation is entirely absorbed by the earth's atmosphere and is not a factor in coating degradation. A small amount of UVB radiation reaches the earth's surface, while about 6% of the total radiant energy reaching the earth's surface falls in the UVA region. Nevertheless, this is sufficient to cause substantial damage to organic coatings.

Photodegradation of organic coatings is caused by the absorption of energetic UV light, which either breaks chemical bonds directly or results in the formation of free radicals, which can subsequently result in bond breaking or bond formation. Photodegradation can result in chain scission and a loss of molecular weight, or cross-linking and an increase in molecular weight. Either process will result in changes in the physical properties of the coating, usually for the worse.

Most organic bonds do not absorb UV light at frequencies above approximately 280 nm and therefore are not directly damaged by sunlight. Unfortunately, most resins used in coatings contain small amounts of peroxide and ketone impurities, which can absorb sunlight to form free radicals [15] and initiate a process of photo-oxidation. The problem is encountered particularly in polymers with conjugated carbon–carbon double bonds (such as many of the natural oils), which stabilize the formation of free radicals. At the other end of the scale, silicone resins are highly stable towards photo-oxidation, which accounts for the greater exterior durability of silicone-modified alkyds.

Two important coating types which show significant absorption of UV light at wavelengths above 280 nm are aromatic urethanes and epoxies. Sunlight can directly break bonds in these coatings. Urethane coatings made from aromatic precursors yellow badly in sunlight, while epoxy coatings will chalk rapidly.

2.2.5 Osmotic Blistering

Another mechanism by which coatings can fail is permeation. It is debatable as to whether or not permeation should be considered as being a stress, but it can certainly lead to coating failures.

Permeation of a coating by water and oxygen can lead to corrosion of steel substrates, and entire texts have been written on the role of coatings in corrosion science. The subject will not be treated here except to say that coatings mitigate corrosion either by acting as a physical barrier to moisture and oxygen, or by incorporating anti-corrosive pigments which function by passivation of the steel substrate.

In general, inorganic salts, acids and bases have difficulty passing through most coatings, due both to their physical size and the fact that, being ionic, they are usually incompatible with the organic polymers making up the coating. Of course, this situation can change if these chemicals can actually react with the coating, or if physical defects such as pinholes are present.

Water, however, is a relatively small molecule, and under normal conditions, only a tiny fraction of it is ionized. Water can permeate through all present-day coatings, although much more slowly for some than others. Under normal circumstances, this would not cause premature failure. However, if there is a water-soluble species behind the coating, or within a lower coat of a multi-coat system, the situation can change dramatically. When a normal, usually miniscule, amount of water permeates the coating and comes in contact with the water-soluble species beneath, it will dissolve it. If this species is absent in the water on the outside of the coating, an osmotic cell has been created.

An osmotic cell consists of a semipermeable membrane (the coating) separating a solution of high concentration (the submicroscopic droplets of dissolved material) from a solution of low concentration (e.g. relatively pure potable water in an elevated water tank). This is a non-equilibrium condition and the two solutions have different values of free energy or chemical potential. There will be a strong driving force to make these two concentrations, and hence the chemical potential of the solvent (water in this example), equal. In practice, more water than 'normal' will permeate the coating in an attempt to dilute the concentration of the soluble species within or behind the coating film. The driving force for this is the concentration gradient and hence chemical potential gradient, across the coating membrane. The consequence is a water-filled blister. Such blisters are fairly common, and are referred to as osmotic blisters.

It is often stated that in osmotic blistering, the water is 'forced' through the coating under high 'osmotic pressure'. This statement is *incorrect*. If there really was an external pressure normal to the coating film, then the pressure would serve to push against the coating and keep it attached to the substrate. The term 'osmotic pressure' refers to the pressure that is building up inside or behind the coating as more and more water accumulates there. The water is not accumulating because it is being forced through the coating under pressure; it is accumulating in an attempt to eliminate the concentration gradient. In fact, it will continue to do so until the pressure being exerted inside the forming blister is energetically equivalent to the chemical potential difference across the coating due to the concentration gradient.

It is logical to ask as to the identity and source of these water-soluble species which can cause osmotic blistering. One such species is salt. There are several ways in which salt could get on steel or other substrates. It could originate from road salt which gets on new steel during transport, or on old steel by breeching old, cracked or peeling coating. It could contaminate steel in a marine environment, either by direct contact with salt water or via salt-laden moisture in the air. Concerns have also been expressed [16, 17] that otherwise clean steel can be contaminated with salts by blast cleaning with abrasives which themselves have high salt content. Such salts could originate naturally as a consequence of the production of the abrasive or could result from contamination of recycled abrasive. While this seems plausible, little quantitative research has been published in this area. One paper, however [18], concluded that only small amounts of salts (approximately 0.15–0.25% of the salt content of the abrasive) were actually transferred from the abrasive to the steel surface during blast cleaning. However, since it is best to err on the side of safety, high salt-containing abrasives would best be avoided when blast cleaning steel surfaces.

In addition to salt, another common agent responsible for osmotic blistering is water-soluble or partially water-soluble solvent. These solvents could be part of the coating formulation or could originate from thinner (either authorized or unauthorized) added to the coating to improve application properties.

It has been the author's experience that small amounts of solvents can be retained in coatings for months if not years. Solvent, depending on its evaporation rate and the temperature and air flow during application, initially evaporates rapidly from a coating. However, as the coating begins to dry and harden, solvent evaporation drops off rapidly. It has been reported [19] that up to 10–20% of the original solvent may be retained within the coating, and that in some cases, up to 5–10% of it may be retained for years. James [20] and Funke and Haagen [21] both demonstrated the existence of osmotic blistering due to water-soluble solvents. Storfer and Yuhas [22] also showed a strong correlation between the water solubility of the tail solvent used in paint and its tendency to blister when exposed to water. Schilling [23] also pointed out the danger of osmotic blistering as a result of water-soluble solvent.

2.2.6 Electroendosmotic Blistering

In the 1940s, Kittleberger and Elm noticed that blisters on painted panels immersed in sea water developed only after other areas of the panels had begun to corrode due to flaws in the coating. This led to a study [24] where the authors conducted testing which led to the conclusion that such blistering was due to electroendosmosis.

Definitions of electroendosmosis, or electro-osmosis, sometimes define it is the passage of water through a semipermeable film by an electrical potential gradient. While this is the effect, this definition is not strictly correct. What is actually happening is that the electrical potential gradient (associated with the corrosion observed by Kittleberger and Elm) causes ions to migrate through the coating. These ions have what is known as a hydration sheath associated with them, which is basically a few molecules of water clinging to and hydrating the ions. As the ions migrate through the semipermeable coating under the influence of the potential gradient, the water moves right along with them [25]. Another interpretation of this effect [26] pictures the water as being pulled along by the migrating ions via viscous drag and concludes that 'the effect of liquid flow under the action of an electrical potential difference is called electroosmosis'.

It should be noted that many of the studies or demonstrations of electroendosmosis in the literature employ such things as capillaries of glass and silica, porous plugs of finely divided solids, ion-exchange membranes and even layers of soil. Indeed, in the field of environmental science, electroendosmosis has even been proposed as a means of removing certain contaminants from saturated clay [27]. Even the study by Kittleberger and Elm was on permeable thin film linseed oil coatings, and much more work would have to be done to determine how effective a blistering mechanism electroendosmosis is on dense, modern, thick film coatings such as epoxies.

Blistering immediately at or very close to sites of corrosion can also be due to a type of blistering known as cathodic blistering [28]. As part of the corrosion process under basic or neutral conditions, hydroxyl ion (OH^-) is produced at the cathode. This results in localized, strongly alkaline conditions where the hydroxyl ions are trapped beneath the coating film. It is thought that this strongly alkaline environment can cause localized film delamination (blistering), either through attack of the coating binder or the metal oxide surface. This type of blistering would be more prevalent with coatings which have poor resistance to high pH, such as alkyds or polyesters.

REFERENCES

1. P. E. Pierce and C. K. Schoff, *Coating Film Defects*, Federation of Societies for Coatings Technology, Blue Bell, Philadelphia, PA, USA, 1988, p. 8.
2. C. H. Hare, *Protective Coatings*, Technology Publishing Company, Pittsburgh, PA, USA, 1994, p. 386.
3. D. G. Weldon, A. Bocham and M. Schleiden, *J. Prot. Coat. Linings*, **4**, 46 (1987).
4. B. R. Appleman (Ed.), *Systems and Specifications*, 7th Edn, Vol. 2, The Society for Protective Coatings (SSPC), Pittsburgh, PA, USA, 1995, pp. 9–66.

5. H. J. Roper, R. E. Weaver and J. H. Brandon, *J. Prot. Coat. Linings*, **22** (6), 52 (2005).
6. M. A. Kuehner, *Met. Finish.*, **83**, 15 (1985).
7. W. J. Wittke, *Met. Finish.*, **87**, 24 (1989).
8. W. Rausch, *The Phosphating of Metals*, Finishing Publications Ltd, Teddington, Middlesex, UK, 1990, p. 153.
9. B. N. McBane, *Automotive Coatings*, Federation of Societies for Coatings Technology, Blue Bell, Philadelphia, PA, USA, 1987, p. 11.
10. B. M. Perfetti, *Met. Finish.*, **81**, 57 (1983).
11. J. M. Hutchinson, *Prog. Polym. Sci.*, **20**, 703 (1995).
12. A. Skaja, D. Fernando and S. Croll, *J. Coat. Technol. Res.*, **3** (1), 41 (2006).
13. W. D. Cook, M. Mehrab and G. H. Edward, *Polymer*, **40**, 1209 (1999).
14. L. C. E. Struik, *Physical Aging in Amorphous Polymers and Other Materials*, Elsevier, Amsterdam, the Netherlands, 1978.
15. Z. W. Wicks Jr, S. N. Jones and S. P. Pappas, *Organic Coatings: Science and Technology*, Vol. 2, John Wiley, New York, 1994, p. 134.
16. R. Garrabant, J. Bieseinger, G. Jacinto and M. Melampy, *J. Prot. Coat. Linings*, January, **16** (1), (1999).
17. W. C. Johnson, *J. Prot. Coat. Linings*, September, **7** (9), (1990).
18. C. L. Fairfull and D. G. Weldon, *Salt Contaminated Abrasives and their Effect on Coating Performance*, Proceedings of the SSPC Conference, Atlanta, GA, USA, 2001.
19. C. M. Hanson, *Ind. Eng. Chem. Prod. Res. Dev.*, **9** (3), 282 (1970).
20. D. M. James, *J. Oil Colour Chemists Assoc.*, **43**, 653 (1960).
21. W. Funke and H. Haagen, *Ind. Eng. Chem. Prod. Res. Dev.*, **17** (1), 50 (1978).
22. S. J. Storfer and S. A. Yuhas, Corrosion 87, *Paper No. 232*, National Assoc. of Corr. Eng., 1987.
23. M. Schilling, *J. Prot. Coat. Linings*, February, **20** (2), 24 (2004).
24. W. W. Kittleberger and A. C. Elm, *Ind. Eng. Chem.*, **39** (7), 876 (1947).
25. C. Hamann, A. Hamneet and W. Vielstich, *Electrochemistry*, Plenum Press, New York, 1998.
26. P. Reiger, *Electrochemistry*, Chapman & Hall, 1994, p. 73.
27. A. Shapiro and R. Probstein, *Environ. Sci. Technol.*, **27** (2), 283 (1993).
28. C. Hare, *J. Prot. Coat. Linings*, March, **15** (3), 17 (1998).

3

Pigments

Although coating failures caused by the wrong choice of pigment are relatively rare, a basic understanding of the types and characteristics of various pigments is important in gaining insight into how coatings work. Pigments can be either inorganic or organic.

3.1 INORGANIC PIGMENTS

In the field of failure analysis, inorganic pigments will be encountered much more often than organic ones. This is because most failure investigations deal with heavy-duty coatings that are intended to provide protection from the environment, rather than brightly coloured coatings that are used for aesthetic reasons. The subject will be subdivided into inorganic coloured pigments, extender pigments and corrosion-resistant pigments.

3.1.1 Inorganic Colour Pigments – White

Titanium dioxide (TiO_2) is the most common white pigment in use today. It has a clean white colour and a high refractive index. This latter means that it is very good at bending light rays, and consequently has very good hiding power. This is an important feature if one wishes to hide the colour of the underlying paint in a single coat. Titanium dioxide is relatively hard, and coatings formulated with it tend to have improved abrasion resistance.

There are two common crystal forms of titanium dioxide, namely, rutile and anatase. The rutile form is much more resistant to the combined effects of ultraviolet radiation and moisture, and generally will not chalk. It is particularly suited

Failure Analysis of Paints and Coatings, Revised Edition Dwight G. Weldon

to exterior use. The anatase form will chalk badly and is generally not suitable for exterior use.

Zinc oxide is occasionally used as a white pigment, but usually in combination with titanium dioxide. It has the advantage of inhibiting mould growth if present at high levels [1], but is alkaline and can react with certain types of coatings, such as alkyds, to produce soaps.

White lead ($2PbCO_3{\cdot}Pb(OH)_2$), also known as basic lead carbonate, was once used extensively as a white pigment, but is not used today because of the toxic nature of lead. The same is true of basic lead sulfate ($2PbSO_4{\cdot}PbO$). Both pigments tend to darken if exposed to sulfur in polluted atmospheres.

3.1.2 Inorganic Colour Pigments – Yellow

The yellow iron oxides ($Fe_2O_3{\cdot}H_2O$) are one of the more common families of inorganic yellow pigments. They can range in colour from a dull, dirty yellow to a dark yellow brown and are naturally occurring. They are resistant to alkaline and organic acids but can be attacked by mineral acids [2]. At temperatures above about 350 °F (177 °C), they can lose their water of hydration and turn red [2].

Lead chromate pigments ($PbCrO_4$) and chrome yellow (mixtures of lead chromate and lead sulfate) are available in a wide range of hues, and generally produce brighter, cleaner colours than the iron oxides. They have good lightfastness (resistance to fading) and opacity. However, they suffer from poor chemical resistance. Atmospheres polluted with sulfur can cause them to darken, and if used in an alkaline environment or formulated with alkaline vehicles, they can redden [2]. They are also toxic.

Cadmium yellows (CdS) are synthetic pigments that range in colour from yellow through primrose to orange. They are heat resistant, lightfast and alkaline resistant, but have poor acid resistance. Although a very useful pigment, there is some concern about the toxic effects of cadmium.

Bismuth vanadate is a brilliant greenish-yellow pigment with excellent lightfastness and good hiding power. Although expensive, it is an excellent replacement for chrome yellow [2].

3.1.3 Inorganic Colour Pigments – Orange

Current inorganic orange pigments contain lead, cadmium or chromium, all of which are considered toxic. As a result, their use is declining. One of the most popular alternatives is molybdate orange, which gives a clean, clear, brilliant orange with excellent hiding, good lightfastness and good chemical resistance. Two other inorganic oranges include chrome orange and cadmium orange.

3.1.4 Inorganic Colour Pigments – Red

The best high-performance red pigments for aggressive environments are the iron oxides [3]. While not as rich and brilliant as some of the organic reds, they have good heat resistance, help stabilize the coating vehicle against ultraviolet light, and are resistant to alkalis and organic acids. However, harsh mineral acids can affect them. They have good opacity and are available in a variety of shades.

3.1.5 Inorganic Colour Pigments – Blue

Prussian blue ($Fe_4[Fe(CN)_6]_3$), less commonly known as iron blue, and (as its potassium derivative) potassium ferric ferrocyanide is a synthetic bright blue pigment with high tinting strength but low opacity. It has good lightfastness and is reasonably resistant to acids, although it is very sensitive to alkaline materials.

Ultramarine blue ($3Na_2O_3 \cdot 3Al_2O_3 \cdot 6SiO_2 \cdot 2Na_2S$) is a synthetic pigment, but also occurs naturally as lapis lazuli. The shade can be altered by varying the amount of silica [1]. This material can be difficult to disperse, although it has good lightfastness, heat resistance and alkali resistance. However, it displays poor tint strength and is decomposed by acids.

3.1.6 Inorganic Colour Pigments – Green

The most widely used inorganic green pigment is chromium oxide (Cr_2O_3), also known as chromium green oxide. This is a dull green synthetic pigment that has good stability to heat, light, acids and alkalis, is relatively inexpensive and has good opacity. Since the chromium is in the +3 oxidation state, it is not toxic like many of the +6 (hexavalent) chromium pigments. It cannot, however, produce bright greens with a high chroma.

Chrome green pigments, containing both hexavalent chromium and lead, are seldom used anymore, largely as a consequence of toxicity concerns.

3.2 EXTENDER PIGMENTS

Extender pigments are relatively low-cost inorganic pigments which provide little colour or corrosion resistance, but which are used to provide bulk and to adjust density and flow properties. They can also affect physical properties such as hardness, permeability and gloss. They consist, for the most part, of silica/silicates, calcium carbonate and barytes.

3.2.1 Silica/Silicates

Silicas fall into two broad types, that is, natural and synthetic. The synthetic silicas are used in small amounts and are perhaps better considered as additives. The three major natural silica pigments are crystalline silica, amorphous silica and diatomaceous silica. All are chemically inert, hard, and have low specific gravities and refractive indexes.

Crystalline silica is very durable, but is also very hard. This can make it difficult to disperse, and pigment settling can then become a problem. It is very useful in floor and traffic paints, however, because of its good abrasion and skid resistance [4].

Amorphous silicas are slightly softer and finer than crystalline silica, and are easier to disperse in most formulations. They are inert and weather resistant, and contribute to good abrasion resistance.

Diatomaceous silicas are seldom used in coatings. They are very light and fluffy, and have very high oil absorption values, which essentially means that they can absorb large amounts of resin. Even small amounts can rapidly decrease a coating's critical pigment volume concentration (*CPVC*).

There are a variety of silicate extender pigments, such as talc (magnesium silicate, $3MgO{\cdot}4SiO_2{\cdot}H_2O$), china clay (kaolin, $Al_2O_3{\cdot}2SiO_2{\cdot}2H_2O$), mica ($K_2O{\cdot}2Al_2O_3{\cdot}6SiO_2{\cdot}2H_2O$) and several other miscellaneous materials.

Talc may be the most commonly used extender pigment. It is a naturally occurring mineral available in platy or fibrous structures. The fibrous grades can both provide a reinforcing effect and improve flexibility, while the platy version can reduce water permeability and improve corrosion resistance. Talcs wet and disperse easily in most binders, do not display settling problems and impart good brushability.

China clay is another very common extender pigment. This is a naturally occurring platy pigment of low specific gravity and moderate oil absorption, which tends to reinforce and add toughness to films. Calcined, or anhydrous, versions are bright and can actually impart significant hiding power, especially when employed with titanium dioxide, and are popular for interior latex coatings.

Mica is a naturally occurring pigment having the form of thin platelets. This lamellar structure results in a leafing effect that can significantly reduce the rate of water permeation. This pigment also has good resistance to ultraviolet light, and improves a coating's resistance to cracking and checking. As a consequence, it is often used in exterior coatings.

3.2.2 Calcium Carbonate

Calcium carbonate ($CaCO_3$), also known as whiting or chalk, is another very common extender pigment, particularly in masonry coatings such as block fillers.

It is available both naturally and as a synthetic. Dolomite, a mixture of calcium and magnesium carbonates, is also available but is of generally poorer quality.

Calcium carbonate pigments have low oil absorptions and low specific gravities. They can take up a considerable volume in a coating with little demand for the vehicle, accounting for their widespread use in coatings such as masonry block fillers. They are, however, alkaline pigments, and should therefore not be used in conjunction with alkaline-sensitive pigments.

3.2.3 Barytes

Barytes or barium sulfate can be obtained naturally from the mineral baryte or can be produced synthetically. It is hard, very dense, and is resistant to both acids and alkalis. Its density, however, can sometimes lead to settling problems. It has a very low oil absorption and therefore has little effect on viscosity.

3.3 CORROSION-RESISTANT PIGMENTS

Before discussing the specific qualities of various corrosion-resistant pigments, a basic introduction to the theory of corrosion and the mechanism of corrosion inhibition is necessary. This discussion will also prove valuable in understanding the mechanisms of certain coating failures. It has been estimated that corrosion-related damage in the United States alone amounts to over $15 billion annually [5].

The corrosion, or rusting, of iron is simply an oxidation–reduction reaction. In chemical terms, oxidation is the release of electrons, or an increase in the oxidation state, while reduction is a gaining of electrons, or a reduction in the oxidation state.

The oxidation of iron to form rust can be written as follows:

$$2Fe + O_2 + 2H_2O \rightarrow 2Fe(OH)_2. \tag{3.1}$$

Once formed, the iron hydroxide often reacts further to form various oxides and hydrated oxides. The various 'blends' of these oxidation products account for the range of colours observed in rust, which can vary from black through yellow to orange red.

The anodic or oxidation reaction is as follows:

$$Fe \rightarrow Fe^{2+} + 2e^-. \tag{3.2}$$

The cathodic or reduction reaction can vary depending on the environment. In neutral or basic solutions, it is as follows:

$$2H_2O + O_2 + 4e^- \rightarrow 4OH^-. \quad (3.3)$$

In the absence of oxygen, or in acidic solutions, the following reaction occurs:

$$2H^+ + 2e^- \rightarrow H_2. \quad (3.4)$$

While in acidic solutions when oxygen is present, we have the following:

$$2H^+ + \frac{1}{2}O_2 + 2e^- \rightarrow H_2O. \quad (3.5)$$

For corrosion to occur, several conditions must be met. There must be a corrosion cell consisting of an anode (where oxidation occurs), a cathode (where reduction occurs) and an electrical conductor to facilitate transfer or movement of ions. In the corrosion of iron, the iron is the anode; water with small amounts of dissolved salts from the environment is the electrolyte, and dissolved oxygen is often the electron acceptor.

Furthermore, the surface of steel is not homogeneous. There are microscopic variations in the crystalline nature of neighbouring areas, and there are subtle variations in the type and/or amount of surface contaminants and oxide layers. The end result is a plethora of extremely small anodes and cathodes on the steel surface itself, just waiting for the proper environment for rust formation.

There are three ways in which coatings can provide corrosion protection, namely, barrier protection, inhibitive pigment protection and cathodic protection.

Protecting steel from rusting by coating it with a more or less impermeable barrier is perhaps the oldest technique, going back to the days of wiping tools and weapons with animal fat. If one could truly isolate steel from all access of water and oxygen, the corrosion process would cease. Barrier protection is still a popular technique, now involving greatly improved technology when compared to medieval animal fat.

Barrier coatings are simply relatively thick coatings (usually approximately 10–40 mil) composed of relatively impermeable moisture-resistant resins, sometimes pigmented with flake- or platelet-type pigments. Epoxies, coal tars, coal tar epoxies and, more recently, moisture-cured urethanes are common types of resins. Pigments that add to the barrier properties include aluminium flake, glass flake, mica and micacious iron oxide. The latter is a special flake-like form of hematite (Fe_2O_3). These pigments all function by forming an interleafing, lamellar structure, much like the shingles on a roof. In this case, water and dissolved materials have a much more tortuous path to reach the steel substrate than if the coating were pigmented with amorphous or spherical pigments.

It is important to note that all coating barriers will eventually be breached, either through normal permeation or through defects such as pinholes or scratches. If the coating is intended for atmospheric service, the corrosion protection can be bolstered by the addition of small amounts of inhibitive pigments. However, this would not be a good idea for coatings intended for immersion service.

Inhibitive pigments function by reacting with the steel substrate to produce extremely thin films that serve to passivate the metal surface. Passivation refers to a lowering of the rate of metal ion production. This can be accomplished by forming a physical barrier, or by forming an 'electrochemical barrier' by reducing the potential difference between adjacent cathodes and anodes on the steel surface.

The most common example of a physical passivation barrier is the extremely thin layer of aluminium oxide that forms on aluminium sheet under normal atmospheric conditions. This is essentially a layer of aluminium 'rust' which forms extremely rapidly, and which essentially hinders the formation of any additional aluminium corrosion product by sealing off the underlying surface from additional oxygen.

There are a variety of inhibitive pigments for use in primers over steel. These include chromates, lead-based pigments (including lead chromate), zinc phosphate, zinc hydroxy phosphite, calcium borosilicate, barium metaborate and various molybdate pigments.

Chromate and lead-containing pigments are seldom used today because of their toxicity. Two of the more common lead-based pigments are red lead (Pb_3O_4) and basic lead silicochromate. The latter is produced by the reaction of litharge (PbO), silica and chromic acid at high temperatures. This results in a combination of tribasic lead silicate and monobasic lead chromate on a silicon core, with a typical composition as follows [6]:

$3PbO{\cdot}PbSiO_3$, 25%
$PbO{\cdot}PbCrO_4$, 29%
SiO_2, 46%.

Three of the more common chromate-based inhibitive pigments include zinc potassium chromate, also called zinc chromate ($4ZnO{\cdot}K_2O{\cdot}4CrO_3{\cdot}3H_2O$); zinc tetraoxychromate, also called basic zinc chromate ($ZnCrO_4{\cdot}4Zn(OH)_2$); and strontium chromate ($SrCrO_4$). Unfortunately, all of these contain hexavalent chromium and are therefore not as popular as they once were.

There are several lead- and chromate-free inhibitive pigments on the market, with one of the more common being calcium borosilicate, a mixture of CaO, B_2O_3 and SiO. This is a basic pigment with a water solubility of $3.5\,g\,l^{-1}$. However, because of its basic nature, it should not be used in acidic vehicles. It is a white pigment of low hiding power and can be formulated into a wide range of colours.

Zinc phosphate ($Zn_3(PO_4)_2{\cdot}2H_2O$) is an anti-corrosive pigment of very low reactivity, thus making it compatible with both basic and acidic binders. It is distinguished from most inhibitive pigments by having a platy, or lamellar, structure, thus allowing it to also function as a barrier pigment. There is some debate as to whether zinc phosphate actually functions as an inhibitive pigment, since its solubility in water is only 0.0008 g l^{-1}. For a pigment to be inhibitive, it must be slightly water soluble in order to react with the substrate and form a protective, passivating film.

Zinc hydroxy phosphite (at 0.04 g l^{-1}) is considerably more soluble than zinc phosphate, and is used in both solvent- and water-borne primers. It has low tint strength, allowing it to be formulated into numerous colours. It is thought to reduce corrosion by forming a passivating film of iron phosphites and phosphates, as well as neutralizing various acidic corrosive species as a consequence of its basic nature [6].

Barium metaborate ($BaB_2O_4{\cdot}H_2O$), due to its excessive water solubility, is modified with silica in order to make it a suitable inhibitive pigment. It is highly alkaline, and is not useful in acidic binders or in coatings which are cured by acid catalysts. It suppresses corrosion by fostering an alkaline environment. The metaborate ion may also passivate the anodic areas of the substrate.

Molybdate pigments include basic zinc molybdate, basic calcium zinc molybdate and zinc molybdenum phosphate. It is thought that they function by producing a thin film of iron (III) molybdate on the steel surface, which is insoluble at neutral and basic pH [7].

Another way of interfering with the corrosion reaction is to alter the corrosion cell such that some species other than iron becomes the anode. This is the basis of cathodic, or sacrificial, protection, and is achieved by using primers containing very high levels of metallic zinc dust. These coatings are referred to as zinc-rich primers.

Zinc is a more active metal than iron. When the two are in electrical contact in the presence of oxygen and an electrolyte, zinc becomes the anode and hence oxidizes, while iron is the cathode. Not only is the steel protected electrochemically, but the white zinc corrosion products quickly form a protective film over the steel, thus providing barrier protection as well.

Zinc-rich primers typically have 70–90 wt% metallic zinc dust pigment and are often formulated by using special inorganic binders based on silicates. However, urethane and epoxy zinc-rich primers also exist and are referred to as organic zinc-rich primers. The zinc loading must be above the *CPVC* in order for the zinc particles to achieve electrical contact with the steel.

3.4 ORGANIC PIGMENTS

In contrast to inorganic pigments, the role of organic pigments is primarily that of aesthetics, with little contribution to the physical properties of the film. Some organic pigments are partially soluble in the coating's solvent systems, but most are dispersed as minute particles. The organic pigments span a much broader range of hues than do the inorganic pigments and typically give brighter, cleaner colours. Other than their colour or hue, the important properties of organic pigments include their tinting strength and their resistance to fading (lightfastness).

There are a very large number of organic pigments on the market. A few of the more common ones will be briefly discussed in the following.

3.4.1 Organic Red Pigments

Three of the most common organic red pigments include barium lithol red, BON red and toluidine red.

Barium lithol red (Figure 3.1) falls into the general category of a metallized azo red. The azo group (–N=N–) is known as a chromophore. Chromophores are the groups within a molecule which absorb visible light, thus imparting colour.

Most, if not all, of the organic pigments have relatively complicated molecular structures, and most also incorporate benzene rings and are therefore considered aromatic pigments. Two fused benzene rings are referred to as a napthalene structure.

Barium lithol red is primarily used in the ink industry due to its relatively poor lightfastness and poor resistance to both acids and alkalis. This pigment is bright red in colour and is suitable for interior use, or occasionally at high concentrations (masstones) in exterior applications where lightfastness is not critical.

Calcium BON red (Figure 3.2) is a very clean, bright red pigment. Simply substituting manganese for calcium results in 'blue shades of red', which have better lightfastness than the calcium variety. Manganese BON red is suitable for exterior exposure.

Toluidine red (Figure 3.3) is one of the more popular non-metallized azo reds. Because non-metallized azo pigments do not have a metal cation that can be displaced by acids and alkalis, they have better chemical resistance than their metallized counterparts. Toluidine red is a bright red of moderate lightfastness and good hiding power. In tints, however, the lightfastness declines, so it is usually used at high strength for farm or lawn equipment.

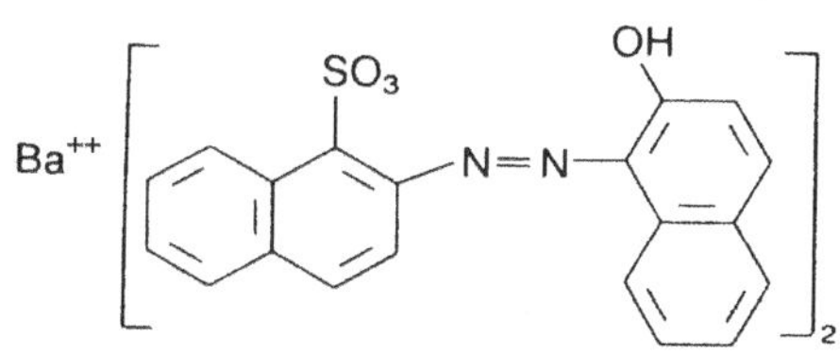

Figure 3.1 Structure of barium lithol red.

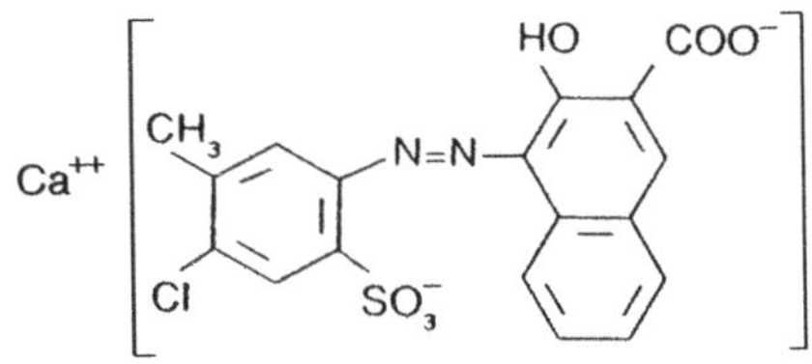

Figure 3.2 Structure of calcium BON red.

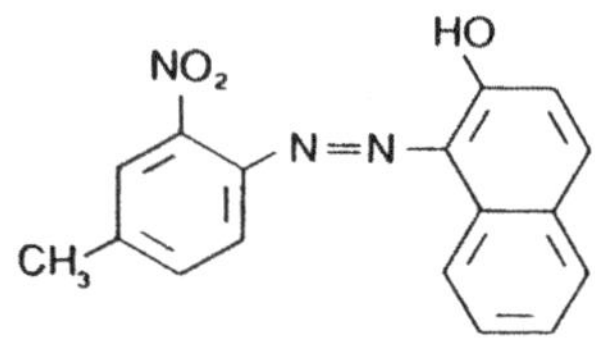

Figure 3.3 Structure of toluidine red.

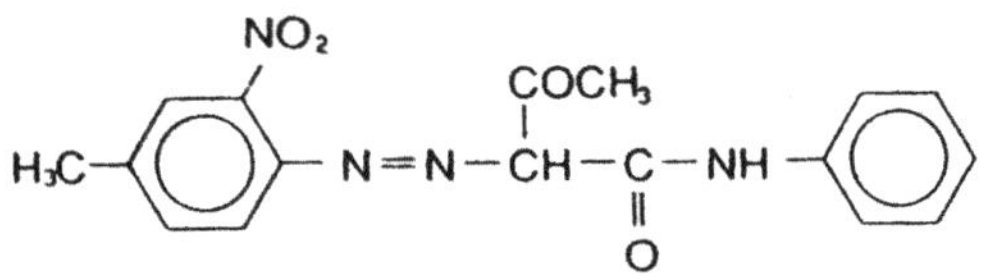

Figure 3.4 Structure of Pigment Yellow 1.

The red pigments discussed above would be considered low-to-medium-grade organic reds, and could serve many uses. For critical applications where appearance and longevity are primary concerns, such as automotive paints, a number of high-performance red pigments exist. These include the quinacridone reds and those based on anthraquinone.

3.4.2 Organic Yellow Pigments

The four main classes of organic yellow pigments are monoarylide yellows, diarylide yellows, benzimidazolone yellows and heterocyclic yellows.

Pigment Yellow 1 (Hansa Yellow 6) (Figure 3.4) is a bright monoarylide yellow often used in trade sales and emulsion paints. However, it does have a tendency to bleed and has poor lightfastness in tints.

Several diarylide yellow pigments are on the market. They are of low cost and have reasonable heat and chemical resistance, and improved bleed resistance

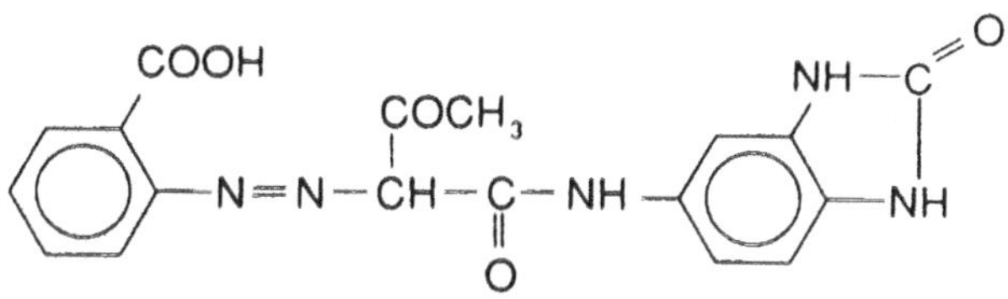

Figure 3.5 Structure of Pigment Yellow 151.

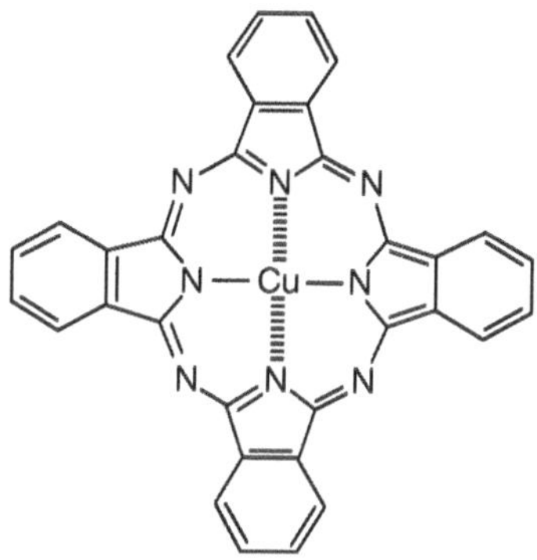

Figure 3.6 Structure of copper phthalocyanine blue.

when compared to the monoarylide yellows. However, they have poor lightfastness.

Pigment Yellow 151 (Figure 3.5) is an example of a benzimidazolone yellow pigment. These pigments have very good heat resistance, good solvent resistance (little tendency to 'bleed' into other coats of paint) and very good lightfastness. These properties make them a popular choice for high-quality industrial coatings.

There are numerous heterocylic yellow pigments of widely varying complex structures. They are high-cost, high-performance yellow pigments with excellent lightfastness.

3.4.3 Organic Blue Pigments

The most common organic blue pigment in the coatings industry is copper phthalocyanine blue (Figure 3.6). This is a bright, versatile pigment. There are numerous grades of such pigments, which vary in hue primarily as a result of different crystal sizes and structures. The pigments have good tint strength and excellent lightfastness.

3.4.4 Organic Green Pigments

The most common organic green pigments are very similar structurally to copper phthalocyanine blue and are called phthalocyanine greens. They differ from copper phthalocyanine blue principally by replacing many of the hydrogen atoms on the six-membered aromatic rings with either chlorine or bromine atoms. The pigments can go from a blue green to a yellow green, depending on the ratio of bromine to chlorine.

The phthalocyanine greens are economical and have good lightfastness. They also possess excellent resistance to heat and solvents.

REFERENCES

1. J. Boxall and J. A. Von Fraunhofer, *Paint Formulation, Principles and Practice*, Industrial Press, New York, 1981, pp. 16–18.
2. C. H. Hare, *Protective Coatings*, Technology Publishing Co., Pittsburgh, PA, USA, 1994, pp. 321, 322.
3. C. W. Fuller, 'Colored Iron Oxide Pigments, Synthetic', in *Pigment Handbook*, Vol. I, T. C. Patton (Ed.), John Wiley, New York, 1973, p. 333.
4. W. J. Polestak and T. D. Thompson, *J. Coat. Technol.*, **60**, 31 (1988).
5. O. M. Considine and G. D. Considine (Eds), *Van Nostrand Reinhold Encylcopedia of Chemistry*, 4th Edn, Van Nostrand Reinhold, New York, 1984, p. 294.
6. A. Smith, *Inorganic Primer Pigments*, Federation of Societies for Coatings Technology, Blue Bell, PA, USA, 1988, pp. 13–16.
7. M. S. Vukasovitch and J. P. G. Farr, *Mater. Perform.*, **25**, 9 (1986).

4

Additives and Solvents

A basic understanding of solvents is important in order to appreciate some of the failure modes encountered with protective coatings. For diagnosing certain types of defects, an understanding of the role of additives is also important, although problems due to additives, or the lack thereof, are more commonly of the aesthetic variety (craters, sagging, etc.) as opposed to actual performance deficiencies.

4.1 ADDITIVES

Modern coatings are complicated systems where a wide variety of both inorganic and organic materials of widely varying molecular weight and physical properties are blended together and expected to act as a homogeneous whole. Since this is almost too much to expect, a number of additives exist which can be viewed as molecular 'band-aids' to make up for some of the irritating limitations of the natural laws of physics and chemistry. These additives are usually added at small (less than 1%) levels, but can have a significant impact on the coating system.

The following is a brief summary of additives based on the properties that they confer to a coating or on the type of problems that they are designed to prevent.

4.1.1 Anti-settling Agents

Pigments, particularly some of the inorganic ones such as barytes, have much higher densities than the resins and solvents in which they are dispersed. Consequently, problems can develop with hard settling of these pigments, making re-incorporation into the body of the coating a difficult task. For pigments to be

Failure Analysis of Paints and Coatings, Revised Edition Dwight G. Weldon

effective, and for coatings to be processed economically, the pigments must be readily dispersed in the resin/solvent system of the liquid coating. Large agglomerations of pigment must be broken down, and air and moisture must be displaced so that the pigment particles can be wetted by the resin/solvent system. This process is sometimes more difficult in water-based coatings than in solvent-based ones.

Several anti-settling agents have been developed to alleviate this problem. Many of their formulas are proprietary, although some are waxes, chemically treated clays (silicates) or low-molecular-weight polyethylenes. Such materials can both change the surface tension of the pigmented particle and, by increasing its effective volume, also reduce its density.

The agents used to disperse pigments are usually surface-active agents, or surfactants. These function by modifying the surface chemistry, and hence surface tension, of the pigment particles such that the surface tension of the pigment more closely matches that of the resin/solvent phase in which it is dispersed. Structurally, they include such compounds as sulfonated oils, fatty acid salts, quarternary ammonium compounds and non-ionic surfactants such as ethoxylated castor oil.

4.1.2 Viscosity Modifiers

It is possible to formulate a coating with excellent protective properties that will not sell in the marketplace because it is too difficult to apply, either because it is too viscous or because it is too thin and tends to sag or run. The problem here is one of viscosity, or more precisely, the relationship between viscosity and shear.

Viscosity falls under the science of rheology, or the study of the flow of materials under stress. The mathematical definition of viscosity is the ratio of shear stress to that of shear rate. Shear stress can be envisioned as the force per unit area pulling one imaginary layer of fluid across another. Shear rate is the velocity at which these two layers move relative to one another. We can therefore write the following:

$$\text{viscosity} = \frac{\text{shear stress}}{\text{shear rate}}. \tag{4.1}$$

Some liquids, such as water, have a nearly constant viscosity, such that as one or the other of the above quantities change, the other changes in the same proportion, and the viscosity stays the same. Such liquids are said to be Newtonian in behaviour.

Most coatings, however, display non-Newtonian behaviour, and their viscosities are different at different rates of shear. This is probably fortunate for coatings formulators.

From an application point of view, a house paint should have a high viscosity on setting (very low shear rate) since this will reduce the tendency of pigment particles to settle to the bottom of the can during storage. As the shear rate increases, such as during stirring of the paint or applying it by brush or spray, it is desirable for the viscosity to decrease. This facilitates mixing of the ingredients and improves application properties – the coating will not drag or snap back from the brush, and it will flow out enough to eliminate brush marks. As the shear rate decreases shortly after application, the viscosity will begin to increase such that excessive flow – which would result in sags and runs – does not occur. Liquids that show a reduced viscosity at high shear rates are called thixotropic, and agents that impact this characteristic are referred to as thixotropic agents or thixotropes.

Several compounds are available which can increase viscosity or impart thixotropy. These include cellulose ethers, micronized silica and many natural or modified clays (bentonites).

4.1.3 Surfactants and Emulsifying Agents

Many of the additives already mentioned are, technically speaking, surfactants. A surfactant is a compound that alters the surface tension of a liquid or solid. Many additives, such as the dispersing agents described above, are really just surfactants, but are more conveniently classified based on their specific end use, rather than on their chemical properties.

There are specific additives termed either surfactants or emulsifying agents, formulated for the express purpose of altering the surface tension of resins or polymers such that they become compatible with water. Such surfactants are crucial in the formulation of stable water-based latex or emulsion paints. These materials generally consist of a fairly long molecule, where one end is hydrophobic and the other end is hydrophilic. The hydrophobic end (usually a long hydrocarbon tail) is compatible with and orients itself towards the resin, while the hydrophilic end orients itself outward, that is, towards the water phase. This hydrophilic sphere or shield around the resin makes it compatible with water.

Surfactants or emulsifying agents fall into one of three categories, namely, anionic surfactants, cationic surfactants and non-ionic surfactants.

As the name implies, anionic surfactants ionize in solution to produce a negatively charged group on the hydrophilic end. Common anionic surfactants include the salts of fatty acids, such as palmitic or stearic acid, and molecules which have sulfate groups.

Cationic surfactants ionize in solution to produce a positively charged end group. They are often based on amine salts and quarternary ammonium compounds.

Non-ionic surfactants do not form ions in solution. As with the ionic surfactants, the hydrophobic end usually consists of a fairly long hydrocarbon chain. The hydrophilic end consists of several relatively polar organic moieties, such as hydroxyl or ether groups. The non-ionic surfactants do not usually have the emulsifying power of the ionic surfactants, and sometimes the hydrophilic end must be physically longer when compared with the ionic surfactants.

4.1.4 De-foaming and Anti-foaming Agents

The production of water-based coatings, and to a lesser extent solvent-based coatings, is sometimes accompanied by the formation of excessive foaming or air bubbles. Foam can result in under-filled cans, or in craters in the applied film. De-foaming and anti-foaming agents are specially formulated surfactants that reduce surface tension and break up the formation of foams.

4.1.5 Driers

Coatings that ultimately cure by air oxidation, such as alkyds, are usually formulated with catalysts known as driers to speed up this process. Many of these driers are organometallic compounds such as cobalt and manganese napthenates. Some, such as the cobalt complexes, act primarily on the surface of the coating and are referred to as 'surface driers'. Others aid in the curing of the bulk of the coating and are referred to as 'through driers'. The latter include compounds of lead or zirconium. An imbalance in the selection of driers can cause wrinkling if the surface of the coating dries significantly faster than its interior.

4.1.6 Plasticizers

Plasticizers are additives that increase the flexibility of coating films. They are usually relatively low-molecular-weight, non-volatile organic compounds, although some resins, such as nitrocellulose, can be plasticized by flexible, high-molecular-weight resins such as coconut oil alkyds. While most additives are used at levels of 1% or less, plasticizer contents may range from 4 or 5% to as high as 50%.

For plasticizers to be effective, they must be compatible with the resin system, non-volatile and must not impart undesirable properties such as colour formation. There are usually optimum plasticizer levels below which the desired degree of flexibility has not been achieved, and beyond which a reduction in physical properties occurs. It is possible for too much plasticizer to reduce such properties as the tensile strength, hardness, permeability, chemical resistance and adhesion.

There are a wide variety of plasticizers available. Common types include phthalates such as dioctyl phthalate, chlorinated paraffins, organic phosphates such as tricresyl phosphate and non-drying fatty oils or the alkyds formulated from them.

4.1.7 Ultraviolet Stabilizers

Although ultraviolet light makes up only about 6% of the radiant energy of sunlight at the earth's surface, it is one of the major factors responsible for the degradation of coatings.

The energy of light, or of any electromagnetic radiation, is directly related to its frequency, according to the following:

$$E = h\nu, \tag{4.2}$$

where E is energy; ν is the frequency of light in cycles per second (or hertz), and h is a fundamental constant known as the Planck constant. The latter has a value of 6.626076×10^{-34} J-s. Since the frequency of light is inversely proportional to its wavelength, it is also correct to say that lower-wavelength light is more energetic than longer-wavelength light.

Because of the screening effect of the earth's atmosphere, much of the short-wavelength radiation emitted from the sun never reaches the surface. The lowest-wavelength (highest-energy) light reaching the surface has a wavelength of approximately 300 nm, which falls into what is called the B region of the ultraviolet spectrum (see earlier). Radiation in the B region is, however, still energetic enough to break the covalent bonds of many coating materials.

The additives used to protect coatings from ultraviolet light fall into two classes, based on their mechanism of protection, that is, ultraviolet absorbers and ultraviolet stabilizers.

Ultraviolet absorbers function in much the same way as sunscreen lotions – they absorb ultraviolet light before it can do any damage. These additives are usually complex, substituted aromatic compounds such as benzophenones. Once they absorb the ultraviolet light, they then convert it into heat energy, which is harmlessly dissipated by the coating.

Ultraviolet stabilizers function by an entirely different mechanism. Instead of absorbing the ultraviolet light, they protect the coating binder by scavenging free radicals. Since the production of free radicals is one of the first steps in the degradation of coatings by ultraviolet light, stabilizers interrupt the mechanism of degradation. They are often complex amine compounds known as hindered amine light stabilizers (HALS). It is common to employ a combination of an absorber and a stabilizer to achieve optimum protection from ultraviolet light.

4.1.8 Anti-skinning Agents

Coatings that cure by air oxidation, such as alkyds and other oil-based paints, can sometimes skin over in the can as the surface partially oxidizes and cures. In order to prevent this, additives known as anti-skinning agents are used. These are relatively mild, volatile antioxidants such as methyl ethyl ketoxime and butyraldehyde oxime. Strong, non-volatile antioxidants, such as the ultraviolet stabilizers, cannot be used because they could actually inhibit the normal drying, or curing, of the coating. By using small amounts of volatile oxidizing agents, skinning in the can is prevented, while cure is not retarded because the anti-oxidants quickly evaporate during solvent evaporation.

4.1.9 Biocides

Fungi and other microorganisms can prove detrimental to certain types of coatings. Water-based coatings can be subject to the growth of microorganisms while in the can, resulting in objectionable odours and even changes in the rheological properties of the coating. Certain biocides can prevent this by interfering with the organism's metabolism, thus preventing the synthesis of the enzymes that are actually responsible for the degradation of the coating [1]. Currently used biocides include complex phenols, formaldehyde compounds and substituted oxazolidines.

Although solvent-based coatings are not susceptible to biological deterioration while in the can, coatings containing natural products such as linseed oil can be attacked by the growth of fungi, such as mildew, once they have been applied. This usually results only in appearance problems, but occasionally inter-coat delamination can result from excessive mildew growth.

There are several additives and some pigments which are effective in controlling fungal growth. Zinc oxide and barium metaborate are two pigments with mildewcidal properties, although to be effective, relatively high (20–40%) levels must be used. These levels can be reduced if other, more efficient, mildewcides are used.

For many years, phenyl mercury and tributyl tin compounds have been used as mildewcides. These are toxic to humans, however, and are therefore no longer used. The phenyl mercury compounds tended to turn black in sulfur-containing environments. Currently used mildewcides include carbamates, benzothiols and other complicated organic compounds.

4.1.10 Flow-Modifying Agents

There are several appearance defects in coatings that are related to their ability to wet the substrate and to subsequently flow or level adequately. Some of these defects, such as brush marks or runs and sags, have been previously discussed and can often be addressed by the incorporation of viscosity modifiers. Other defects related to flow and levelling problems include 'orange peel', floating, crawling and cratering. Such defects require a different type of remedial action.

Orange peel is an objectionable appearance due to shallow surface wrinkling, thus resulting in a texture similar to that of the skin of an orange. This defect is typically caused by a too rapid increase in viscosity, which does not allow the coating sufficient time to flow out and eliminate imperfections. Sometimes, it can be cured simply by adjusting the solvent blend. Other times, it can be reduced by using flow-control additives, such as those based on polyether-modified alkylpolysiloxanes.

Floating is an effect whereby the various pigments in a coating segregate into small, localized areas. This results in a mottled or blotchy appearance. Thixotropic additives that reduce the mobility of the pigment can often alleviate this problem. Another approach is the use of low-surface-energy additives such as silicone fluids.

Crawling is a de-wetting phenomenon where the coating pulls away from areas on the substrate which have low surface tension, where the latter is usually the result of a poorly cleaned or contaminated surface. Cleaning the surface is the most desirable means of solving crawling. Low-surface-tension additives, such as silicones and certain other surfactants, can mitigate this problem.

Cratering is crawling on a minute level. This is usually caused by a small speck of a low-surface-energy material that could be a contaminant. Sometimes, it is caused by seeds or gel particles in the coating itself. Again, surface-tension modifiers such as silicones can alleviate cratering, so can good application practices such as thorough cleaning of the substrate and perhaps filtering of the coating.

However, there can be some unwanted side effects with indiscriminate use of low-energy-flow additives such as silicones. While they can frequently cure the immediate problem, they can often make re-coating difficult by causing excessive cratering in subsequently applied topcoats.

4.2 SOLVENTS

In order for a coating to be applied by conventional methods such as brush, roller or spray, it must have a reasonable viscosity. One of the simplest and most effective ways to control viscosity is through the use of solvents. Not only do solvents aid in viscosity adjustment, but they can also strongly influence the appearance of the dried film, and to some extent the performance properties as well. Usually, solvents are thought of as low-molecular-weight organic compounds, but it should be pointed out that the major solvent in a latex coating is water.

The wrong choice of solvent can result in both coating defects (appearance or cosmetic deficiencies) and coating failures (a breakdown in protective properties). Too low a viscosity can result in objectionable sags and runs, and in a reduced film build at edges and vertical surfaces. In turn, the reduced film build may lead to failures such as premature corrosion. Too high a viscosity may result in defects such as orange peel and may hinder a coating's ability to wet the substrate, thus resulting in a catastrophic adhesion failure.

One of the primary properties of a solvent is its effect on viscosity. A somewhat related property is its evaporation rate. The concepts of viscosity and evaporation rate are related because the amount and type of solvent remaining in the applied film affects the viscosity of the drying coating. If the evaporation rate is too slow, drying times will be extended and defects such as sagging may occur. Conversely, if the evaporation rate is too fast, the film may not flow adequately and orange peel defects may result.

There are other consequences to a poor choice of solvents. In oxidizing coatings that cure very slowly, such as alkyds, there must be a relationship between the evaporation rate of the solvent and the drying or curing of the surface of the coating. If this surface dries and cures too fast compared to the release of solvent from the interior of the coating, wrinkling can result. The proper choice of driers also influences wrinkling of alkyds and oil-based coatings.

The presence of large numbers of voids in the interior of a coating can greatly reduce its ability to function as a protective barrier. Sometimes, voids are caused by air entrapment and sometimes by a poor choice of solvent. If the viscosity of the coating increases at a much greater rate than the evaporation of the solvent, relatively large quantities of solvent are trapped in the viscous coating, which can result in the formation of voids. Oddly, this is usually more of a problem with volatile solvents due to their higher vapour pressures. A slowly evaporating solvent may become more readily trapped in a curing coating, but its vapour pressure is low enough that instead of 'boiling off' and causing voids, it simply slowly migrates out of the film over time.

One particular aspect of solvents that needs to be closely considered, at least for coatings intended for immersion service, is their water solubility. The presence of a water-soluble solvent in a coating which is immersed in water can result in blistering due to an osmotic process (see Section 2.2.5).

Some solvents can actually react with components of the coating, thus seriously hindering the curing process and the development of physical properties. Two-component urethanes rely on a reaction between a hydroxyl functional component (a polyol) and an isocyanate. If an alcohol were used as a solvent in such a system, its hydroxyl group would also react with the isocyanate, thus interfering with the normal curing mechanism. A mistake like this would almost never be made by the formulator, but could be caused by a contractor adding the wrong thinner at a job site.

Apart from such practical considerations as cost and toxicity, solvents are generally selected based on their ability to efficiently dissolve the resins in the coating and to evaporate at the proper rate. In some instances, formulators rely on the old rule of thumb that 'like dissolves like', but a somewhat more quantitative approach involves the use of solubility parameters.

The solubility parameter is generally considered to be a measure of the attractive forces between molecules of a particular solvent. This parameter is defined as the square root of the energy of vaporization of the solvent, divided by its molar volume.

The usefulness of such a concept is that resins are most easily dissolved or are compatible with solvents that have solubility parameters similar to the solubility parameter of the resin itself. While the solubility parameter of a solvent can be accurately measured, approximate solubility parameters for resins can only be determined indirectly. The solubility parameters of many solvents and solvent blends are listed in ASTM D3132 (Standard Test Method for Solubility Range of Resins and Polymers) and are shown in Table 4.1.

In addition to solubility parameters, Burell [2] recognized the importance of hydrogen bonding in solubility, and subdivided solvents and resins into three hydrogen bonding categories, that is, poor, moderate and strong. Hydrogen bonding refers to a molecule's ability to 'share' a hydrogen atom with electronegative groups on a neighbouring molecule. Water is perhaps the ultimate hydrogen bonding solvent, as illustrated in Figure 4.1.

Not too surprisingly, alcohols fall into the strong hydrogen bonding category (*n*-butanol has a hydrogen bonding parameter of 18.7), while simple alkanes fall into the poor hydrogen bonding category (heptane has a hydrogen bonding parameter of 0). Table 4.1 presents two values for the hydrogen bonding parameters because there are two methods for calculating it.

Dipole moments – a measure of localized electron density – can also play a role in solubility, but are seldom considered as much as solubility parameters and hydrogen bonding in this context.

Coating formulators sometimes refer to 'true solvents' and 'diluents'. A true solvent actually dissolves the resin. Its solubility parameter and hydrogen bonding parameters are similar to that of the resin. A diluent does not actually dissolve the resin, but, when present along with a true solvent, can serve to reduce the viscosity of the system.

Solubility parameters are additive. An interesting consequence of this is the possibility of blending two diluents together to obtain a solubility parameter which matches that of the resin. The blended diluents now constitute a true solvent for that particular resin. Whether anything practical is gained by this is problematical. One potential downfall is that unless both diluents have the same evaporation rate, the solubility parameter of the blend will change as the more volatile diluent evaporates. The remaining blend may not have enough solvency to keep the resin in solution, with undesirable consequences. The evaporation rates of various solvents are shown in Table 4.2. The rates shown here are relative to that of *n*-butyl acetate, which is assigned a value of 1.0.

Not counting water, most modern solvents fall into one of two broad categories, that is, hydrocarbon solvents and oxygenated solvents.

Hydrocarbon solvents fall into four basic subcategories, namely, aliphatic hydrocarbons, aromatic hydrocarbons, chlorinated hydrocarbons and terpenes (which are rarely used anymore).

Aliphatic hydrocarbon solvents are usually produced by the distillation of crude oil, and are subsequently treated to improve odour and colour. Common aliphatic hydrocarbon solvents include mineral spirits, hexane and VM&P naphtha. They are commonly used for alkyds and other oil-based coatings.

Aromatic hydrocarbon solvents also originate from the distillation of petroleum. The major aromatic solvents include toluene, xylene and various aromatic naphthas. The aromatic hydrocarbon solvents are stronger solvents than the

Table 4.1 Solubility parameters of a selection of solvents and solvent mixtures (blends), from Standard Test Method for Solubility Range of Resins and Polymers, ASTM D3132, 1990. Copyright American Society for Testing and Materials (ASTM). Reprinted with permission of the ASTM.

Solvent or solvent mixture[a]	Solubility parameter, δ	Hydrogen bonding γ^b	Hydrogen bonding γ^c	Dipole moment,[d] μ
Diisopropyl ether	6.9	11.7	6.6	1.3
n-Pentane	7.0	0.0	2.2	0.0
n-Heptane	7.3	0.0	2.2	0.0
50% Diisobutyl ketone/50% diisopropyl ether	7.4	10.1	5.9	2.0
Diethyl ether	7.4	13.0	6.9	1.2
50% *n*-Heptane/50% diisobutyl ketone	7.6	4.2	3.7	1.4
25 *n*-Heptane/75% diisobutyl ketone	7.7	6.3	4.5	2.0
Methyl cyclohexane	7.8	0.0	2.2	0.0
66.7% *n*-Heptane/33.3% *n*-butyl acetate	7.8	2.9	3.3	0.6
Diisobutyl ketone	7.8	8.4	5.2	2.7
57% Diethyl ether/43% *n*-butyl acetate	7.9	11.2	6.3	1.5
66.7% Diethyl ether/33.3% 2-ethylhexanol	8.1	14.9	7.6	1.4
Cyclohexane	8.2	0.0	2.2	0.0
40% *n*-Heptane/60% *n*-butyl acetate	8.2	5.3	4.1	1.1
71% Cyclohexane/29% toluene	8.4	1.3	2.7	0.1
n-Butyl acetate	8.5	8.8	5.4	1.9
40% Cyclohexane/60% toluene	8.6	2.7	3.2	0.2
50% Cyclohexane/50% benzene	8.7	0.0	2.2	0.0
50% *n*-Butyl acetate/50% toluene	8.7	6.7	4.6	1.2
50% EGMBE[e]/50% *n*-butyl acetate	8.7	10.9	6.2	1.8
33.3% Diethyl ether/66.7% 2-ethylhexanol	8.8	16.8	8.2	1.5
Toluene	8.9	4.5	3.8	0.4
2-Butoxyethanol	8.9	13.0	6.9	1.6
Benzene	9.2	0.0	2.2	0.0
Propylene oxide	9.2	10.0	5.8	1.8
56% EGMBE[e]/44% 2-ethylhexanol	9.2	15.5	7.8	1.6
50% Methylene chloride/50% toluene	9.3	3.0	3.3	1.0
50% Toluene/50% dioxane	9.4	7.1	4.8	0.4
2-Ethylhexanol	9.5	18.7	8.9	1.7
Methyl acetate	9.6	8.4	5.2	1.7
Methylene chloride	9.7	1.5	2.7	1.5
66.7% Methylene chloride/33.3% dioxane	9.8	4.2	3.7	1.1
50% Dioxane/50% methylene chloride	9.8	5.6	4.2	1.0
Dioxane	9.9	9.7	5.7	0.4
Cyclohexanone	9.9	11.7	6.4	2.7
2-Ethoxyethanol	9.9	13.0	6.9	1.6
75% EGMEE[f]/25% methyl isobutyl carbinol	9.9	14.4	7.4	1.6
Carbon disulfide	10.0	0.0	2.2	0.0
30% EGMEE[f]/70% methyl isobutyl carbinol	10.0	17.0	8.2	1.7
Methyl isobutyl carbinol	10.0	18.7	8.8	1.7

Table 4.1 *Continued*

Solvent or solvent mixture[a]	Solubility parameter, δ	Hydrogen bonding γ^b	Hydrogen bonding γ^c	Dipole moment,[d] μ
66.7% Methylene chloride/33.3% nitroethane	10.2	1.8	2.8	2.2
66.7% Dixane/33.3% nitroethane	10.3	7.3	4.8	1.5
50% Toluene/50% acetonitrile	10.4	5.4	4.2	2.2
50% EGMEE/50% *n*-amyl alcohol	10.4	15.9	7.9	1.7
2-Ethylbutanol	10.5	18.7	8.9	1.7
33.3% Methylene chloride/66.7% nitroethane	10.6	2.2	3.0	2.9
75% Nitroethane/25% dioxane	10.8	4.3	3.8	2.8
75% Dioxane/25% propylene carbonate	10.8	8.5	5.3	1.6
Dimethylacetamide	10.8	12.3	6.6	3.8
80% EGMEE[f]/20% methanol	10.8	14.1	7.3	1.6
33.3% EGMEE[f]/66.7% *n*-butanol	10.9	16.8	8.2	1.7
n-Amyl alcohol (pentanol)	10.9	18.7	8.9	1.7
48% Benzene/52% nitromethane	11.0	1.3	2.7	1.8
50% Dioxane/50% dimethylformamide	11.0	10.7	5.6	2.1
Nitroethane	11.1	2.5	3.1	3.6
25% *n*-Butyl acetate/75% acetonitrile	11.1	6.9	4.5	3.4
n-Butanol	11.4	18.7	8.9	1.7
50% Nitroethane/50% acetonitrile	11.5	4.4	3.8	3.8
65% EGMEE[f]/35% methanol	11.5	15.0	7.6	1.6
66.7% *n*-Butanol/33.3% dimethylformamide	11.6	16.4	8.2	2.4
50% Dioxane/50% dimethyl sulfoxide	11.7	8.7	5.4	2.2
50% Nitroethane/50% nitromethane	11.9	2.5	3.1	3.5
Acetonitrile	11.9	6.3	4.5	3.9
30% *n*-Butanol/70% dimethylformamide	11.9	13.8	7.2	3.2
n-Propanol	11.9	18.7	8.9	1.7
70% Acetonitrile/30% dimethylformamide	12.0	8.0	5.1	3.9
68% Dimethylformamide/32% acetonitrile	12.0	10.0	5.8	3.8
Dimethylformamide	12.1	11.7	6.4	3.8
50% *n*-Propanol/50% ethanol	12.3	18.7	8.9	1.7
60% Nitromethane/40% acetonitrile	12.4	4.0	3.7	3.6
50% Ethanol/50% dimethylformamide	12.4	15.2	7.7	2.8
23% Dioxane/77% propylene carbonate	12.5	6.0	4.4	3.9
40% Dimethylformamide/60% dimethyl sulfoxide	12.6	9.3	5.6	3.9
70% Ethanol/30% dimethylformamide	12.6	16.6	8.2	2.3
80% Dimethylformamide/20% methanol	12.6	13.1	6.9	3.4
Nitromethane	12.7	2.5	3.1	3.4
Ethanol	12.8	18.7	8.9	1.7
72% Dimethyl sulfoxide/28% ethanol	12.9	10.8	6.1	3.4
Dimethyl sulfoxide	13.0	7.7	5.0	4.0
46% Methanol/54% dimethylformamide	13.2	14.9	7.6	2.8
Propylene carbonate	13.3	4.9	4.0	5.0
80% Dimethyl sulfoxide/20% methanol	13.3	9.9	5.8	3.5

Table 4.1 *Continued*

Solvent or solvent mixture[a]	Solubility parameter, δ	Hydrogen bonding γ^b	Hydrogen bonding γ^c	Dipole moment,[d] μ
70% Dimethylformamide/30% monomethylformamide	13.3	11.8	7.0	3.8
50% Ethanol/50% methanol	13.6	18.7	8.9	1.7
66.7 Methanol/33.3% dimethylformamide	13.7	16.4	8.1	2.4
57% Methanol/43% dimethyl sulfoxide	13.9	14.0	7.8	2.7
Methanol	14.5	18.7	8.9	1.7
Propylene glycol	15.0	20.6	9.4	2.2
63% Methanol/37% monomethylformamide	15.1	16.2	8.8	2.5
Ethylene glycol	17.1	20.6	9.6	2.3

[a] Mixture compositions given as vol%.
[b] Crowley, J. D. *et al.*, "A Three Dimensional Approach to Solubility," *Journal of Paint Technology*, Vol 38, No. 496, 1966, p. 269; Vol 39, No. 504, 1967, p. 19.
[c] E. I. du Pont de Nemours & Co., Bulletin PA 12-770, "Solvent Formulating Maps for Elvacite Acrylic Resins," Serial A-70562, July 1970.
[d] McLellan, A. L., *Tables of Experimental Dipole Moments*, W. H. Freeman & Co., San Francisco, 1963.
[e] EGMBE, ethylene glycol monobutyl ether (2-butoxyethanol).
[f] EGMEE, ethylene glycol monoethyl ether.

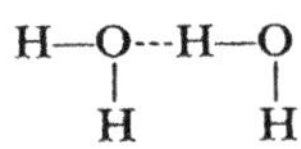

Figure 4.1 Illustration of hydrogen bonding in water.

aliphatic hydrocarbon solvents and can dissolve a greater variety of resins. They are used in a variety of coatings such as alkyds, acrylics, silicone resins, epoxies, phenolics and urethanes.

Chlorinated hydrocarbon solvents, such as methylene chloride and trichloroethane, have previously been seldom used in coatings. This has changed somewhat recently due to environmental concerns, as these solvents are considered exempt from volatile organic content (VOC) regulations.

A number of solvents fall into the family of oxygenated solvents, including ketones, alcohols, esters and glycol ethers and acetates. While it is common in the coatings industry to refer to these solvents as stronger solvents than the hydrocarbon solvents, this is only true because their solubility and hydrogen bonding parameters more closely match those of many of the commonly used polymers. A so-called 'weaker' solvent might actually prove to be a better solvent

Table 4.2 Relative evaporation rates of various solvents. Reproduced by permission of the Federation of Societies for Coatings Technology, from W. H. Ellis, *Solvents*, 1986.

Solvent	Evaporation rate[a]
Ketones	
Acetone	19.0
Methyl ethyl ketone	7.0
Methyl *n*-propyl ketone	2.7
Methyl isobutyl ketone	1.5
Methyl isoamyl ketone	0.4
Methyl *n*-amyl ketone	0.3
Diisobutyl ketone	0.2
Cyclohexanone	0.3
Esters	
Methyl acetate	1.4
Ethyl acetate	7.0
Isopropyl acetate	4.3
n-Propyl acetate	2.3
Isobutyl acetate	1.3
n-Butyl acetate	1.0[a]
n-Amyl acetate	0.4
Alcohols	
Methyl alcohol	10.0
Ethyl alcohol	4.0
Isopropyl alcohol	3.3
n-Propyl alcohol	1.5
Isobutyl alcohol	0.9
n-Butyl alcohol	0.4
Amyl alcohol	0.3
Glycol ethers	
Ethylene glycol monomethyl ether	0.6
Ethylene glycol monomethyl ether acetate	0.2
Ethylene glycol monoethyl ether	0.4
Ethylene glycol monoethyl ether acetate	0.2
Ethylene glycol monobutyl ether	0.06
Ethylene glycol monobutyl ether acetate	0.03
Propylene glycol monomethyl ether	0.1
Propylene glycol monomethyl ether acetate	0.05
Hydrocarbons	
Lacquer diluent	7.0
Toluene	2.9
VM&P naphtha	1.3
Xylene	0.7
Mineral spirits	0.4

[a]Expressed relative to *n*-butyl acetate, which has a value of 1.0.

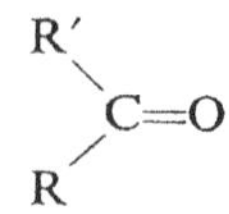

Figure 4.2 General structure of a ketone.

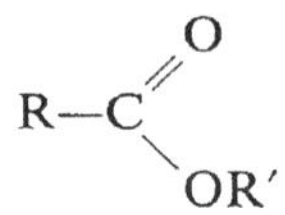

Figure 4.3 General structure of an ester.

for a certain polymer or resin, if the solubility and hydrogen bonding parameters match up well.

Ketones, of which methyl ethyl ketone (MEK) is the most common, have the general structure shown in Figure 4.2, where the R (and R′) groups are usually relatively short-chain alkyl groups. Ketones are relatively polar solvents and are often used with nitrocellulose, acrylic, vinyl, urethane and epoxy coatings. Alcohols all have a hydroxyl (—OH) group attached to a carbon atom and are usually very polar in nature. Alcohols are used on a limited basis in the coatings industry, primarily for shellac and nitrocellulose lacquers, and sometimes for the ethyl silicate vehicles used in certain inorganic zinc-rich primers.

Esters have the general structure shown in Figure 4.3. Butyl acetate is perhaps the most common ester solvent used in the coatings industry. Such solvents are usually of moderate polarity and can be used with many coating types.

Glycol ethers and glycol acetates vary in polarity from moderate to high, with some actually being water soluble. They can be used in numerous coating types and have relatively low evaporation rates. Glycol ethers contain both ether (—C—O—C—) and alcohol (C—OH) groups, while the glycol acetates are made by converting the alcohol group to an acetate group.

REFERENCES

1. J. Boxall and J. A. Von Fraunhofer, *Paint Formulation, Principles and Practice*, Industrial Press, New York, 1981, p. 31.
2. H. Burrell, *Official Digest of the Federation of Societies for Paint Technology*, **27**, (no. 369), 726 (1955).

5

Coating Types and Common Failure Modes

Previous chapters in this volume have discussed some of the basic mechanisms by which coatings can fail, as well as the basic coating ingredients such as pigments, additives and solvents. However, the single most important ingredient in a coating is its binder or resin. The type of resin or polymer is so important that this is how coatings are named and marketed. One often hears of an 'alkyd' coating or an 'epoxy' coating. In contrast, one never hears of a 'titanium dioxide' coating or a 'methyl ethyl ketone' (MEK) coating.

This chapter will describe many of the common resins or polymers used in coatings as well as their particular strengths and weaknesses. It must be noted, however, that even in a single coating family, products of widely varying physical and chemical properties can exist. One of the intriguing characteristics of polymer science is its diversity. Slight changes in the chemistry of the monomeric building blocks of an acrylic, for instance, can result in a final polymer with dramatically different characteristics. Epoxies can be formulated hard and brittle, or relatively soft and almost flexible. While it is, to some degree, necessary to talk in generalities, one should always be aware of the existence of exceptions.

5.1 NATURAL RESINS AND OILS

Natural resins and oils have been used for centuries in the manufacture of decorative and protective coatings. They are derived from natural sources such as plants, animals and fossilized remains. Although oils such as linseed oil or fish oil are certainly 'natural resins,' they are usually considered separately in most discussions of the subject. This convention will be adhered to here.

Failure Analysis of Paints and Coatings, Revised Edition Dwight G. Weldon

5.1.1 Natural Resins

Many natural resins come from the tapping of trees and are exuded from the tree in order to protect it from cuts or wounds. Many of them have a carboxyl functionality and are acidic. The acidity of a resin is often expressed as its acid number, which is the number of milligrams of potassium hydroxide required to neutralize the acid in one gram of resin. The natural resins are often of relatively high molecular weight, and in the pure state are solids at room temperature. They are often used in the production of varnish, where they are heated together with varying amounts of vegetable oils. The natural resins impart hardness, gloss, moisture resistance and improved drying time, while the vegetable oil imparts flexibility and durability.

Examples of natural resins from trees, plants and fossilized materials include elemi, copal, dammar and rosin.

Elemi is a soft resin with a melting range of 176–248 °F (80–120 °C) and a low acid number (20–35). It is derived from a tree in the Philippines, is compatible with many resins and solvents, and imparts flexibility to lacquers and varnishes.

In contrast to elemi, copal is a hard, high-molecular-weight material with melting points as high as 300 °F (149 °C) and acid numbers ranging from 50, for the New Zealand variety, to as high as 140 for deposits in Central Africa. It consists of fossilized material from various tropical trees, and is of little use unless distilled under pressure and heat. Once treated, it can produce varnishes of excellent durability and weather resistance.

Manilla copal, obtained from the Philippines or the East Indies, is not fossilized but is rather obtained from the tapping of live trees. It is somewhat softer than fossilized copal and is soluble in alcohol. Harder, fossilized varieties are also available.

Dammar is a relatively soft resin with a melting range of 158–235 °F (70–113 °C) and a low acid number (20–30). It is obtained from certain trees in the East Indies and is soluble even in weak hydrocarbon solvents. It is used in varnishes, nitrocellulose lacquers and even as a modifier in certain alkyd coatings. It tends to improve gloss and colour retention.

Rosin, also known as colophony, is one of the most widely used natural resins. It has a melting point of approximately 176 °F (80 °C), is highly acidic (acid number of 150–180), and is soluble in alcohols and hydrocarbons. It is an exudate of pine trees and can be obtained domestically. Chemically, it consists of derivatives of phenanthrene (a polynuclear aromatic hydrocarbon), such as abietic acid (Figure 5.1).

Rosin usually has poor resistance to water and alkalis, tends to oxidize over time and is slightly tacky. Because of these deficiencies, rosin is often used as a precursor in the manufacture of other resin types. When reacted with glycerol or other polyhydric alcohols, rosin is converted to a product known as ester gum, a hard material used in varnish production. The reaction occurs between the car-

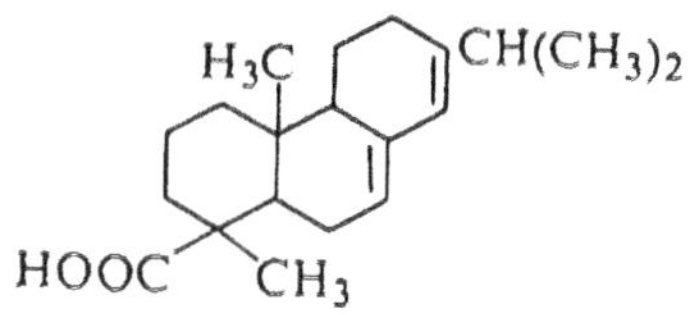

Figure 5.1 Structure of abietic acid.

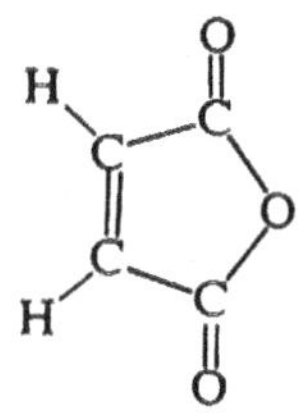

Figure 5.2 Structure of maleic anhydride.

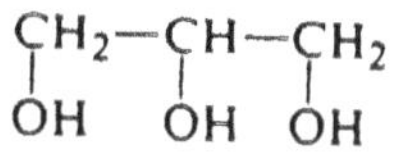

Figure 5.3 Structure of glycerol.

boxylic acid group (–COOH) on the rosin and the hydroxyl groups on the glycerol to produce a higher-molecular-weight ester.

When reacted with maleic anhydride (Figure 5.2) and subsequently with glycerol (Figure 5.3), so-called rosin maleic esters are produced. The pattern of conjugated unsaturation in the original rosin is eliminated, thus resulting in improved colour retention, and the higher-molecular-weight product does not have the tackiness of rosin. Rosin maleic esters are used along with vegetable oils to produce non-yellowing binders, to improve the gloss and hardness of alkyds and in the manufacture of varnishes.

Shellac is a natural resin of a distinctly different origin from those discussed above. Shellac is an excretion of the lac insect, which is native to India and Thailand. The dried excretion is crushed and washed, is then melted and dried in sheets which are broken up into flakes, and subsequently dissolved in alcohol. It is used for such products as knot sealers and lacquers.

The primary use of most natural resins is in varnish manufacture. In the making of a varnish, the natural resin is 'cooked' together with a drying oil, such as linseed oil, to obtain a homogeneous solution, which is then thinned to obtain a workable viscosity. Such varnishes are referred to as oleoresinous varnishes. Although these vanishes are still in use for certain specific applications, they have been widely replaced by synthetic resins that offer improved performance and which contain less environmentally regulated solvent. Although the term 'varnish' is still used to describe these transparent coatings, few of them are varnishes in the original sense of the word.

5.1.2 Oils

Vegetable oils, and to a much lesser extent, fish oils, are by far the most widely used class of natural resins, and have been used for centuries in the production of decorative and protective coatings. In the nineteenth and early twentieth centuries, they were by far the predominant paint binder. While coatings based solely on oil binders still exist, they basically constitute a niche market as wood preservatives and varnish-type films. However, large quantities of oils are still used in the production of more sophisticated coatings such as alkyds and epoxy esters.

Chemically, the naturally occurring oils are triglycerides. Triglycerides are the triesters of glycerol and a variety of fatty acids. Fatty acids are simply carboxylic acids that have a saturated or unsaturated aliphatic hydrocarbon group, usually from 15 to 17 carbon atoms in length, such as oleic acid (Figure 5.4). A typical oil (triglyceride) has a structure shown in Figure 5.5, where R can be derived from a variety of fatty acids.

The formulas of some of the more important fatty acids are as follows:

Stearic acid: $CH_3(CH_2)_{16}COOH$
Palmitic acid: $CH_3(CH_2)_{14}COOH$
Oleic acid: $CH_3(CH_2)_7CH{=}CH(CH_2)_7COOH$
Linoleic acid: $CH_3(CH_2)_4CH{=}CH\ CH_2CH{=}CH(CH_2)_7COOH$
Linolenic acid: $CH_3CH_2CH{=}CH\ CH_2\ CH{=}CH\ CH_2\ CH{=}CH(CH_2)_7COOH$
Ricinioleic acid: $CH_3(CH_2)_5\ COH\ H\ CH_2\ CH\ CH{=}CH\ (CH_2)_7\ COOH$
Eleosteric acid: $CH_3(CH_2)_3CH{=}CHCH{=}CHCH{=}CH(CH_2)_7\ COOH$

$$CH_3(CH_2)_7CH{=}CH(CH_2)_7COOH$$

Figure 5.4 Structure of oleic acid, a typical fatty acid.

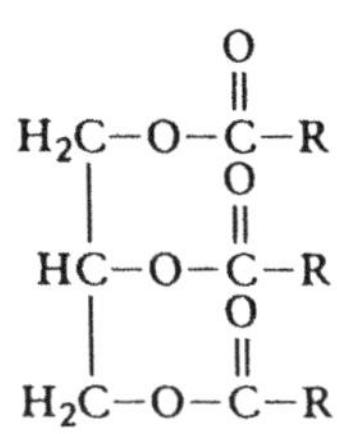

Figure 5.5 General structure of a triglyceride, where the R groups can be any combination of fatty acids.

Table 5.1 Typical compositions of some selected oils[a].

	Fatty acid				
Oil	**Saturated**[b]	**Oleic**	**Linoleic**	**Linolenic**	**Other**
Linseed	10	22	16	52	
Safflower	11	13	75	1	
Soybean	15	25	51	9	
Tung	5	8	4	3	80[c]
Tall oil	8	46	41[d]	3	2[e]
Castor	3	7	3		87[f]
Coconut	91	7	2		

[a] Values expressed as %.
[b] Mainly stearic and palmitic acids, although coconut oil contains several other saturated acids.
[c] Eleosteric acid.
[d] Linoleic and various isomers.
[e] Rosin.
[f] Ricinoleic acid.

The natural oils do not consist of a single triglyceride made from glycerol and a single fatty acid, but rather consist of mixtures of triglycerides composed of a variety of different fatty acids. Table 5.1 gives the composition of some of the more commonly used oils [1].

The composition of a single oil type can vary significantly based on such factors as climate, soil and the particular strain of the actual plant.

The sources of the oils are rather obvious from their names, with the possible exceptions of linseed oil, which comes from flax, and tall oil, which is a by-product of the manufacture of Kraft paper from wood pulp.

Oils are classified as drying oils, semi-drying oils or non-drying oils. Drying oils form relatively hard, solid films upon exposure to air; semi-drying oils form tacky films, while non-drying oils essentially remain as viscous liquids. The category to which an oil belongs basically depends on the amount of carbon–carbon double bonds, or unsaturation, in the fatty acids making up the oil. The mechanism by which oil dries or cross-links is quite complex, involving oxidation at the double bonds, which is initiated by free radical formation. A free radical is a molecule that has an unshared electron. Peroxides and hydroperoxides (ROOH), which are naturally present in small amounts, are particularly prone to forming free radicals, as shown in Figure 5.6. These free radicals abstract hydrogen atoms on methylene (CH_2) groups adjacent to double bonds, thus producing a free radical on one or more of the fatty acid groups on the oil. Two free radicals formed on separate triglyceride molecules can combine to form a neutral species, and therefore a chain reaction of propagation and condensation (cross-linking) is initiated, resulting in a gradual increase in molecular weight and in the production of a relatively hard, solid film. The entire process is referred to as autoxidation.

$$ROOH \longrightarrow RO\cdot + \cdot OH$$

Figure 5.6 Illustration of the formation of free radicals from a peroxide.

The rate of the autoxidation reaction can be quite slow. In order to speed things up, catalysts are often used. These catalysts, or 'driers', are often metal salts of octanoic or napthenic acids. Some, such as cobalt and manganese salts, primarily catalyze cross-linking at the coating surface and are referred to as 'surface driers'. Others, such as zirconium salts, catalyze cross-linking throughout the thickness of the coating and are known as 'through driers'. Although properly formulated oil-based coatings can dry to the touch in a few hours, the process of oxidative cross-linking can continue for years. The free radical reaction can also result in the cleavage of bonds, thus resulting in low-molecular-weight by-products. Therefore, oil-based coatings based on drying oils can embrittle and discolour over a period of many years.

It is relatively rare to encounter coatings consisting of binders composed solely of natural oils. When used alone, oils form coatings which are relatively soft (but which can embrittle on ageing) and have poor impact, abrasion and chemical resistance. Oils are, however, extensively used in the next category of coatings to be discussed – alkyds and epoxy esters.

5.2 ALKYDS AND EPOXY ESTERS

5.2.1 Alkyds

Alkyds have been one of the mainstays of the coatings industry for much of the twentieth century. Despite the proliferation of more sophisticated, higher-performance coatings such as epoxies and urethanes, alkyds continue to be used in both the architectural and industrial maintenance markets. However, the regulatory drive towards lower and lower volatile organic contents (VOCs) may eventually result in a significant decline in their use.

Alkyds are basically oil-modified polyesters. They are prepared by reacting together polyols, dibasic acids and fatty acids (or oils composed of fatty acids).

Although several polyols can be used, two of the most common are glycerol (see Figure 5.3) and pentaerythritol (Figure 5.7). The most common dibasic acid is actually used in its anhydride form and consists of phthalic anhydride (Figure 5.8). Isophthalic acid (Figure 5.9) is also used fairly commonly.

There are two processes used to produce alkyds, namely, the monoglyceride process and the fatty acid process.

In the monoglyceride process, the oil (such as linseed oil) is first 'cooked' with a polyol (typically glycerol), which results in a transesterification reaction to form a monoglyceride (Figure 5.10). Once the monoglyceride has formed, the dibasic

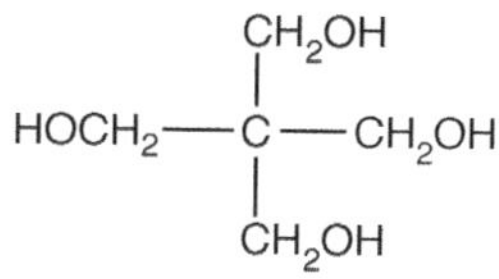

Figure 5.7 Structure of pentaerythritol.

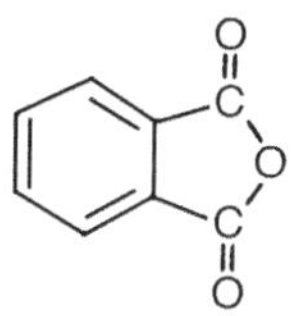

Figure 5.8 Structure of phthalic anhydride.

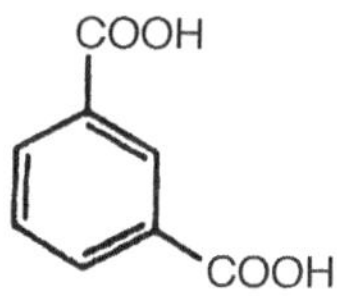

Figure 5.9 Structure of isophthalic acid.

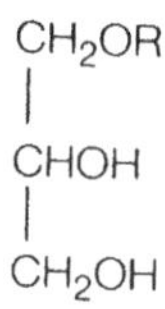

Figure 5.10 General structure of a monoglyceride, where the R group is a fatty acid.

acid, such as phthalic anhydride, is added. The acid (or anhydride) groups of the phthalic anhydride can then react with the hydroxyl groups of the monoglyceride, thus forming the oil-modified polyester, or alkyd.

If it is desirable to use a polyol other than glycerol, the above process cannot be used. In this case, the fatty acid process is used whereby fatty acids are employed rather than oils (triglycerides), and the fatty acids, polyols and dibasic acids are heated together in the reactor in a single step.

There are two overlapping ways of classifying alkyds. They can be classified as drying or non-drying (sometimes referred to as oxidizing or non-oxidizing), or by a characteristic known as oil length.

Alkyds basically cure by the same autoxidation reaction previously described for the natural oils from which they are made. Therefore, alkyds based on drying oils (oils whose fatty acids have conjugated carbon–carbon double bonds) such as linseed oil dry or oxidize to solid films and are therefore appropriate for use as coatings. In contrast, alkyds based on saturated, non-drying oils do not cure but basically remain as viscous liquids. They are not suitable for coatings, but are often used as plasticizers.

Oil length refers not to the length of the fatty acid chain (i.e., C_{16} versus C_{18}) but to the actual amount, in wt%, of oil (or fatty acid) used. Alkyds with an oil length greater than approximately 60 (60% oil by weight) are called long oil alkyds. Alkyds containing approximately 40–60% oil are considered to be medium oil alkyds, and those with less than 40% oil are known as short oil alkyds.

All other things being equal, long oil alkyds have lower surface tensions and hence better wetting properties than short oil alkyds. They are more tolerant of marginally prepared surfaces, and are better able to lift and displace small amounts of soluble organic contaminants such as oil or grease. They are also better able to penetrate and partially encapsulate minor amounts of rust or scale than are short oil alkyds. Initially, at least, they also result in softer, more flexible films than short oil alkyds. The latter are often used as 'shop primers', which dry rapidly such that steel can be shipped quickly to job sites for erection and subsequent topcoating.

However, since the oxidative curing process of an alkyd resin can continue for years, the final properties of the coating can be very much different from the initial ones. A long oil alkyd containing a large fraction of drying oils can, over many years, become as brittle, or even more so, than a short oil alkyd containing a relatively small fraction of drying oils. This is one example of why it can be dangerous to generalize too much about the properties of a specific class of coatings. Formulation variables can sometimes outweigh the so-called 'distinctive characteristics' often ascribed to a particular coating type.

Alkyds are good general-purpose coatings for a wide variety of applications. However, primarily because the backbone of the coating contains numerous relatively reactive ester linkages, they have poor resistance to alkaline environments. The reaction between an acid and an alcohol to form an ester is a reversible one, particularly so under moist, alkaline conditions. The reversible degradation of an ester to an alcohol and an acid is called hydrolysis, or saponification. Since the reaction occurs under alkaline conditions, the salt of the acid is formed rather than the free acid. The acid salt can usually be readily detected by analytical methods, thus making identification of saponification quite straightforward.

In addition to saponification, alkyds are also susceptible to chemical and sunlight-induced attack at the remaining double bonds. Certain chemicals, such as mineral acids, can add across double bonds. The unsaturation also provides sites for oxidation and free radical reaction.

Another failure occasionally encountered with alkyds is wrinkling. Wrinkling occurs when the surface of the alkyd dries considerably faster than the interior. The surface then contracts, and since the interior is still 'mushy', it is pulled along by the shrinkage of the surface, and a wrinkle results. This is usually a conse-

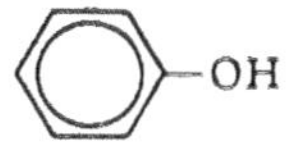

Figure 5.11 Structure of phenol.

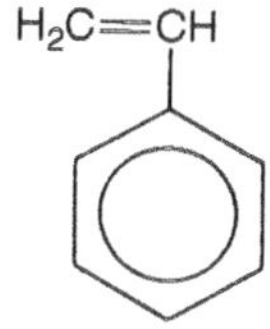

Figure 5.12 Structure of styrene.

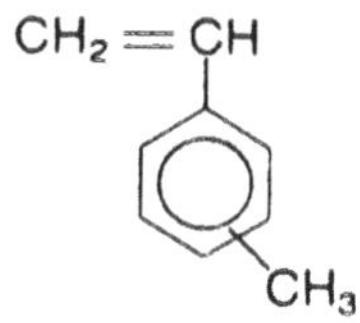

Figure 5.13 General structure of vinyl toluene

quence of the coating being applied too heavily, or an improper balance of surface and through driers.

There are numerous ways to modify alkyds to improve certain properties. Common modifications include phenolic, styrene or vinyl toluene, and silicone resins.

Phenolic resins are made by reacting phenol (Figure 5.11) or substituted phenols with formaldehyde. They can be used in relatively small quantities to upgrade the adhesion, hardness and corrosion resistance of alkyds. Phenolic alkyds are sometimes termed 'universal' metal primers, since they may be recoated with a large variety of solvent-based coatings without the fear of lifting and wrinkling that can sometimes occur with the relatively solvent-sensitive unmodified alkyd coatings.

Alkyd resins can also be modified via an extra 'cooking' step where they are reacted with styrene (Figure 5.12) or vinyl toluene (Figure 5.13). This typically results in the formation of higher-molecular-weight resins, which translates to faster drying of the coating, along with a somewhat improved resistance to moisture and alkali. As might be expected of a 'vinyl'-type modification, the resistance to solvents and aliphatic hydrocarbons such as grease and oils suffers.

Alkyds are also modified with silicone resins. The principal advantages are improved durability and heat resistance.

5.2.2 Epoxy Esters

Technically, epoxy esters could simply be considered as another type of modified alkyd. However, since they are usually referred to as 'epoxy esters' rather than 'epoxy-modified alkyds', it is felt that they deserve their own heading here.

Coatings based on epoxy resins constitute one of the most important classes of high-performance coatings. Epoxy esters, however, are not epoxy coatings and should not be confused with them. The coatings industry would probably be better off if indeed they were referred to as epoxy-modified alkyds.

Chemically, an epoxy-functional group (also called an oxirane group) is a three-membered ring consisting of two carbon atoms and an oxygen atom. Most epoxy resins are based on the diglycidyl ether of bisphenol A (DGEBA), which has the structure shown in Figure 5.14.

The commercially available epoxy resins are condensation products of DGEBA, with the structure shown in Figure 5.15, where *n* is usually from 1 to 12.

The resins contain hydroxyl groups and oxirane groups, both of which can react with the carboxylic acid groups of the fatty acids present in an alkyd, thus resulting in the formation of the so-called epoxy ester. The epoxy resins can, to some extent, react with themselves (the secondary hydroxyl group of one molecule can react with the oxirane group on a second molecule), although this can be minimized by proper 'cooking' procedures.

The properties of epoxy esters can vary widely, depending to a large extent on the oil length of the alkyd. Long oil epoxy esters are more 'alkyd-like' and are soluble in aliphatic solvents such as mineral spirits. Short oil epoxy esters are more 'epoxy-like' and require stronger aromatic solvents such as xylene. However, even short oil epoxy esters are more like alkyds than epoxies.

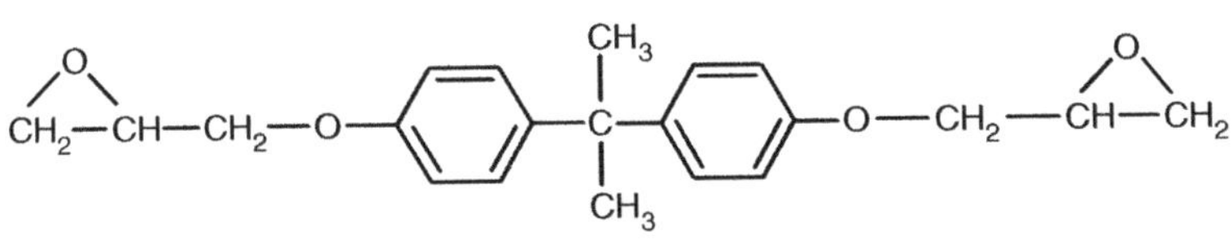

Figure 5.14 Structure of the diglycidyl ether of bisphenol A (DGEBA).

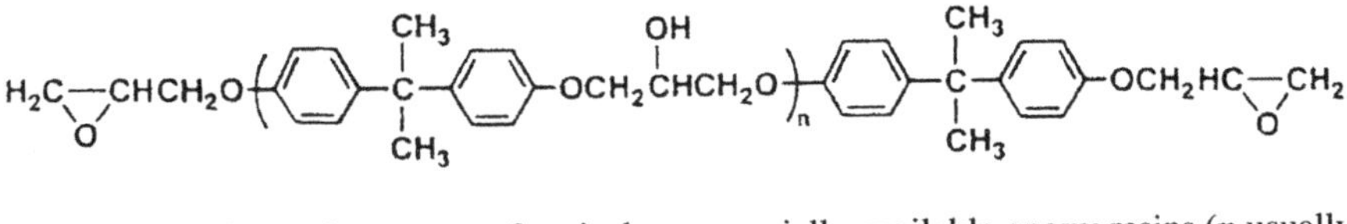

Figure 5.15 General structure of typical commercially available epoxy resins (*n* usually has values of 1–12).

Epoxy esters can be formulated with numerous advantages (and concomitant disadvantages) when compared to alkyds. They can dry faster and produce harder films with better adhesion and improved chemical resistance (although they are still susceptible to saponification). Although inferior to true epoxies in terms of most chemical and physical properties, they are easier to use and are more forgiving of minor surface contaminants. Unfortunately, like epoxies, they also tend to chalk badly when exposed to sunlight. Keeping in mind the various advantages and disadvantages of epoxy esters versus alkyds, the modes of coating failure are generally similar.

5.3 EPOXIES

When epoxy resins first became commercially available in the 1940s, they were used primarily in adhesives. They now constitute perhaps the single most important class of high-performance thermoset coatings. Epoxies can be formulated as two-component ambient-cured coatings, as high-temperature baking systems, as 100% solids fast-cure plural spray systems and even as powder coatings. Compared to the alkyds and epoxy esters previously described, they have superior adhesion, corrosion resistance, chemical resistance, and mechanical and physical properties.

The molecular weights of epoxy resins vary from approximately 400 ($n = 1$ in Figure 5.15) to as high as approximately 4000 ($n =$ approximately 12). As the molecular weight increases, so too does the equivalent weight (the number of grams of resin containing 1 equivalent of epoxy groups). In some very high-molecular-weight epoxy resins (molecular weight of 15 000–50 000), there is so little epoxy (the oxirane groups simply make up the two ends of the long molecule) that the resin is really just a polyfunctional alcohol. Such resins are termed phenoxy resins and do not require curing agents. When used as single-pack primers, they have good chemical resistance and adhesion, and in some cases their impact and abrasion resistance is superior to two-component thermoset epoxies.

When n is 1 or less, the resins are viscous liquids. At n values of approximately 2, the resins are amorphous solids with epoxy equivalent weights of approximately 450–500 (molecular weights of approximately 900–1000), which 'melt' or soften at roughly 65–75 °C. They are usually sold as high-solids (70% or so) solutions in aromatic- or ketone-type solvents. These resins are used to produce the bulk of the conventional, two-component, ambient-cured epoxy coatings currently in use.

Since epoxy resins are of low molecular weight and are basically viscous liquids, they only produce useful coatings when they are cross-linked to form higher-molecular-weight material. Although some of these thermosetting reactions rely on the secondary hydroxyl groups on the epoxy resin backbone, most of them take advantage of the highly reactive terminal epoxy groups. These groups are capable of reacting with almost any molecule which has an active hydrogen atom, including amines, carboxylic acids, phenolics, amino resins, anhydrides and isocyanates. The properties of the cured films depend to a large degree on the type of curing agent used.

5.3.1 Amine and Amide Curing Agents for Epoxy Resins

Amines and amine-functional amides are the most common curing agents for high-performance, ambient-cured industrial maintenance epoxy coatings. Aliphatic and cycloaliphatic primary amines react with epoxy resins via the mechanisms shown in Figure 5.16.

The secondary amine thus formed can now react, at a slower rate due to steric hindrance, with another molecule of epoxy resin.

Simple, unmodified aliphatic amines such as diethylene triamine (DETA) (Figure 5.17) are very reactive and result in coatings with short pot lives. Because DETA has two primary amine hydrogens and three secondary amine hydrogens for a total amine functionality of five, it forms very highly cross-linked coatings with excellent chemical and heat resistance and high cohesive strength. However, because of the high degree of cross-linking, flexibility and impact resistance tend to suffer.

Another disadvantage of amine cross-linking agents is their low equivalent weight, as well as significant toxicity problems. The equivalent weight of DETA, for example, is only 21. If used with a typical epoxy resin having an equivalent weight of 500, the stoichiometric mix ratio is 25 : 1 by weight. Such a large mismatch makes it difficult for a painting applicator to accurately mix the components in the correct ratio, thus increasing the possibility that the coating may not obtain its optimum chemical and physical characteristics. To complicate matters further, there is a very large difference in the viscosity of the epoxy resin component compared to the amine component. This difference in viscosity makes uniform mixing of the two components difficult, even if the proper amounts of each component are used.

Figure 5.16 Illustration of the reaction between an amine and an epoxy resin.

$$H_2N{-}CH_2{-}CH_2{-}NH{-}CH_2{-}CH_2{-}NH_2$$

Figure 5.17 Structure of diethylene triamine (DETA).

Another disadvantage of low-molecular-weight amine curing agents is the fact that they are usually somewhat water soluble, and also somewhat volatile. Particularly when exposed to high humidity or condensation shortly after application, they have a tendency to exude or 'blush' to the surface. In severe cases, this blush is visible to the unaided eye, but even in less severe cases it can sometimes interfere with the adhesion of subsequently applied coats of paint.

In the presence of moisture and carbon dioxide, the amine blush can be converted to a carbonate blush.

One common way to minimize blush formation is to allow the mixed coating to sit in the can for 15–45 minutes prior to applying it. This so-called induction or 'sweat-in' time allows the amine–epoxy reaction time to proceed, such that little or no free amine is available to blush. This procedure works reasonably well, but it does shorten an already short pot life.

This is a good time to point out another semantic flaw in the coatings industry. All too often, cross-linking agents such as amines are referred to as catalysts. This is a major error in nomenclature, since by definition a catalyst is an agent that speeds up a chemical reaction without actually taking part in it. Clearly, an amine is a cross-linking agent, or a co-reactant, and *not* a catalyst. True catalysts are, however, widely used in coatings.

One way to avoid some of the problems associated with amine curing agents is to use amine adducts instead. They are produced by reacting low-molecular-weight epoxy resins with an excess of an amine, such as DETA, to form an amine-terminated adduct.

The amine adducts have higher equivalent weights, higher viscosities, lower vapour pressures and reduced toxicities than their amine counterparts, and therefore avoid many of the problems associated with amine cross-linkers. Coatings produced from amine adducts generally have somewhat longer pot lives, and hence somewhat slower cure rates, than amine-cured epoxies. As a whole, their chemical resistance and solvent resistance, while still quite good, is somewhat less than that of amine-cured coatings. They generally have better flexibility and impact resistance. Blushing problems, while significantly reduced, are not necessarily eliminated through the use of amine adducts.

Other important alternatives to simple amine cross-linking agents are the so-called polyamides. These are prepared by reacting an amine, such as DETA, with dimer fatty acids to form amine-terminated polyamides. Dimer fatty acids are relatively high-molecular-weight carboxylic acids, obtained by reacting two C_{18} fatty acids. When these dimer acids are reacted with DETA or some other multifunctional amine, amine-functional polyamides are produced.

Although these cross-linking agents do have amide functional groups, it is somewhat confusing to refer to them as polyamide cross-linkers since the actual reactive groups are still the amine groups (especially the primary amine groups at the two ends of the molecule).

As is the case with amine adducts, there are a wide variety of polyamide cross-linkers on the market, all imparting somewhat different properties to the final epoxy product. Since all of the polyamides are much larger molecules than the simple amine cross-linkers, and since there is a much greater distance between the highly reactive terminal primary amine groups, the cross-link density of a polyamide-epoxy coating is much less than that of an amine-cured epoxy. Reaction rates are slower; pot lives are longer; mix ratios are usually 1 : 1 or 2 : 1 by volume, and toxicity, blushing and mix ratio errors are reduced (but not eliminated).

In terms of the properties of the cured films, polyamide epoxies generally form less densely cross-linked, more flexible films than do amine-cured epoxies. Flexibility and impact resistance are improved, as is adhesion. However, chemical and solvent resistance suffer, although these properties are still good when compared to many other generic coating types. The combination of relative ease of use and good chemical and physical properties has made polyamide-epoxy coatings one of the workhorses in the area of general industrial maintenance coatings, as well as in several speciality coating applications such as food processing, marine and organic zinc-rich primers.

Cross-linking agents known as amido-amines are prepared by reacting monofunctional fatty acids, rather than dimeric fatty acids, with low-molecular-weight polyamines. Compared to the simple, unmodified amines, amido-amines have higher molecular weights and lower amine functionalities. The cross-linking reactions proceed mainly at the single primary amine group at the amine-terminated end of the molecule, with a much lower reaction rate associated with any secondary amines that may be present in the body of the molecule. The cross-link density is very low when compared to a simple amine-cured epoxy, and is even lower than that of a polyamide epoxy. Such coatings have very long pot lives, cure slowly, and have a better flexibility and impact resistance than even the polyamide epoxies. While their chemical resistance is comparable to the polyamide epoxies (and somewhat inferior to amine-cured epoxies), their solvent resistance is significantly poorer.

Ketimines are a rather specialized type of curing agent, where long pot lives and low application viscosities are especially important. Ketimines are considered to be 'blocked' curing agents, and are made by reacting a low-molecular-weight ketone such as MEK with an aliphatic primary amine. The structure of a typical ketimine formed from MEK is shown in Figure 5.18.

Since ketimines are tertiary amines, they have a very low reactivity towards epoxy groups. They decompose upon exposure to atmospheric moisture, forming

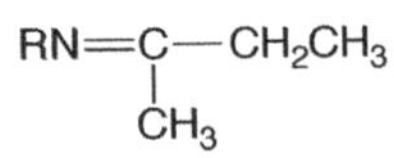

Figure 5.18 Structure of a ketimine formed from a primary amine and methyl ethyl ketone (MEK).

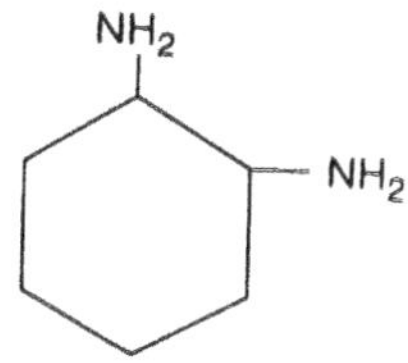

Figure 5.19 Structure of a diamine cyclohexane.

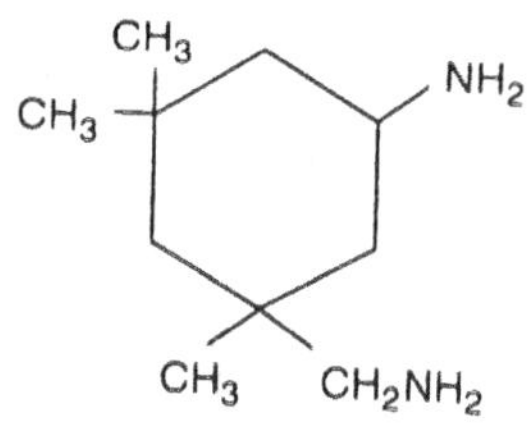

Figure 5.20 Structure of isophorone diamine.

the primary amine and, in the above example, MEK. The MEK volatilizes from the film, thus leaving the primary amine free to react with the epoxy resin.

The cycloaliphatic amines are a special class of aliphatic amines. Two examples of cycloaliphatic amines are diamine cyclohexane (Figure 5.19) and isophorone diamine (Figure 5.20). When cycloaliphatic amines are modified with organic acids as accelerators, satisfactory curing can be achieved at near-freezing temperatures. Although they are less volatile and hence less irritating than their non-cyclic counterparts, they are still often reacted with small amounts of very low-molecular-weight epoxy to further reduce volatility and blushing. In addition to low-temperature cure, they also have good adhesion to damp surfaces, along with the superior chemical, solvent, heat and moisture resistance associated with epoxies. They do, however, tend to be brittle.

The Mannich bases are another type of curing agent used to obtain low-temperature curing and adhesion to damp surfaces. Mannich bases are formed by the reaction between methyol phenol and a multifunctional primary amine. The phenolic hydroxyl is a powerful accelerator of epoxy ring opening reactions.

Powder coatings are a special class of epoxy coatings. These are solventless coatings provided in powder form, where the epoxy resin, pigments and curing agents are mixed, compounded by melting in extruders, cooled and then ground into a homogenous powder. The virgin powder has a high enough glass transition temperature (T_g) such that it will not partially fuse during storage. The coatings are applied by specialized equipment either to hot substrates or to room-temperature articles that are then subsequently baked, in order to achieve both flow and cure and also to complete the cross-linking. A curing agent commonly used with epoxy powder coatings is dicyandiamide (Figure 5.21). Powder coatings are used in

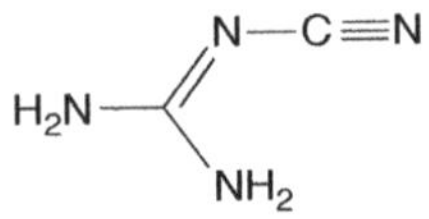

Figure 5.21 Structure of dicyandiamide.

numerous original equipment manufacture (OEM) applications such as electrical boxes and appliances, and in particular, for buried underground pipelines.

5.3.2 Epoxy Failure Modes

There are numerous types of epoxy resins that can be reacted with an even greater number of curing agents. There are also a large number of pigments, catalysts, additives and solvents to choose from. It is difficult, and sometimes risky, to generalize about epoxy coatings (or any class of coatings, for that matter). This author has often heard statements such as 'that doesn't appear normal for an epoxy'. Such observations are often difficult to support unless one has had previous experience with the particular product in question, or unless one has access to published data concerning the specific coating's chemical and physical characteristics.

There are, however, some general, if sometimes tentative, conclusions that can be made about the performance of specific classes of epoxies and about their failure mechanisms.

Compared to thermoplastic coatings, as well as alkyds, epoxies have better adhesion properties, as well as a greatly improved resistance to chemicals, solvents, water and elevated temperatures. Abrasion resistance, flexibility and impact resistance can vary widely among the various classes of epoxies. In some cases, brittleness can be a major shortcoming. The corrosion resistance of properly formulated epoxies is generally quite good, and usually dramatically better than the thermoplastics and alkyds.

One of the major drawbacks of most, if not all, epoxies, is a poor resistance to sunlight. Because of the aromatic structure of epoxies, they absorb ultraviolet light. In addition to causing yellowing, the absorption of ultraviolet light results in a breakdown of the surface of the epoxy resin, resulting in a thin layer of loose, chalky material. The depth of this degradation is very thin and does not detract from the coatings chemical and physical properties. It can be objectionable if aesthetics are important, and can also cause serious problems with the adhesion of subsequently applied coats of paint if not removed. It is sometimes possible for an epoxy to chalk sufficiently in only a week of outdoor exposure such that certain types of coatings will not adhere well to them. In such instances, the degree of chalking is so slight that it is not readily observable.

In 1980, Croll [2] investigated the phenomenon that thick epoxy coatings appeared to cure faster than thin epoxy coatings. Croll confirmed that this was indeed the case by measuring the indentation hardness of coatings applied at varying thickness. By conducting his experiments under different atmospheric conditions, he was able to show that the effect was essentially non-existent when the coatings were cured under an inert atmosphere of dry nitrogen, but was quite pronounced when they were cured in moist air, and even more so when cured in a humid carbon dioxide-rich atmosphere. Croll concluded that the hardening of an epoxy coating is influenced by absorption of moisture, or moisture and carbon dioxide, from the atmosphere, and that thin coatings were affected the most because the water and gas can more completely diffuse through their entire thickness. He suggested that the cure of the coating was being disrupted by the reaction of the amine curing agent with carbon dioxide, likely catalyzed by moisture, to form a carbamate or an ammonium bicarbonate. This suggests another problem unique to epoxy coatings, that of 'amine blush'.

Many of the lower-molecular-weight amine curing agents used in epoxies are somewhat water soluble. When exposed to high humidity or condensation shortly after coating application, or when the coating has been applied under relatively cool, damp conditions, the curing agents have a tendency to exude to the surface of the coating, resulting in what is usually termed an 'amine blush'. The possibility of the formation of an amine blush on the surface of epoxy coatings has been widely recognized [3–5]. In many cases, the blush is visible to the naked eye, often as a thin, greasy or whitish layer. While it typically does not affect the properties or performance of the epoxy coating itself, it can sometimes cause significant adhesion problems with subsequently applied coats of paint.

Upon initial formation, the blush is, of course, the amine curing agent. However, as time goes by, this amine can react with carbon dioxide and/or water to form either an ammonium carbamate (sometimes referred to as an amine carbamate or as a carbamate salt) or an ammonium bicarbonate (sometimes referred to as an amine bicarbonate) [3–5]. These reactions are shown in Figures 5.22 and 5.23.

The problem of amine blush can be minimized by using higher-molecular-weight cross-linkers such as amine adducts and polyamides, or by observing

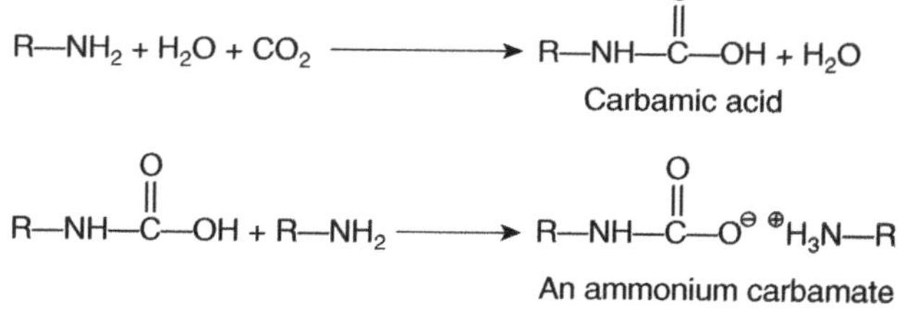

Figure 5.22 Formation of an amine carbamate (a carbamate salt) via the reaction of an amine curing agent with water and carbon dioxide.

Figure 5.23 Formation of an ammonium bicarbonate via reaction of an amine curing agent with water and carbon dioxide.

induction times. However, it is doubtful that it can be entirely eliminated. Often the problem can be mitigated by water washing of the coating prior to the application of succeeding coats of paint. If the coating is top-coated within a reasonable length of time with another coating that is compatible with the blush (such as another coat of epoxy), there may be no problems with adhesion. This is an inexact science, however, since this author has occasionally seen large amounts of blush successfully top-coated, while small amounts of blush can sometimes cause serious failure. As pointed out by Dinnissen [5], the problem of blush formation might be exacerbated during indoor applications during cold weather, when direct fired gasoline or kerosene heaters are being used, since they can considerably increase the carbon dioxide content of the air and can also produce significant amounts of water vapour.

Totally apart from problems with blushing and chalking, certain epoxies and other thermoset coatings can be difficult to top-coat simply because they are smooth and hard. The problem will be compounded if the topcoat has a relatively weak solvent system that cannot penetrate or soften the surface of the cured epoxy. Manufacturers often refer to a 're-coat window', beyond which time the coating will be difficult to top-coat. Often it is recommended that the surface be lightly scarified by rough sanding or abrasive blasting. Occasionally, the applicator may be directed to rub the surface with a strong solvent, or to spray apply a solvent on the epoxy, in order to slightly soften it prior to topcoating.

Another problem relating to certain types of epoxies is that of stress. To one degree or another, all coatings have a tendency to shrink upon curing, and this is especially true with two-component thermoset coatings such as epoxies. If the coating, for whatever reason, is constrained from shrinking, it can develop considerable internal stress, and may be unable to resist the added stress imparted by the environment, and cracking or other types of failure may occur.

Commonly, however, the epoxy does shrink to relieve its internal stress. This can be a problem if it is applied to substrates (including certain coatings) which have poor cohesive strength, since the relaxation of internal stress in the epoxy results in the transfer of stress to the underlying coating or substrate. This so-called 'shrinkage stress' can sometimes result in massive cohesive failure (splitting) within the underlying layer. A classic example of this type of failure is the use of two-component solvent-based epoxies as topcoats over vinyl latex block fillers used on masonry walls. Not only does the highly filled (pigmented) block filler have poor cohesive strength, but it is also susceptible to the relatively strong solvents used in the epoxy.

Another problem encountered with the more densely cross-linked epoxies (or any densely cross-linked coating) is their inherent brittleness. There is usually no problem on thick, immobile substrates such as heavy structural steel, provided that there is no impact damage. However, if the substrate flexes in high wind, or is subject to expansion and contraction by thermal forces, the coating will be unable to flex with it. In this event, one of two things is likely to happen, depending on the degree of adhesion to the substrate.

If the coating has very good adhesion to the substrate, the stress may be relieved by cracking. Depending on the environment, this may lead to the ingress of water or other harmful materials to the substrate.

If the coating has marginal adhesion to the substrate, the stress may be relieved simply by detachment of the coating from the substrate. If the coating is thin, it may come off in small flakes. If it is thick (over about 5 mils), it may come off in large pieces. Neither of these options is desirable.

Another potential problem with epoxies, or with any coating that has a pot life of only a few hours or less, is the application of the coating near the end of, or slightly after, its pot life. It is important to remember that the reaction between the epoxy resin and the cross-linker begins immediately upon mixing the two components. The chemical composition of the paint in the can is continually varying, and the paint that is being applied 2 hours after mixing is, on a molecular level, not the same paint that was applied 2 minutes after mixing. The biggest differences between the '2-minute' paint and the '2-hour' paint are molecular weight and viscosity. These two interrelated factors can have significant consequences for both the flow and wetting of the coating. Although the incipient formation of a cross-linked, three-dimensional film will probably result in an increase in surface tension, if applied to a metallic substrate, the surface tension of the coating will still be lower than that of the substrate and wetting can still occur, at least theoretically. However, opposing the ability of the coating to wet the substrate will be the resistance to flow as a result of increased viscosity. It is entirely possible that localized areas of poor adhesion can occur simply because of the time at which a coating was applied after mixing.

The above discussion of failure modes assumes that there are no applicator errors. It is widely believed that applicator errors are responsible for the majority of coating failures, an opinion which is shared by this author. However, this opinion has been gleaned by piecemeal observations and individual experience, and any attempt to assign an actual percentage to the number of applicator-related failures would be pure guesswork. Based on the author's experience, it is safe to say that more than half of all coating failures are applicator related.

Epoxies can fail by the same type of applicator errors that affect the performance of other types of coatings. Modern high-performance coatings are typically less forgiving of applicator error than simple coatings such as oil-based alkyds. Poor surface preparation, over-thinning, poor spray gun technique, contamination or painting when it is either too hot, too cold, or too damp can all contribute to the failure of an epoxy, as can applying the coating too thinly or too thickly. In addition, there are certain applicator-related errors that are specific to two-component coatings.

One of the more common of these errors is 'mis-mixing' of the epoxy. This is more likely to happen with amine-cured epoxies where the mix ratio between the two components is disparate, such as 10 parts of component 'A' to 1 part of component 'B'. The chances for mixing errors with coatings that have simple 1 : 1 or 2 : 1 ratios are not as great. Nevertheless, this author has seen more than one coating failure caused by the failure of the applicator to add any of the cross-linker (or converter) component.

When analyzing an epoxy coating for potential errors in mixing, two things should be pointed out. Firstly, since coatings are mixed at job sites and not in laboratories, small deviations from the ideal mix ratio should be expected. If the coating in question has to be mixed exactly at a ratio of 1.00 gal of component 'A' to 1.00 gal of component 'B', or else failure will occur, it is highly questionable as to whether or not it is a merchantable product.

The second factor to consider if mis-mixing is being investigated is whether or not the specific batch in question was produced properly at the factory. The finding that a coating contains less epoxy resin than expected does not necessarily mean that the applicator is at fault. The error could have occurred at the factory.

Other epoxy application errors include failure to observe induction times (which could cause blushing and possibly inter-coat adhesion failure), application of the coating at or beyond its pot life (which could cause adhesion failures) and topcoating beyond the re-coat window (possibly causing inter-coat adhesion failure).

5.4 MODIFIED EPOXIES

Epoxy chemistry is very versatile, and the types of curing agents and reaction pathways are too numerous to be adequately covered in this present text. However, a few important classes of modified epoxies will be included here, as the investigator of coating failures will likely encounter most of these at one time or another.

5.4.1 Acrylic Epoxies

The term acrylic epoxy may appear incongruous, as acrylics are often thought of as soft, water-based coatings and epoxies as hard, solvent-based coatings. As the three-membered epoxy ring is highly reactive towards functional groups containing active hydrogens, it is possible to react it with carboxylic acids.

If the carboxylic acid group is attached to an acrylic resin, one has formed an acrylic epoxy. The coatings usually require both a catalyst and a relatively high-temperature bake to achieve full cure. Such coatings have properties which are intermediate between an acrylic and an epoxy. They tend to have better hardness, adhesion and chemical resistance than an acrylic, and better light stability and flexibility than an epoxy.

5.4.2 Coal Tar Epoxies

Coal tar epoxies are an important class of epoxy coatings. They are used extensively in the water and sewage industries due to their thick barrier properties, reasonably good chemical and moisture resistance, and relatively low cost.

Refined coal tar pitches are used in ratios of approximately two to three times the amount of epoxy/cross-linker. The coal tar and cross-linking agents are combined into one component, with the epoxy resin as the second component. Many of the curing agents discussed previously can be used. Higher levels of coal tar result in softer, more slowly curing coatings with reduced chemical and solvent resistance, while higher levels of epoxy result in harder, faster curing, more brittle coatings.

The characteristics imparted to non-modified epoxies by the various curing agents are also observable in the coal tar epoxies. Amines give faster curing coatings and harder, more abrasion-resistant films with better chemical resistance, while polyamide systems are generally more flexible and impart better wetting properties. In waste-water service, an additional advantage of amines is a better resistance to hydrogen sulfide and bacteria.

Coal tar epoxies generally come in one colour, that is, black. They have poor resistance to ultraviolet light and will tend to develop a bronze colour on the surface. They are also extremely difficult to re-coat, and therefore usually must be applied as a single thick coat.

5.4.3 Epoxy Phenolics

Phenolic resins are produced by reacting phenol (Figure 5.11), or a substituted phenol, with formaldehyde (Figure 5.24) to give the structure shown in Figure 5.25.

The phenolic hydroxyl group is relatively reactive, and when reacted with epichlorohydrin (Figure 5.26), one of the precursors of epoxy resins, epoxy functionality can be grafted onto the backbone of the phenolic resin. Coatings made

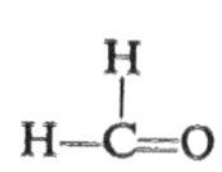

Figure 5.24 Structure of formaldehyde.

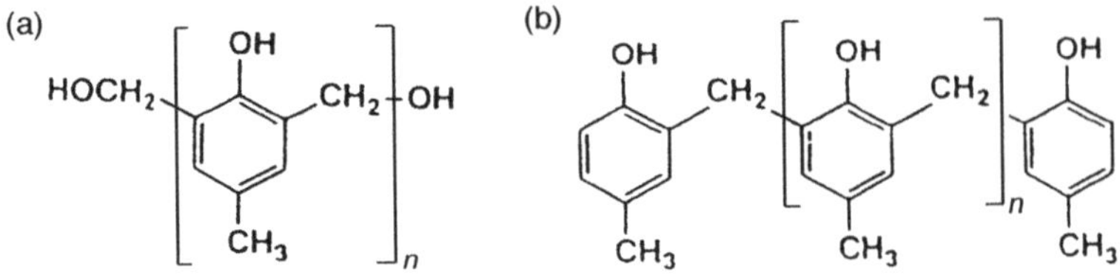

Figure 5.25 Structures of (a) a resole phenolic and (b) a novolac phenolic resin.

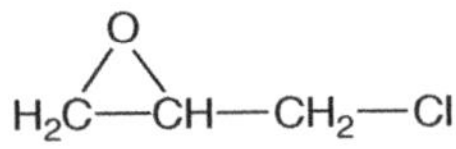

Figure 5.26 Structure of epichlorohydrin.

from these resins have excellent adhesion and chemical resistance and a high cross-link density.

In another type of epoxy phenolic coating, the phenolic resin may be considered a cross-linking agent and reacts via its phenolic hydroxy group (in the case of phenolic novolacs) with the epoxy ring on an epoxy resin. This type of coating is one of the most chemically resistant systems available, with excellent adhesion. It usually requires a bake in the 300–400 °F (149–204 °C) region to attain optimum properties, and is often used where extreme chemical exposure is required, such as in the lining of railroad tanker cars.

5.5 PHENOLICS

Phenolic resins are produced by reacting formaldehyde with phenol or substituted phenols. When the reaction is carried out under alkaline conditions, the resins are termed resole phenolics, and when carried out under acidic conditions, novolac phenolics are formed.

5.5.1 Resole Phenolics

The general structure of a resole phenolic prepared from cresol is shown in Figure 5.25.

Other commonly used substituted phenols include methyol phenol and *p*-tertiary butyl phenol. Depending on the reaction conditions, some of the phenolic monomers can be cross-linked by ether bridges (–CH_2–O–CH_2–) rather than by methylene bridges.

Depending on the raw materials used and the extent of reaction, two different subcategories of resoles can be produced, namely, those that are soluble in oxygenated solvents (termed 'alcohol-soluble phenolics') and those that are soluble in hydrocarbon solvents (termed 'oil-soluble phenolics').

Alcohol-soluble resole phenolic resins require baking to achieve cure, usually in the presence of an acid catalyst. Under these conditions, the resins self-condense via the reaction between the phenolic hydroxyl groups, or between these hydroxyl groups and methyol groups. In this process, ether linkages are formed, and either water or formaldehyde is liberated.

Because all of the linkages in a baked resole phenolic coating are either of the methylene or ether variety, there is nothing to hydrolyze. Such coatings have excellent resistance to moisture and alkali. Because of the stability of these linkages combined with high cross-link density, the coatings have excellent chemical resistance and low permeability, which imparts superior corrosion resistance. Adhesion is also usually quite good. Such coatings are used in tank linings in severe chemical service, and also for can linings in the food and beverage industries. However, because of their high cross-link density, they tend to be brittle and often require plasticization. Because of their aromaticity, they can darken significantly, both on baking and on exposure to sunlight.

One of the main uses for oil-soluble resole phenolics is in the production of high-quality varnishes. The phenolic resin is cooked with linseed or tung oil, sometimes after first being modified with rosin.

5.5.2 Novolac Phenolics

Novolac phenolics have the general structure shown in Figure 5.25.

Novolac phenolic resins contain little or no methyol functionality, and will generally not self-condense. Such resins are thermoplastic. They are usually soluble in alcohol or other oxygenated solvents, but if prepared from *p*-tertiary butyl phenol, solubility in hydrocarbons can be achieved. Alone, they have little use as coatings. However, they are used to some extent in varnish production.

5.5.3 Phenolic Failure Modes

Two of the biggest problems with phenolics are their brittleness and baking-related problems.

The brittleness is a consequence of both their rigid aromatic structure and their high cross-link density. This can be ameliorated somewhat by external plasticization, or to a lesser extent by the use of larger hydrocarbon groups on the substituted phenolic monomer.

In a controlled factory environment, heat-related problems of thin-film can coatings are relatively rare. However, this can sometimes be a problem when higher-build coatings are applied to large storage vessels or railroad tank cars. Portable heat sources are often used in such situations, and constant, accurate temperature control can sometimes be a problem. Since phenolics darken upon baking, the degree of colour change is often used to determine whether adequate cure has been achieved.

5.6 AMINO RESINS

The resins discussed in this current section, sometimes referred to as amino resins, do not have suitable physical properties with which to make coatings. However, they are widely used as cross-linkers with many other types of resins. Rather than discussing them piecemeal in later sections, they will be introduced as a group here.

Amino resins usually amount to perhaps 30% or less of the non-volatile vehicle content of various thermoset coatings. They are easily the major cross-linking agent for such coatings, and can be used with acrylics, polyesters, alkyds, epoxy esters and even epoxies. Unlike the thermoset epoxy coatings discussed in the previous section, thermoset coatings cross-linked with amino resins are almost always single-component, rather than two-component, coatings. This is possible because their reaction rates are extremely slow at low temperatures, and they must be baked to achieve cure.

The most commonly used amino resins are based on melamine, or 2, 4, 6-triamino-1, 3, 5-triazine (Figure 5.27).

The three primary amino groups on melamine make it a very versatile precursor. Pure melamine itself is not used as a cross-linking agent. It is reacted with formaldehyde and an alcohol. The first step, the reaction with formaldehyde, results in the formation of methyolated melamine. The structure of fully methyolated melamine (hexamethyol melamine) is shown in Figure 5.28.

Depending on the stoichiometric ratio of melamine to formaldehyde, less than fully methyolated melamine can be produced.

The next step in the production of melamine–formaldehyde (MF) resins is termed 'etherification' and involves reaction with a low-molecular-weight alcohol (usually methyl or butyl alcohol). Depending on the degree of methylolation of the initial melamine, the type of alcohol and the ratio of methylated melamine to alcohol, numerous types of useful MF cross-linking resins can be produced. One of the more common ones is hexamethoxymethylmelamine (HMMM) (Figure 5.29).

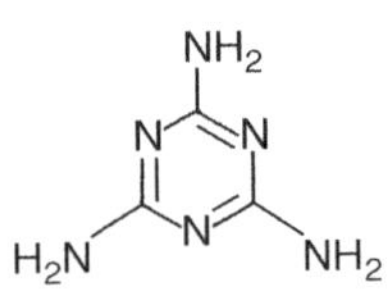

Figure 5.27 Structure of melamine.

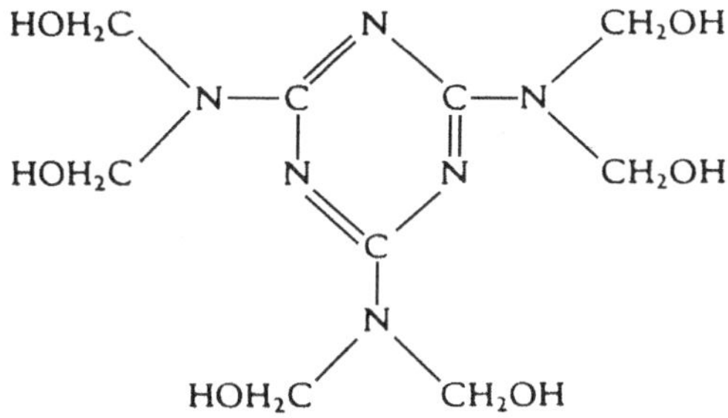

Figure 5.28 Structure of hexamethyol melamine.

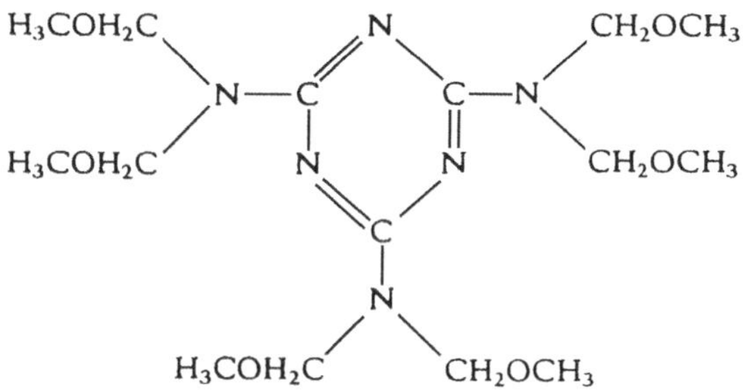

Figure 5.29 Structure of hexamethoxymethylmelamine (HMMM).

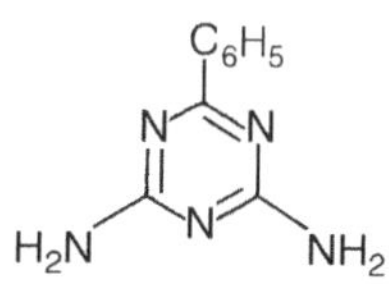

Figure 5.30 Structure of benzoguanimine.

In addition to the etherification reaction between the methyol groups and the alcohol (methanol, in the case of HMMM), it is also possible for limited self-condensation (self-etherification) between two methyol groups on adjacent molecules of methyol melamine. It is not unusual for a small amount of dimer to be formed during the production of HMMM or other MF resins.

Benzoguamine resins (Figure 5.30) are perhaps the next most widely used type of amino cross-linking agent. Like melamine, benzoguanime is also based on triazine.

Benzoguamine differs from melamine only in the replacement of one of the $-NH_2$ groups with a phenyl group. It undergoes the same types of reactions with

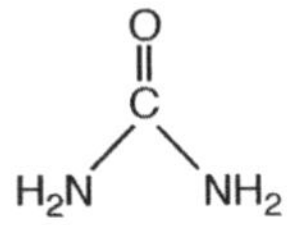

Figure 5.31 Structure of urea.

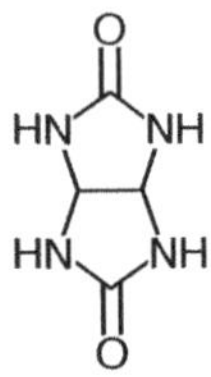

Figure 5.32 Structure of glycoluril.

formaldehyde and alcohols as melamine, resulting in similar cross-linking agents but with lower functionality.

Urea (Figure 5.31) can also be reacted with formaldehyde to form methylol derivatives. These derivatives, like its melamine and benzoguamine counterparts, can be subsequently etherified with alcohol to make urea–formaldehyde cross-linking resins. Like melamine and benzoguanimine resin, some amount of self-condensation also occurs to produce dimers.

Yet another class of amino resins is based upon glycoluril (Figure 5.32).

Glycoluril can react with formaldehyde to form tetramethylolglycoluril (TMGU). Depending on the pH of the reaction medium, lower methylolated analogues, as well as self-condensation products, can be formed.

The etherification reaction previously described between other methylolated amino resins and alcohols also occurs for TMGU, resulting in the formation of glycoluril–formaldehyde resins. Tetramethoxymethyl glycoluril is a solid used as a cross-linker for certain powder coatings. Mixed ether resins such as dimethoxy-methyldiethoxymethyl glycouril are liquids, and are used in coil and can coatings to provide greater flexibility than many of the other amino resins [1].

5.7 ACRYLICS

The most accurate definition of an acrylic resin is any resin formed from derivatives of acrylic acid (CH_2=CH–COOH). This simple definition belies the variety and complexity of acrylic resins, which are used in such diverse applications as plastics, fabrics, caulks and sealants, latex house paint, and high-performance thermoset coatings.

Table 5.2 Structures of some common acrylic monomers.

Monomer	Structure
Ethyl acrylate	$CH_2{=}CHCOOCH_2CH_3$
Methyl methacrylate	$CH_2{=}\overset{\overset{CH_3}{\mid}}{C}COOCH_3$
Acrylamide	$CH_2{=}CHCONH_2$
Hydroxyethyl acrylate	$CH_2{=}CHCOOCH_2CH_2OH$
Acrylic acid	$CH_2{=}CHCOOH$
Styrene	$CH_2{=}CHC_6H_5$

Table 5.2 shows the structures of some of the common acrylic monomers used in coatings. Although stryene may not be considered a derivative of acrylic acid, it is included because of its widespread use in many acrylics.

Acrylic resins are produced from acrylic monomers via a free radical addition reaction occurring at the carbon–carbon double bond site. An almost endless variety of polymers can be built depending on the monomers used, their ratios and the reaction conditions.

Acrylic coatings can be considered as belonging to one of three subgroups, namely, solution acrylics, acrylic latex coatings and thermosetting acrylics.

5.7.1 Solution Acrylics

Solution acrylics, also referred to as acrylic lacquers, constitute the earliest use of acrylic polymers as coatings. Since they are single-component, thermoplastic coatings that dry and cure only by solvent evaporation, they are also referred to as acrylic lacquers. They are used on furniture, and at one time huge quantities were used for automotive coatings. Their use in the automotive industry is now restricted primarily to the refinish market. They are also widely used in acrylic aerosols for consumer use.

Since solution acrylics do not cross-link, their chemical and physical properties must be 'built in' at the time of manufacture. They have relatively high molecular weights (usually 75 000 or higher), and consequently are rather low in solids content to achieve workable viscosities.

In addition to molecular weight, another fundamental property of a solution acrylic is its glass transition temperature (T_g).

The glass transition temperature has several different definitions, all of which are approximately equivalent. The most common definition in the coatings industry is that the T_g is that temperature below which a polymer is in a hard, glassy state, and above which it is in a softer, rubbery state. Another useful definition is that the T_g is that temperature where there is a sudden, discontinuous increase in the polymer free volume. The latter may be roughly described as the amount of space available for polymer segments or chains to freely rotate, or to move about, in.

A furniture coating provides a simple example of why the T_g is an important consideration for an acrylic lacquer. If the coating's T_g is too low (more than a few degrees below room temperature), the coating will be flexible (a good thing) and soft (a bad thing), and heavy objects such as bookends might actually stick to the coating. Conversely, if the T_g is too high, the coating will be hard (a good thing) and brittle (a bad thing), and a heavy object (the bookend is a good example) dropped on the coating might very well crack it. With a thermoplastic coating such as an acrylic, the T_g can be controlled to some extent by molecular weight, but to a much greater extent by the choice of monomers. For instance, the T_g of a homopolymer of methyl acrylate is 10 °C, while it is 105 °C for a homopolymer of methyl methacrylate [1]. Generally, the T_g is roughly an additive property characteristic of the relative proportions of the monomers used. Thus, a polymer made up of a 50/50 blend of methyl acrylate and methyl methacrylate would be expected to have a T_g of roughly 58 °C.

An examination of the structure of acrylic resins shows that they contain ester groups. Such groups are subject to alkaline hydrolysis or saponification. However, the ester groups do not constitute the backbone of acrylic coatings but are instead pendant to it. Although saponification of an acrylic lacquer can occur, the consequence is usually not as disastrous as saponification of an alkyd.

Some advantages of acrylic lacquers include a good resistance to ultraviolet light, gloss retention, flexibility and a reasonably good resistance to alkali. Disadvantages include difficulty in obtaining high-gloss pigmented coatings, relatively poor adhesion, very poor solvent resistance and high VOC levels. Acrylic lacquers are in much lower demand now than they were even 10 years ago. VOC restrictions will likely end up outdating them entirely.

5.7.2 Acrylic Latex Coatings

Acrylic latexes are used in huge amounts in the coatings industry. They make up the bulk of architectural coatings, and are now gaining a significant market share in industrial maintenance areas as well. They have the notable advantage of being very low in VOC (their principal solvent is water), without having to sacrifice molecular weight.

A latex is basically a dispersion of polymer particles (the discontinuous phase) in water (the continuous phase). The latex is usually made by a process called emulsion polymerization. In this type of polymerization process, relatively high levels of surfactants are mixed and agitated in hot water.

When the concentration of surfactants exceeds what is known as the 'critical micelle concentration' (*CMC*), they will associate into discrete spheres or particles known as micelles. Their hydrophobic ends will be oriented inward, while their hydrophilic tails will be oriented outward. The inside of the sphere, then, is a localized hydrocarbon-rich environment conducive to the solubilization of organic molecules. When acrylic monomers are added to this blend of surfactants and water, they, along with polymerization initiators, will migrate into the interior of the micelles, and polymer formation will begin.

One of the advantages of this type of system is that its viscosity is, for the most part, dependent on the continuous (aqueous) phase. Very high-molecular-weight polymers (several hundred thousand) can be formed in micelles with little or no increase in the viscosity of the system. As was the case with solution acrylics, the properties of the final polymer can be controlled by judicious selection of monomers and reaction conditions.

How latex coatings form films is still the focus of considerable research. It is commonly believed that it is a multi-step process.

The first step is the evaporation of water, allowing the discrete latex particles to come into physical contact. The next step, which is assisted by the incorporation of small amounts of organic co-solvents, or 'coalescing aids', is a melting or fusion of latex particles into one another. The final step is the evaporation of the residual water and, over time, the evaporation of the co-solvents.

The coalescence of the latex particles is critical in achieving the desired properties of the coating. Two of the main factors governing this are humidity and the glass transition temperature (T_g) of the acrylic emulsion.

For coalescence to occur, the latex particles must first come into close contact with one another, which will not happen until most of the water evaporates from the coating. The rate of evaporation of water will be much slower under humid conditions than under dry conditions.

Once the latex particles are in close proximity to one another, effective coalescence will still not occur if the T_g of the latex is significantly higher than the ambient temperature. In such a situation, the latex will be in a hard, glassy state, with little free volume and limited molecular motion. The films formed under such circumstances will have little cohesive strength and high permeability.

Paradoxically, it is generally desirable to have a final coating with a T_g several degrees higher than the expected ambient temperature. A T_g that is lower than ambient would cause the coating to be soft and sticky, and objectionably porous. Windows would stick to their window sills; house paint would pick up and retain dirt; and coatings for metals would allow moisture and possibly other deleterious materials easy ingress.

One way in which this conundrum can be solved is by the coalescing aid serving in the dual role of a volatile plasticizer. During film formation, the coalescing aid, or organic co-solvent, helps to solubilize and plasticize the coalescing latex particles, thus depressing their T_g to below ambient. Within a day or two after application, however, the co-solvent has largely evaporated, and the T_g of the dried film is now higher than the prevailing ambient temperatures.

Two monomers which are routinely used to elevate the T_g are methyl methacrylate (T_g of the homopolymer is 105 °C) and styrene (T_g of the homopolymer is

100 °C). Methyl methacrylate imparts exterior durability, and the methyl group makes hydrolysis more difficult due to steric hindrance. Styrene also imparts good resistance to hydrolyis because it has no ester linkages to hydrolyze, although it can yellow and degrade when exposed to sunlight. It is often employed in primers to impart hardness, water resistance and lower permeability.

Acrylic latex coatings make up the bulk of the house paint and architectural coatings market. Two big advantages to the consumer are low odour and an easy clean-up with water. A big advantage in the regulatory arena is an extremely low VOC. When first introduced to the marketplace, latex coatings were greatly inferior to the tried-and-true oil-based paints, but rapid advancements in technology have quickly closed the gap. In addition to the application and VOC advantages of acrylic latexes, they also tend to be more 'breathable' than oil-based coatings. They allow the moisture inside a structure to more easily diffuse through the coating and therefore have less tendency to blister than do oil-based products.

In addition to breathability, acrylic latexes are also usually more flexible than oil-based paints, especially after the latter have oxidized and cured for a year or two. Acrylics will flake and 'disbond' less readily from wood that has undergone repeated expansion/contraction due to changes in humidity and temperature.

However, acrylic latex coatings do not penetrate as effectively into the pores of the wood as do oil-based coatings, somewhat compromising initial adhesion. Latex coatings often contain a plethora of additives required to 'fine tune' the rheological properties of the coating. Many of these are water soluble, thus imparting water sensitivity to the dried films. Some additives are cellulose based, therefore making the coating subject to microbial attack while in the can.

Because of tighter VOC requirements, acrylic latex coatings are beginning to be used more extensively on structural steel. Formulation with fewer water-sensitive additives is required. Because of their higher surface tension and poorer solubilizing ability, surface preparation for such coatings must be quite good.

Many of the failures encountered with acrylic latex coatings are related either to water sensitivity or incompatibility with existing or subsequent coatings.

Water sensitivity can be due to poor cure (coalescence), perhaps because of low temperatures or high humidity. Although the evaporation of water from the films is greatly retarded by high humidity, the evaporation of the co-solvent, or the coalescing aid, is not. When the humidity finally does go down and the water eventually evaporates, the co-solvent may not be around in sufficient quantities to promote good film formation. Water sensitivity can also be due to excessive or non-judicious use of additives.

Acrylic latex coatings have virtually no organic solvents in them, and usually have functional groups of low to medium polarity, with little ability to form covalent bonds. They also have little ability to soften or otherwise bond with hard, smooth coatings. They are generally not recommended for application over thermoset coatings (including aged alkyds). Likewise, solvent-rich, high-stress coatings such as epoxies could soften and lift an underlying coat of acrylic latex.

Related to the acrylic latex coatings are the water-reducible or so-called 'water-soluble' acrylics.

Water-reducible acrylics are made by incorporating a certain amount of carboxylic acid functionality into the pendant groups along the acrylic backbone. This is usually carried out by using a few percent of a monomer such as acrylic acid. By then using a volatile amine such as ammonia to react with the acid groups, salts are formed which allow the resins to be dispersed, or reduced in, water. The pH of such coatings is usually well over 8. Such resins are more often used as part of a baked thermosetting system, rather than as room-temperature lacquers.

5.7.3 Thermoset Acrylics

Another common use of acrylic resins is in baked, thermosetting systems for automotive, appliance and other OEM finishes. The acrylics used in these formulations have functional groups that allow them to react with other cross-linking resins. Since the molecular weight will increase dramatically during the baking process, the initial molecular weight of the acrylic resin can be low (roughly 10 000–20 000), resulting in a relatively high-solids product and a concomitantly low VOC. It is common to use monomers such as hydroxyethyl methacrylate to impart the desired amount of hydroxyl functionality to the acrylic. Most of the other monomers used in thermoset acrylic resins are the same as those used in thermoplastic acrylic lacquers. Such resins are usually cross-linked with MF or benzoguanimine–formaldehyde (BF) resins. Such resins were discussed above in Section 5.6.

Unlike several other types of thermoset coatings, an acrylic-MF or acrylic-BF coating consists of only a single component. The cross-linking reaction is so slow at room temperature that it does not become a factor. However, the reaction rate is not zero, so the choice of catalyst is important if shelf lives of more than 6 months are required. Normally, the applied coatings are baked at 250 °F (121 °C) or higher to achieve satisfactory cure.

The reaction between a hydroxyl-functional acrylic resin (R—OH) and an MF or BF resin occurs at both the hydroxyl and ether groups. The reactions are catalyzed by small amounts of strong acids such as *p*-toluene sulfonic acid (pTSA). As expected, acrylics with primary hydroxyl functionality react faster than those with secondary functionality. The rate of reaction with MF or BF resins will depend somewhat on the volatility of the alcohol formed. The more quickly this reaction product leaves the dried film, then the more the reaction will shift to the right.

Generally, an MF resin is used at levels of perhaps 20–30% of resin solids. Higher levels of MF resin will generally result in brittle coatings, perhaps as a result of excessive self-condensation.

The amount of acid catalyst can affect several properties of the cured film, as well as influence shelf stability. While higher levels of catalysts have the beneficial effect of lowering cure temperature, in some cases they can compromise the durability of the cured coatings. This is because, in addition to cross-linking between hydroxyl groups and MF resins, cross-linking can also occur between carboxylic acid groups on the acrylic and hydroxyl or ether groups on the MF resin. This results in the formation of an ester linkage, which can subsequently degrade by hydrolysis. This hydrolytic degradation can be catalyzed by the residual acid left in the film.

Thermoset acrylic/MF coatings can be formulated that have wide variations in properties, since their final characteristics are derived not only from the specific acrylic and MF resins, but from the type and amount of catalyst, pigmentation, and even curing conditions. Properly formulated and cured coatings have excellent adhesion and chemical resistance, along with superior weathering and ultraviolet light resistance. It is not surprising that they are so widely used in automotive topcoats. They are also used as appliance coatings and as coil coatings.

Potential problems with acrylic/MF coatings include poor application properties if the coatings are near or beyond their storage lives, and potential susceptibility to hydrolysis if too much residual acid catalyst remains in the cured films. Other problems are related to cure problems. Too low a cure can result in unexpectedly soft coatings with low cross-link density and poor chemical and physical properties. Too high a bake can result in brittle coatings and even yellowing.

There are numerous, competing reactions occurring simultaneously when an acrylic-MF coating is curing. The two most obvious competing reactions are that of the acrylic-MF reaction and the self-condensation of the MF resin with itself. Typically, the stoichiometry and catalyst package is optimized for a certain time and temperature. If the actual curing conditions are different, the relative rates of the reactions may change, thus resulting in coatings with less than optimum properties.

BF resins are also commonly used to cross-link with acrylics. Because there are only two $-NH_2$ groups per molecule versus three on an MF resin, their cross-link density will be less. Coatings prepared with BF resins have better alkali and detergent resistance than those prepared from MF resins. They are poorer, however, in terms of their exterior durability. Acrylic resins can also be cross-linked with isocyanates. Coatings produced in this fashion, however, are considered as urethanes, and their chemistry will be discussed below in Section 5.9.

5.8 POLYESTERS

Polyesters are polymers obtained by reacting multifunctional acids or anhydrides with multifunctional alcohols (polyhydric alcohols) to form a polymeric backbone containing ester groups. In this sense, they are similar to alkyds, but without the extra modification due to natural oils. This results in resins or polymers that have a much simpler structure than that of an alkyd.

As a consequence of the lack of oil modification, polyesters do not contain the fatty acid characteristic of alkyds. The acids used in the manufacture of polyester resins do not contain conjugated double bonds. Since these double bonds are the sites for oxidative cross-linking (curing) of alkyd resins, it is apparent that polyester resins must rely on a different curing mechanism than an alkyd. The curing mechanism largely depends on whether or not the polyester resin is saturated or unsaturated. It is convenient to regard polyesters as falling into one of these two classes.

5.8.1 Saturated Polyesters

With the obvious difference that there is no oil modification, saturated polyesters are manufactured much like alkyds. They can be produced from a large number of di- and/or polyfunctional alcohols and acids. Some of the commonly used starting ingredients are shown in Tables 5.3 and 5.4.

Polyester resins equal, or perhaps even surpass, acrylic resins in their ability to be molecularly engineered. With such a large number of combinations of starting materials, polyesters can be produced with a wide range of chemical and physical properties. The characteristics of polyesters can be affected by the resin

Table 5.3 Structures of some common polyols used in the production of polyesters.

Polyol	**Structure**
Neopentyl glycol	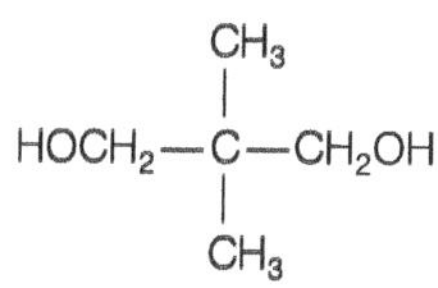
Trimethylol propane	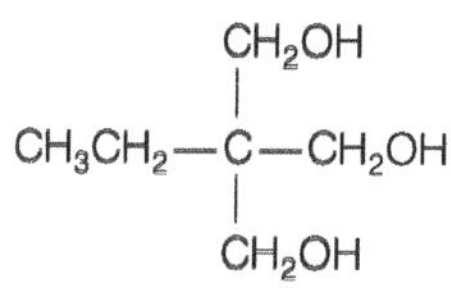
Pentaerythritol	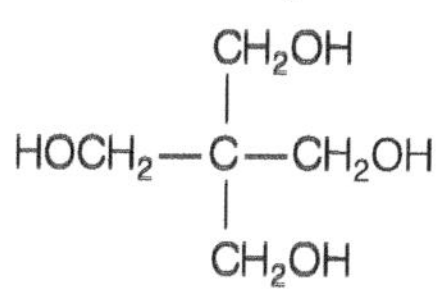
1,6-Hexanediol	

Table 5.4 Structures of some common acids used in the production of polyesters.

Acid	Structure
Phthalic anhydride	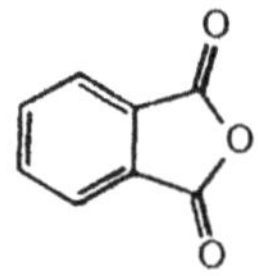
Isophthalic acid	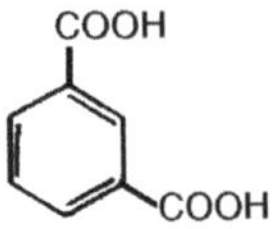
Adipic acid	
Azelaic acid	

used to cross-link them. Therefore, one can see the danger in carelessly 'lumping' all polyesters together.

The type of alcohol, type of acid, stoichiometry and reaction conditions determine the properties of the polyester resin. Four of the most important characteristics are molecular weight, type of end groups, degree of branching and choice of monomers.

Like any other solution resin, viscosity is a function of molecular weight and, therefore, so is the VOC. Most polyester resins have relatively low molecular weights, typically in the 5000–10 000 range.

Resins produced by reacting a diol, such as ethylene glycol, with a diacid, such as isophthalic acid, are linear. If a trifunctional (or higher) ingredient is used, branching will occur. Branched polymers are usually of higher viscosity than linear polymers of comparable molecular weight, and also ultimately result in cross-linked films of greater density and chemical resistance, and less flexibility and impact resistance. All other things being equal, a polyester containing a high level of pentaerythritol (hydroxyl functionality of 4) will result in a much harder, more chemically resistant coating than one based on 1, 6-hexanediol. The latter will perform much better where flexibility and impact resistance are required.

The distance between the functional groups also has an effect on the final properties of the coating. Polyesters prepared from succinic acid, where the acid groups are separated by only two carbon atoms, will be less flexible than those prepared from azelaic acid, with five carbon atoms separating the acid groups.

Since polyesters contain ester groups in their backbone, they are susceptible to alkaline hydrolysis (saponification) in the same way as alkyds. However, they have generally been found to be more resistant to hydrolysis than alkyds.

At least in some cases this increased resistance appears to be due to steric effects, which slow down the rate of the hydrolysis reaction [6]. For example, polyesters produced from neopentyl glycol would be expected to be more resistant to hydrolysis than those produced from 1, 6-hexanediol.

Saturated polyester resins have no mechanism to cure, and therefore coatings prepared from them utilize cross-linking resins such as MF resin, BF resin or epoxy resin. Polyesters designed to cross-link with MF or BF resins are made by starting with an excess of polyol, such that the terminal end groups have hydroxyl functionality, while those designed to cross-link with epoxies utilize an excess of acid to impart carboxylate functionality.

The cross-linking reaction of a hydroxyl-terminated polyester with an MF or BF resin is much the same as that of the hydroxyl-functional acrylics previously discussed. Acidic catalysts are usually used, and baking temperatures are typically in the 250–300 °F (121–149 °C) range. The characteristics imparted by the MF or BF resin are similar to those imparted to the acrylic resins discussed above in Section 5.7.3.

Polyesters produced with carboxylic acid end groups can cross-link at the oxirane (see Figure 1.1) site in epoxy resins. This reaction is somewhat more sluggish than the hydroxyl–MF reaction, and temperatures closer to 400 °F (204 °C) are needed.

Coatings produced from saturated polyesters find a wide range of uses, such as flexible coil stock, can coatings, automotive base coats, appliance coatings, metal furniture coatings and exterior metal siding. For many applications, except those where aesthetics are highly desirable, polyesters can be formulated and applied as powder coatings.

It is also possible to formulate water-based saturated polyester coatings. This is achieved by using carboxylic acid-terminated polyesters, which are neutralized with volatile amines to make them water dispersible. Again, they are cross-linked and baked with appropriate curing agents. Unsaturated polyesters can also be reacted with isocyanates to form urethanes.

Aside from any problems associated with inadequate cleaning or pretreating of the metallic substrates to which these coatings are applied, potential problems include misformulation and curing errors.

Actual misformulation, especially where the end use of the product is known, should be rare. There is always the chance that a batching error can occur. Sending a product out the door with only half of the amount of MF resin will result in a final product with poor durability and inferior chemical resistance.

With any product that requires baking, cure should be one of the first things considered when a coating failure occurs. It is not uncommon for manufacturers to use ovens whose temperatures are not well monitored, or where the equipment used to monitor the temperature are not regularly calibrated. Variations in line speed can also affect cure.

5.8.2 Unsaturated Polyesters

Unsaturated polyester resins may be prepared by using an unsaturated acid in place of some of the saturated acid. Typically, the acids used are maleic anhydride (Figure 5.33) or fumaric acid (Figure 5.34). When the maleic anhydride ring opens, such as upon reaction with water, maleic acid is formed, which is simply the *cis*-isomer of fumaric acid.

In theory, numerous diacids and diols can be used to make unsaturated polyester resins. In practice, the predominant choices are isophthalic acid, maleic anhydride or fumaric acid, and propylene glycol. Polyesters made with fumaric acid tend to be harder than those made from maleic anhydride. More flexibility can be imparted by substituting an aliphatic acid such as adipic acid for some of the isophthalic acid. In some cases, resins are made using epoxy materials for part of the diol component.

Unsaturated polyesters cure by free radical polymerization at the double bond sites. It is common to use peroxides, such as benzoyl peroxide or MEK peroxide, along with a promoter such as cobalt naphthenate, to initiate and catalyze the free radical reaction. It is also common to use stryrene as a reactive dilutant. Styrene is a good solvent for the polyester, and by virtue of its double bond is also a co-reactant with it. It enables coatings to be formulated at near-zero VOC levels.

Oxygen can inhibit free radical reactions by reacting with radicals and by essentially 'capping' the reaction. This 'air inhibition' occurs at the surface of the coating, thus leaving it tacky. One way to overcome this is by formulating a small amount of wax into the coating. The wax will migrate to the surface and partially seal it off from oxygen. Other additives can also be used to reduce air inhibition.

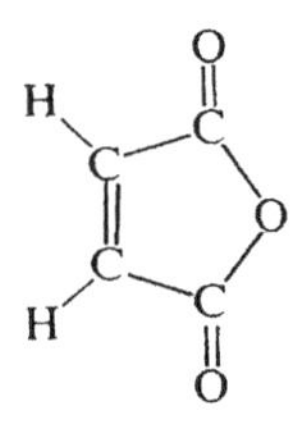

Figure 5.33 Structure of maleic anhydride.

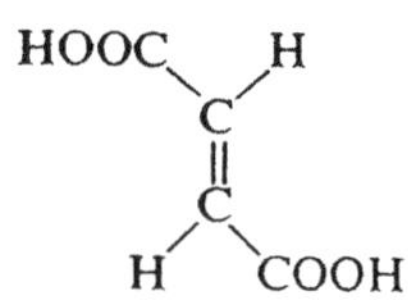

Figure 5.34 Structure of fumaric acid.

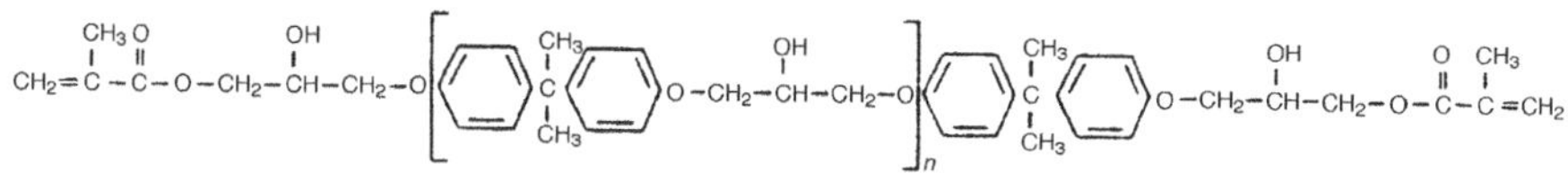

Figure 5.35 Structure of a vinyl ester.

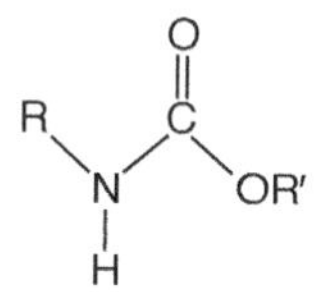

Figure 5.36 Structure of a urethane.

Unsaturated polyesters are much more common in the fibreglass industry than in the coatings industry. As coatings, they are often used in radiation-cured systems, where ultraviolet lamps or electron beam sources replace peroxide as initiators of free radicals. They are also used as high-build linings in severe chemical exposures, such as the interior of chemical storage vessels. In such uses they are often applied by plural spray equipment because of their very short pot lives once catalyzed. Chopped fibreglass may also be one of the ingredients of such coatings. The fibreglass imparts added strength, reduces permeability and allows for high-build applications.

Related to the unsaturated polyesters are the vinyl esters, which could perhaps be more appropriately considered as acrylics. These are typically formed by reacting acrylic acid or methacrylic acid with an epoxy. A typical structure is shown in Figure 5.35.

The only ester groups on these resins are at the end, thus giving them better hydrolytic stability than the typical unsaturated polyester coatings. Like the unsaturated polyesters, they cure by free radical polymerization initiated by peroxide catalysts and also contain styrene as a reactive diluent.

The vinyl esters and unsaturated polyesters typically impart considerable shrinkage stress and are very stiff. Surface cleanliness and surface roughness are thus very important in achieving good adhesion.

5.9 POLYURETHANES

Polyurethanes are one of the most versatile of all coatings, ranging from relatively simple lacquer-type coatings through high-performance automotive coatings to high-build chemically resistant elastomeric linings. They are usually formed by reacting an alcohol (R—OH) with an isocyanate (R—NCO) to form the urethane, or carbamate, structure shown in Figure 5.36.

Like epoxies, they usually consist of two components which must be mixed together shortly before application. One component typically contains a polyol, while the other contains the isocyanate.

The isocyanate group is highly reactive, and di-or polyfunctional isocyanates can be used to cross-link any number of resins that have functional groups containing active hydrogens. Although hydroxyl-functional resins are the mainstay of the industry, isocyanates can also react with aliphatic and aromatic primary amines, phenolics, carboxylic acids, secondary amines, ureas and water. Not all of these reactions result in the formation of urethane linkages, and hence the coatings are technically not urethanes. The classic example of this confusing nomenclature is the so-called 'moisture-cured urethane', which is not a urethane at all but is actually a polyurea.

When a diol is reacted with a diisocyanate, a linear polymer results, based on a backbone of urethane linkages. If water is present, it can react with the isocyanate to form a carbamic acid (Figure 5.37). The carbamic acid is not stable and readily decomposes to form an amine and carbon dioxide. Since amines are very reactive towards isocyanate groups, they will react to form a disubstituted urea (Figure 5.38).

Although the reaction of an isocyanate with water is slower than the reaction of an isocyanate with a primary hydroxyl group, the rate of reaction of isocyanates with primary amines is faster than the corresponding isocyanate–primary hydroxyl reaction. However, the relative reaction rate can change with different concentrations, temperatures and catalysts, which makes the study of such coatings all the more complicated (and interesting).

Once the disubstituted urea is formed, it is possible for the diversity to continue, as urea can also react with isocyanates, to form biurets (Figure 5.39).

The above reactions all occur at room temperature, and can be accelerated by the use of catalysts such as tertiary amines and organotin compounds (specifically dibutyltindilaurate). Isocyanates can also react with carboxylic acids, but at elevated temperatures, to form amides.

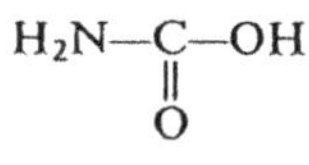

Figure 5.37 Structure of carbamic acid.

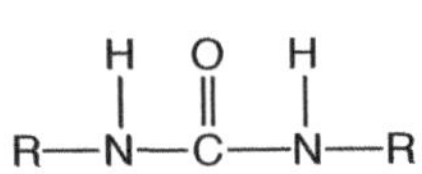

Figure 5.38 General structure of a disubstituted urea.

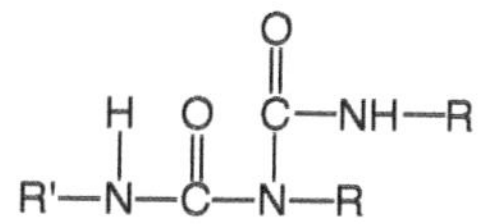

Figure 5.39 General structure of biuret.

The isocyanates used in coatings are usually divided into two types, namely, aromatic and aliphatic. The most commonly used ones are shown in Table 5.5.

Two of the major disadvantages of isocyanates are their toxicity and their volatility. It is common practice to use higher-molecular-weight versions. Thus, both 2,4-toluene diisocyanate (TDI) and 1,6-hexamethylene diisocyanate (HDI) are commonly used as their trimers, which are called isocyanurates. Another way to minimize the problem is to use isocyanate 'pre-polymers', made by reacting excess isocyanate with a polyhydroxy compound. Depending on the polyol used, polyurethanes based on aromatic isocyanates such as TDI and MDI can be formulated to produce hard coatings with excellent chemical and heat resistance. Unfortunately, a weakness of polyurethanes based on aromatic isocyanates is their strong tendency to discolour and chalk upon exposure to sunlight.

Polyurethanes based on aliphatic isocyantes, such as HDI and isophorone diisocyanate (IPDI), are greatly superior to the aromatics in terms of weathering resistance, and also have excellent chemical resistance. They are somewhat slower to cure than the aromatics, however, and are more expensive.

5.9.1 Two-Component Polyisocyanate/Polyol Coatings

The most common type of urethane coatings are the two-component, solvent-borne variety, where one component consists of a polyisocyanate and the other a polyol. The pigmented component usually contains the polyol, where trace amounts of moisture coming from the pigments have no effect on shelf life. Both aliphatic and aromatic isocyanates are used, with the former imparting significantly better colour retention and weathering resistance. The most common types of polyols are hydroxyl-functional acrylics, polyesters and polyethers.

Hydroxyl functionality is typically imparted to acrylics by incorporating a certain amount of 2-hydroxyethyl methacrylate (HEMA) into the monomer blend. HEMA provides primary hydroxyl groups for rapid cross-linking with the isocyanate. Occasionally, 2-hydroxypropyl methacrylate is used. The secondary hydroxyl functionality has the disadvantage of a slower reaction rate with the isocyanate, but the concomitant advantage of longer pot life.

The molecular weight of the acrylic resins used in urethane coatings is usually fairly low, of the order of perhaps 25 000. This is quite low when compared to a conventional single-pack acrylic lacquer, but the molecular weight of the final,

Table 5.5 Structures of some common isocyanates used in the production of polyurethanes.

Isocyanate	Structure
2,4-Toluene diisocyanate (TDI)	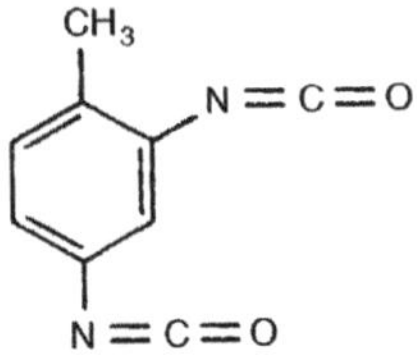
1,6-Hexamethylene diisocyanate (HDI)	$O{=}C{=}N{-}(CH_2)_6{-}N{=}C{=}O$
Isophorone diisocyanate (IPDI)	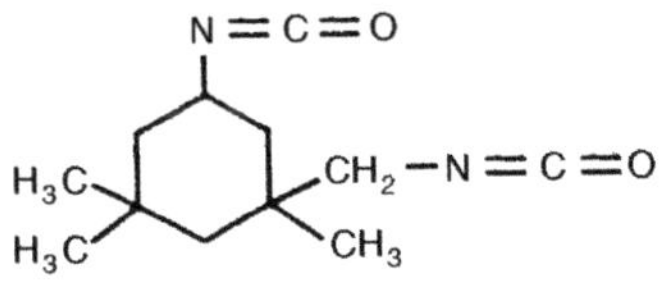
Diphenylmethane diisocyanate (MDI)	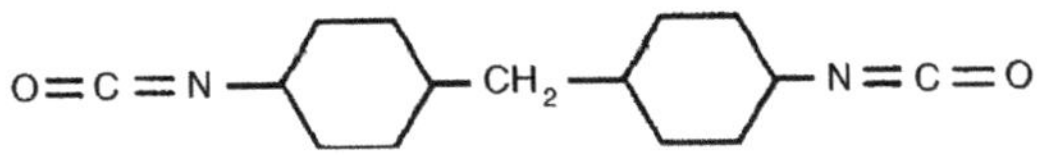
m-Tetramethylxylene diisocyanate (TMXDI)	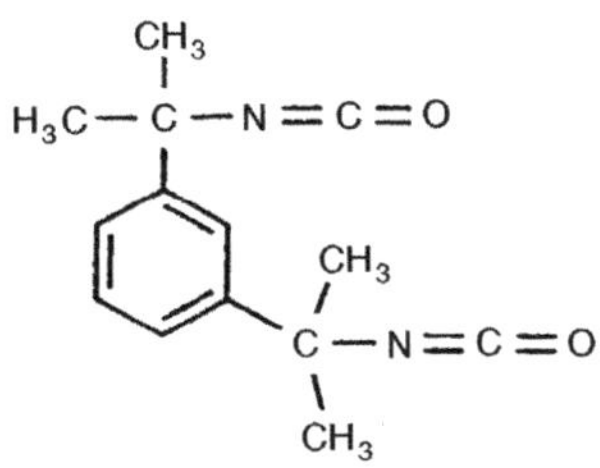
Isocyanurate of toluene diisocyanate	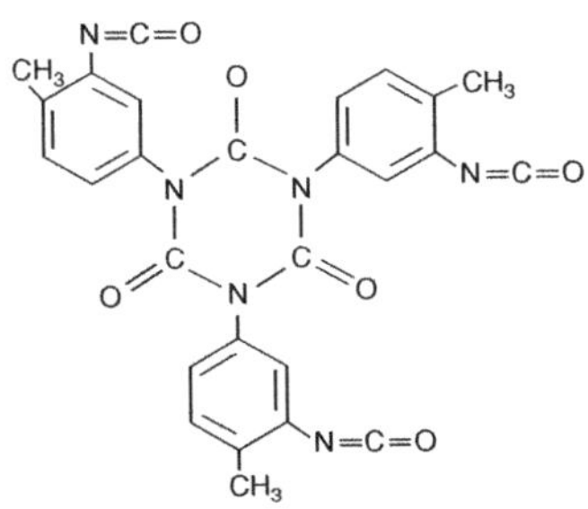
Isocyanurate of hexamethylene diisocyanate	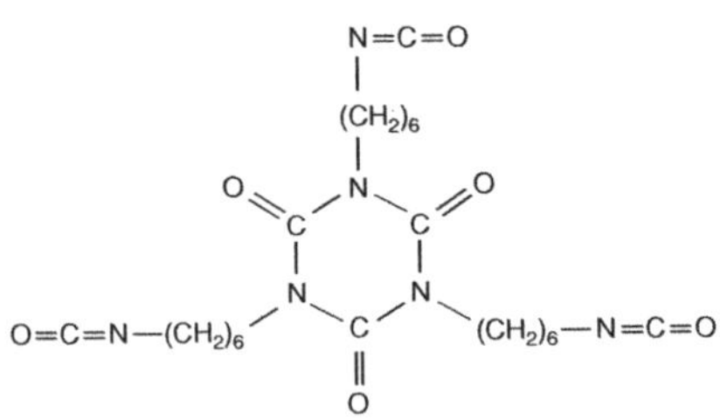

cross-linked film will be extremely high. The low molecular weight of the acrylic resin and of the isocyanate allows high-solids coatings of low to medium viscosity to be easily formulated.

As discussed above in Section 5.7, the variety of monomers allows for a wide latitude in the molecular engineering of acrylic resins. These same monomers are available for 'tailor making' acrylic–polyurethanes with various properties.

One critical factor, however, is the T_g of the final coating. It is desirable to have the T_g of the fully cured coating slightly below the curing temperature (which quite often is room temperature). The T_g of the coating will increase as its degree of cure increases. When the T_g of the partially cured coating becomes equal to or higher than the temperature at which the coating is curing, there will be a sharp decrease in the free volume. The polymer chains will have much less room in which to move, mobility will decrease and eventually cease, and it will not be possible for the reaction to reach completion. One way to avoid this is to cure the coating at an elevated temperature, but this is usually not possible for industrial maintenance coatings.

Hydroxyl-functional polyesters are also widely used in the manufacture of two-component polyurethanes. The chemistry and characteristics of hydroxyl-functional polyesters have already been discussed (see Section 5.8). As in the case of hydroxyl-functional acrylics, the polyesters used to make polyurethanes are of relatively low molecular weight. This allows for the formulation of high-solids, low VOC coatings. The same T_g consideration applies to polyester–polyurethanes as for acrylic–polyurethanes. The polyester–polyurethanes tend to permit higher-solids coatings with somewhat better adhesion to metals compared with the acrylics. The acrylics are often of lower cost, have superior exterior durability and are more resistant to hydrolysis.

Hydroxyl-functional polyethers are also used to make two-component polyurethanes. Since ether linkages are much less restrictive towards molecular rotation than many other functional groups, and since they tend not to have bulky side groups, polyether–polyurethanes tend to be very flexible, with much greater impact resistance than the much harder acrylic and polyester–polyurethanes. Their softness can make them unsuitable for many applications. However, they make superb, high-build elastomeric coatings and linings, and can survive relatively harsh environments.

Although acrylics, polyesters and polyethers are the most commonly used hydroxyl-functional resins used in two-component urethanes, they are not the only ones. Since alkyd resins have unreacted hydroxyl groups, they can also cross-link with isocyanates, as can epoxy resins through their secondary hydroxyl groups.

The two-component, solvent-borne urethanes have become perhaps the premier coating system for applications demanding excellent gloss and colour retention and good chemical resistance. Based on the above discussion, it is apparent that with the wide selection of available polyols and polyisocyanates, it is possible for the polymer chemist to produce urethane coatings that span a broad range of chemical and physical properties. Despite their diversity, there are certain common features among the various urethanes relating to their mode of failure. This is especially true regarding their behaviour with water.

Isocyanates are very reactive and can react with water to form a disubstituted urea. The by-product of this reaction is carbon dioxide, which is a gas at room temperature. Not only does the reaction with water interrupt the designed curing pathway of the coating, resulting in unanticipated and often undesirable physical and chemical properties, but the escaping CO_2 can result in numerous voids in the interior of the cured coating. This 'Swiss cheese' effect can have serious consequences for the permeability and mechanical strength of the coating.

Water can come from many sources, including rain shortly after application, painting during periods of high humidity, using non-urethane-grade solvents which have small amounts of moisture, water in the paint lines and failure to properly seal containers after opening. If the water is present on the substrate as a consequence of a dew point inversion, not only will voids result but likely poor adhesion as well. In order to help guard against the possibility of moisture damaging the curing coating, most urethanes are formulated with a slight excess of isocyanate. Ratios of isocyanate to hydroxyl of 1 : 1 to 1 : 0 are common.

It is not unheard of in the paint industry for a contractor to thin the coating with the wrong solvent. More often than not, this usually has only minor consequences. However, if a two-component urethane is thinned with a solvent containing an alcohol, the monofunctional alcohol will react with the isocyanate and stop the cross-linking process. The consequences will depend on how much alcohol was used, but could be disastrous.

As with any two-component coating, there is always the possibility of a mixing error. Because of the complexity of the various reactions that can occur in a curing urethane, it is not always possible to predict exactly how a certain mixing error will affect the properties of the coating. However, an excess of polyol will nearly always result in a softer coating. This is because there are no side reactions for the polyol to undergo, and it will act as a plasticizer.

In industrial maintenance applications, two-component urethanes are often used as finish coats over epoxy intermediate coats. While not a urethane problem *per se*, they (and other coatings) can be very sensitive towards even small quantities of chalked coating on the surface of the epoxy. In some cases, this extremely thin layer of chalked epoxy, which can sometimes form as quickly as a week or two, can cause peeling of the urethane topcoat from the epoxy.

5.9.2 Urealkyds

Urealkyds, also called urethane alkyds, should not be confused with the two-component urethanes formed by the reaction between an isocyanate and the hydroxyl groups on an alkyd resin. Rather, they are basically modified alkyds where some of the phthalic anhydride is replaced with an isocyanate such as TDI.

Urealkyds are single-component coatings that typically dry faster and harder than conventional alkyds. They also have improved abrasion resistance and resistance to hydrolysis. Unfortunately, they are more expensive and, because of the use of the aromatic TDI, have inferior colour retention. The latter can be improved through the use of aliphatic isocyanates, but at even greater cost. They cure by the same process as alkyd resins, and therefore must be formulated with the same types of driers used in alkyds.

Urealkyds are largely used as varnishes, although technically they are not true varnishes. They are often used as clear varnishes on wood, where hardness and abrasion resistance is important.

5.9.3 Moisture-Cured Polyurethanes

Oddly enough, although water is considered a contaminant with conventional two-component polyurethanes, it is the primary cross-linking agent, along with the isocyanate, in moisture-cured polyurethanes. The source of the water is atmospheric moisture, and the polymer obtained is not a polyurethane, but rather a polyurea.

In the case of a two-component acrylic or polyester urethane, the bulk of the final, cured film is made up of the acrylic or polyester starting material, which initially may have had a molecular weight of 15 000–25 000. However, the hydroxyl 'analogue' to the polyester or acrylic is simply water, with a molecular weight of only 18. Therefore, viable coatings cannot be made simply by reacting water with the same relatively low-molecular-weight isocyanates used in conventional two-pack coatings.

The isocyanates used in moisture-cured urethanes are resins which have two or more isocyanate groups per molecule. They are generally made from hydroxyl-terminated polyesters reacted with an excess of a 'conventional' isocyanate such as TDI or MDI, or with an aliphatic isocyanate if colour retention is important. It is unavoidable that there are also a certain number of urethane linkages formed.

The coatings are single-component products, and when applied cure by reacting with atmospheric moisture. The rate of cure is humidity dependent. Below approximately 30% relative humidity, the cure is rather sluggish, and above 80%, the rate of carbon dioxide evolution from the curing reaction is rapid enough to generate bubbling in the coating. Therefore, humidities in the 30–80% range are preferred.

One of the main advantages of moisture-cured urethanes is the fact that they are single-component products. This saves time for the contractor and eliminates 'mis-mixing' as a potential cause of failure.

Other advantages include the ability of these products to be applied over slightly damp surfaces, rapid cure and a final coating which is hard, tough, somewhat flexible, and which has good adhesion to metal and good chemical resistance. Their light stability is variable, depending to a large degree on the type of isocyanate used.

One of the obvious disadvantages of moisture-cured urethanes is their sensitivity to moisture. They are often packaged under a blanket of nitrogen, and once opened must be used within a few days to avoid gelation. Great care must also be taken in the selection and handling of pigments, since many pigments contain trace amounts of water which can be detrimental to storage stability.

Another potential disadvantage of moisture-cured urethanes is bubbling caused by rapid CO_2 generation under conditions of high humidity. This problem is exacerbated at higher film builds. If high film build is desired, it is sometimes necessary to apply several thin coats.

As was the case with conventional two-component urethanes, alcohols must not be introduced into the coating. If such thinner is used, the alcohol will react with the isocyanate groups and severely interfere with the curing reaction.

5.9.4 Polyurethane Lacquers and Dispersions

Single-component, solvent-based polyurethane lacquers and single-component polyurethane dispersions in water are also available. The lacquers are simply thermoplastic, high-molecular-weight linear polymers prepared by reacting diols with diisocyanates. They are low-solids coatings that dry by solvent evaporation, and are used on flexible substrates such as fabrics and leather.

The polyurethane dispersions are essentially urethane latexes, and various methods are employed in their manufacture. They cure by the same process of water evaporation and coalescence as typical acrylic latex.

5.9.5 Two-Component Water-Borne Polyurethanes

Because of increasingly stringent VOC regulations, two-component, water-borne polyurethanes have been recently introduced. The problems in making such a coating are obvious, since the solvent (water in this case) can react with the isocyanate component. Several approaches have been utilized. These include the use of a blocked isocyanate and the preparation of a polyurethane polyol which is either emulsified in water by using surfactants or is made water reducible by incorporating some carboxylic acid functionality into the polymer and subsequently forming salts via the addition of a volatile amine.

Prior to application, the water-borne polyol component is mixed with a compatible isocyanate, which is not water-borne. The isocyanate then diffuses into the polyol phase, where it is essentially emulsified by the polyol and subsequently cross-links it.

Although the isocyanate reacts faster with the polyol than with the bulk water phase, it is apparent that the water must consume at least some of it. Therefore, NCO : OH ratios approaching 2 : 1 are often used to make up for this. The relative rate of the various reactions will be influenced by, among other things, the temperature and relative humidity at the time of application. If the temperature is high and the humidity is low, the polyol–isocyanate reaction will be greatly favoured. If the temperature is low and the humidity relatively high, the water–isocyanate reaction will proceed more readily. Obviously, it is possible for the same coating used on the same job site to have substantially different properties depending on the weather at the time of application or, if a baked system, to be substantially influenced by oven conditions.

5.10 VINYLS

Strictly speaking, any coating comprised of a polymer produced by the addition polymerization of a vinyl monomer, or a combination of vinyl monomers, could be considered a vinyl. A vinyl monomer is any molecule having the following structure:

$$CH_2{=}CR_1CR_2$$

Based on the above, one could argue that poly(methyl methacrylate) is, therefore, a vinyl. However, in the coatings industry, it is commonly accepted that the term 'vinyl' applies to a relatively small group of monomers, consisting principally of vinyl acetate (VA), vinyl chloride (VC), vinylidene flouride and vinyl butyral.

5.10.1 Solution Vinyls

Solution vinyls are typically based on copolymers of VC and VA. The resulting poly(vinyl chloride) (PVC)–polyvinyl acetate (PVA) copolymers, which contain approximately 14% VA, have molecular weights of 70 000 or higher, and dry by simple solvent evaporation. As a consequence, coatings produced from them are very low in solids, and are being used less and less as VOC regulations continue to tighten.

Although their usage is declining, solution vinyls have several desirable properties, including good resistance to acids, bases and water, and good flexibility and impact resistance. At one time, they were widely used in the water and wastewater industry, in chemical and paper plants, and on submerged structures such as locks and dams.

In addition to their very poor solvent resistance, solution vinyls have relatively poor adhesion. In order to improve their adhesion to metals, a small amount of maleic acid or vinyl alcohol is copolymerized with the VA/VC. The more polar nature of this group greatly improves the adhesion.

5.10.2 Plastisols and Organosols

PVC by itself is not a suitable resin for coatings. It is brittle, difficult to dissolve and has poor adhesion. In the plastics industry, however, hot applied high-molecular-weight PVC particles dispersed in a plasticizer, known as 'plastisols', are used for a variety of applications. Although not a convenient route for a paint or coating, when such plastisols are diluted with solvent, the resulting 'organosols' find considerable use as coil coatings for roofing and siding applications. After application, the coatings are baked to fuse the individual polymer particles into a continuous film.

Cured organosols have many of the desirable properties that were discussed in the previous section on solution vinyls. They have, however, their own set of potential problems.

The most obvious problem is that of poor adhesion, since the resin system is still the very non-polar PVC. Therefore, it is common to use a primer as a 'tie coat' to improve the adhesion of the coating system to the metal, and/or to apply some type of conversion coating to the steel stock to improve the adhesion.

Another potential problem with organosols is the evaporation of plasticizer. Although this can be minimized or even eliminated through proper choice of materials, if the wrong plasticizer is used in a very hot southern environment, embrittlement of the coating can gradually occur over an extended time period.

A third potential problem of an organosol is the fact that PVC is subject to degradation as a result of exposure to heat and light. PVC can undergo dehydrochlorination, resulting in the evolution of hydrochloric acid and the production of conjugated double bonds in the vinyl backbone. Not only can the hydrochloric acid being released have detrimental effects on the pigments and plasticizers in the coating, but the extensive system of conjugation can result in dramatic discolouration of the coating. Furthermore, over time, this conjugated polymer can begin to cross-link via an auto-oxidation process similar to that at work in alkyds, thus resulting in a greater cross-link density and associated embrittlement. This problem can be greatly alleviated by the use of appropriate stabilizing agents.

5.10.3 Vinyl Fluorides

Coatings can also be made from vinyl fluoride polymers, the most important of which is poly(vinylidene fluoride) (PVDF). This polymer is not readily soluble in most common organic solvents and is therefore used as a plastisol-like dispersion in a solution of acrylic resin.

Coatings based on PVDF have an outstanding chemical resistance, solvent resistance, heat resistance and weathering resistance. They are also relatively flexible. Because of the high temperatures (450–500 °F (204–260 °C)) required to fuse the individual particles of PVDF into a cohesive, continuous coating, they are principally used as high-performance coatings on coil stock intended for commercial buildings. Their flexibility allows them to be bent and formed without cracking. They are usually applied by reverse roll coating at thicknesses of 1.5 mil or less.

5.10.4 Poly(vinyl butyral)

Poly(vinyl butyral) is prepared by first partially hydrolyzing PVA to give what could essentially be considered as a copolymer of VA and vinyl alcohol. This 'copolymer' is subsequently reacted with butyraldehyde to form poly(vinyl butyral).

Other poly(vinyl acetal)s can be made by using aldehydes other than butyraldehyde. They are all high-molecular-weight, low-solids coatings, and are therefore used in much smaller quantities than they once were. Because they contain a certain number of secondary hydroxyl groups, they can be reacted with a number of other resins at elevated temperatures to modify their properties. For instance, when reacted with phenolics and epoxies for can linings, they improve formability [6].

In the maintenance coating industry, poly(vinyl butyral)s are best known as 'wash primers'. These are two-package coatings containing an inhibitive chromate pigment, where one component contains a large amount of sulfuric acid. When used as pretreatments over metal surfaces, they actually form primary chemical bonds with the metal, and substantially improve the adhesion of certain subsequently applied coatings such as maleic acid-modified vinyls, alkyds and epoxies. The product is sensitive to formulation error, must be applied in a very thin film (0.1–0.3 mil) and topcoated within a day, or sometimes within only a few hours.

5.10.5 Vinyl Latexes

Perhaps the single greatest use of vinyl coatings is in vinyl latex-based house paints. The high molecular weight of the vinyl resin is not a problem, since the coating takes on the viscosity of the continuous phase, which is water in this case.

The glass transition temperature for pure PVA is too high to achieve good coalescence at room temperature. Although one way to solve this problem is with plasticizers, it is more common to copolymerize the VA monomer with some low-T_g acrylic monomer, resulting in the commonly used 'acrylic-modified vinyl latexes'.

Vinyls and acrylic-modified vinyls are distinctly inferior to the 100% acrylic latexes in terms of both weathering and hydrolysis. Therefore, they are primarily used as interior coatings. Most flat interior wall paints are vinyl latexes. They are also commonly used in latex block fillers, which are highly pigmented coatings, essentially used as a primer to fill the imperfections in rough masonry walls prior to the application of a smoother, glossier topcoat.

Although vinyl latex coatings are inferior to acrylics for exterior application, this does not prevent them being used in this fashion. Furthermore, if used as interior block fillers, they can be subject to premature failure if there is penetration of water through an exterior wall or from a construction defect, or simply from being used in the wrong location, such as shower facilities. Failures are often encountered with such coatings when topcoated with solvent-borne, high-stress coatings such as two-component epoxies. The solvent in the epoxy softens the solvent-sensitive vinyl, and the shrinkage stress of the curing epoxy is adequate to cause a cohesive failure within the weakened vinyl. Such failures are exacerbated by excessive thickness of either the vinyl or the epoxy, and by low curing temperatures of the vinyl block filler, which can often occur on new construction.

5.11 BITUMINOUS COATINGS

The term 'bituminous coating' causes much confusion, primarily because of the poorly defined nature of the materials from which they are made. Basically, a bituminous coating has its origin in either asphalt or coal tar. Although the class also includes coumarone indene resins (obtained from coke ovens) and hydrocarbon resins based on low-molecular-weight unsaturated hydrocarbons obtained from gasoline production, by far the largest quantities of such coatings are the asphalts and coal tars. Both are black or dark brown, due principally to a suspension of fine, elemental carbon particles.

Asphalt can be obtained from both natural sources and as residues from the distillation of certain types of crude petroleum oils. It is basically a semi-solid mixture of numerous hydrocarbons, which for the most part are aliphatic in nature. When obtained from natural sources, they are associated with varying levels of minerals. Asphalt obtained from Trinidad is high in mineral content and is used mostly for road paving material. In contrast, gilsonite, which is obtained from deposits in North America, is a jet black, brittle solid of very low mineral content and is suitable for asphaltic coatings. It is soluble in hydrocarbon solvents and is compatible with many oils and resins. It is used to some extent in high-gloss asphaltic varnishes.

Certain types of crude oil, such as Wyoming sour crude, contain 40% or more asphalt. The asphalt is obtained from the residues of the distillation process. Asphalts obtained from crude oils are generally preferred as coatings since they are of more uniform composition and contain lesser amounts of mineral impurities.

Asphaltic coatings are relatively inert and have extremely good chemical and water resistance. They are also inexpensive. In addition to their well-known use as roofing materials, they are also used as coatings for buried steel and concrete structures. Hot applied asphalt coatings often contain fibrous reinforcing materials, or sometimes even glass cloth, and are applied at thicknesses of 100 mil or more. They are used as barrier coatings on such structures as underground storage tanks and buried pipelines.

Coal tar is a complicated mixture of hydrocarbons, primarily of aromatic character, which constitutes the major part of the liquid condensate obtained from the distillation of coal to coke. As a consequence, the production of coal tar is closely related to the production of steel. Although steel production has steadily risen, the production of coal tar has not, because more efficient blast furnaces require lesser amounts of coke. Although napthalene is the single most prevalent compound in coal tar, it is present at levels of only about 6–10%. It is estimated that coal tar contains from 5000 to 10 000 different compounds, a mere 300 of which have been identified [7]. The compounds range from low-molecular-weight materials to a high-molecular-weight component known as coal tar pitch.

Coal tar is used in thick, hot applied coatings for tanks and pipelines, and as room-temperature, somewhat lower-solids coatings for steel subjected to water immersion. It is often used in the waste-water industry. Like the asphalts, it has good chemical and excellent moisture resistance, is inexpensive and can be applied as thick barrier coatings.

Because of their simple hydrocarbon structure, asphaltic and coal tar coatings have very good chemical resistance and excellent water resistance. They also have some major disadvantages, not the least of which is their aesthetic appeal. In addition to basically coming in only one colour, the coal tar coatings in particular have poor light stability, which is perhaps a consequence of their much greater aromatic content.

Because of their relatively high amount of unsaturation, the upper surface of bituminous coatings tends to oxidize and embrittle, and hence can crack or wrinkle. In addition, because of their poor solvent resistance, there are certain applications where they cannot be used. Not only can solvents from an industrial environment damage them, but they can generally not be topcoated by 'conventional' solvent-borne coatings, since a black-coloured bleed-through may result.

Perhaps the most important class of bituminous coating is the coal tar epoxy. Unlike asphalt and coal tar coatings, these are two-component thermosetting materials. The epoxy and cross-linking chemistries are similar to that already discussed above in Section 5.3, but one or both components are modified with high levels of coal tar. Since there is some evidence that components of the coal tar can slowly react with epoxy resins, many formulations have the coal tar component in with the converter component. Coal tar is used rather than asphalt because of its better chemical and water resistance.

The coal tar undergoes little chemical reaction with the epoxy/converter system, but upgrades the coating by virtue of its improved chemical and water resistance properties, as well as imparting high-build properties. In the same fashion that coal tar imparts its desirable properties to the epoxy, the epoxy imparts properties of its own to the coal tar. This includes improved solvent resistance, high cohesive strength, hardness, abrasion and impact resistance. Although epoxies are not known for their resistance to sunlight, coal tar epoxies have better weathering ability than straight coal tars. They do, however, develop an objectionable surface appearance, aptly described as bronzing.

Coal tar epoxies make very strong, highly impermeable barrier coatings with excellent chemical resistance. As such, they are used extensively in marine application and in the water and sewage treatment industries.

In addition to the surface bronzing of coal tar epoxies, other disadvantages include difficulties in application and re-coating. They are two-component, often highly viscous materials, and as a consequence there is always the possibility of 'mis-mixing'. With other two-component materials, application near the end of their pot life may result in wetting problems and the potential for reduced adhesion. Furthermore, the coatings cure quickly to a dense, hard, relatively smooth finish, which makes for a short (as brief as a day or two) re-coat window. It is often desirable to apply them in a single coat to avoid the possibility of inter-coat delamination.

5.12 INORGANIC AND SILICONE-MODIFIED COATINGS

5.12.1 Silicone Coatings

All of the coating types discussed so far have been based on organic (carbon-based) binders. However, inorganic resins based on silicones and silicates also play important roles in coating technology.

Silicones, which are also called polysiloxanes, have the general structure shown in Figure 5.40. The monomers used to prepare these resins are chlorosilanes, such as trimethylchlorosilane and diphenyldichlorosilane. Chlorosilanes react with water to form silanols, which then condense with the elimination of water to form the silicone (or polysiloxane) resin.

If dichlorosilanes are used, the polysiloxanes are linear. If some trichlorosilane is also used, a three-dimensional branched structure is formed. The incorporation of monochlorosilanes result in chain termination and the production of low-

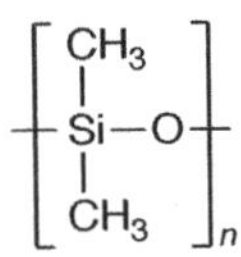

Figure 5.40 General structure of a silicone (polysiloxane) resin.

molecular-weight oligomeric materials. These oligomers are what are commonly referred to as silicone fluids or oils.

Silicone resins used for coatings are based principally on dichlorosilanes, but a certain amount of trifunctional monomer is required to achieve adequate chain branching and cross-linking. These resins are usually of low molecular weight, and once applied have to be baked to achieve full cure and hence the properties expected of them. This thermosetting reaction is not fast, and may take as long as an hour at 400–450 °F (204–232 °C). Certain catalysts can be used to lower either the time or temperature.

Most silicone resins are based on either methyl or phenyl chlorosilanes, or a combination thereof. All are characterized by excellent resistance to weathering and ultraviolet degradation, and display excellent gloss and colour retention. They are also extremely heat resistant. Chemical resistance is better than the alkyds, but not as good as some of the high-performance thermoset organic coatings.

As might be expected, silicones with a high methyl-to-phenyl ratio have greater gloss and colour stability than those containing more phenyl groups, and are more flexible. Those containing more phenyl groups relative to methyl have superior elevated temperature resistance. At very high temperatures, any organic substituents that are present will burn off, and the silicone can continue to cross-link to form high-molecular-weight silicon dioxide (essentially glass). Chimney paints pigmented with aluminium flake have been used for years based on this mechanism. So long as the brittle, glass-like coating is not disturbed, it can provide long-term protection at extreme temperatures.

Pure silicone coatings are primarily used in applications where their temperature resistance is paramount. Typical applications include stacks, mufflers, furnaces, cookware and engine equipment.

Silicone coatings are relatively expensive, and there are numerous applications where extreme temperature resistance is not required. In such cases, the performance of conventional organic coatings can be significantly upgraded by modifying them with a certain percentage of a silicone.

Perhaps the most common of such coatings is the silicone-modified alkyd. In many cases, this is simply a conventional alkyd that has been cold blended, or mixed, with a silicone. This is usually done to improve the exterior durability of the alkyd, although an increase in heat resistance is a beneficial side effect. Levels of 20 or 30% silicone are common. The properties of the coating can be further enhanced if a low-molecular-weight silicone intermediate is chemically reacted with the alkyd. The reaction usually occurs between hydroxyl groups on the alkyd and silanol groups on the silicone.

Silicones can also be used to modify polyesters. The resin is prepared by reacting a silicone with a hydroxy-functional polyester. The product thus obtained must be baked and further cross-linked after application, often with MF resins. These silicone-modified polyesters have an excellent combination of properties such as flexibility, hardness, adhesion, temperature resistance and weatherability.

They are used extensively on coil coated metal for wall panels, as well as on cookware and appliances.

Silicones can also be used to modify acrylics. Again, these are usually intended for the coil coating market, and are baked and cross-linked with a melamine resin. They produce excellent quality finishes.

5.12.2 Silicate Coatings

Binders based on tetraethylorthosilicate (Figure 5.41) are used extensively in the manufacture of one of the most important types of industrial maintenance primers, namely, the inorganic zinc-rich primer. Since occasionally constituents containing other than ethoxy groups are used, the coatings are sometimes referred to as alkyl or alkoxy silicates.

Ethyl silicate binders are prepared by starting with a solution of tetraethylorthosilicate dissolved in ethyl alcohol. When a small amount of water is added to this solution, partial hydrolysis occurs and a certain percentage of the ethoxy groups is replaced by hydroxyl groups, with the liberation of ethyl alcohol (C_2H_5OH). Some of the partially hydrolyzed tetraethylorthosilicate then condenses to form oligomeric pre-polymers, with a concomitant increase in viscosity.

The partially hydrolyzed binders are mixed prior to application with a large quantity of metallic zinc dust pigment (zinc loadings of over 80 wt% are common) to produce the zinc-rich primer. After application, the pre-polymers react with atmospheric moisture and subsequently condense, as shown in Figure 5.42. The chemical term for the binder being formed is polysilicic acid, and it is believed that this also partially reacts with small amounts of zinc salts which are naturally present at small levels in the metallic zinc dust pigment.

Zinc-rich primers have superb corrosion resistance. This is because they are formulated above the critical pigment volume concentration (*CPVC*) with a pigment (zinc dust) that is more reactive than iron. The zinc, which is in electrical contact with steel, will protect it galvanically. Furthermore, as some of the zinc begins to corrode, the oxidation products tend to fill in some of the porosity of

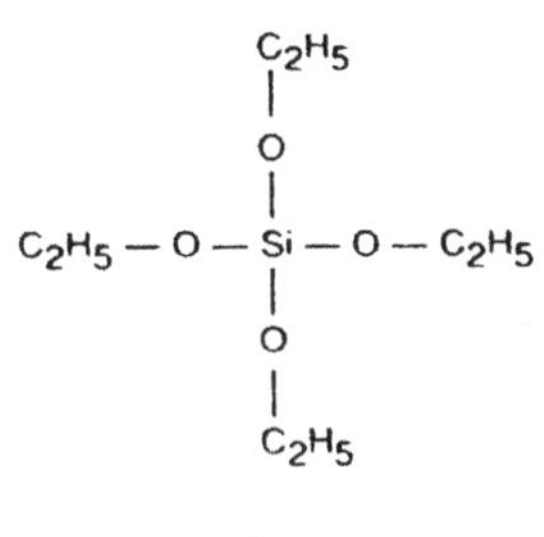

Figure 5.41 Structure of tetraethylorthosilicate.

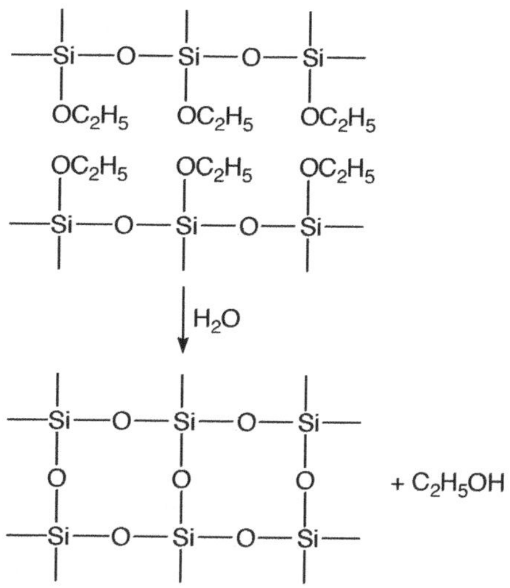

Figure 5.42 Illustration of the curing reaction of the ethyl silicate binder used in inorganic zinc-rich primers.

the high-PVC coating, thus improving its barrier properties. Inorganic zinc-rich primers are also hard and abrasion resistant, and have excellent adhesion to properly blast-cleaned steel and very good resistance to water, solvents and heat. They are used extensively on structural steel for a variety of applications.

However, inorganic zinc-rich primers do have their drawbacks. As might be expected of any coating formulated over its *CPVC*, they are brittle and have poor cohesive strength and impact resistance. Since the curing reaction with atmospheric moisture starts at the surface and works its way down, coatings applied much thicker than 4 mil or so have a tendency to mud-crack. Although inorganic zinc-rich coatings have very good adhesion to steel, they have poor adhesion to themselves. It is generally not possible to apply a second coat if the first coat was too thin. Luckily, a 1.5- or 2.0-mil coat gives excellent corrosion protection, especially when coated with a compatible intermediate or topcoat.

An additional problem occasionally encountered with zinc-rich coatings is the failure to keep the dense zinc dust pigment properly suspended in the pot during spray application. If some arrangement to keep the coating agitated is not observed, a film with inconsistent ratios of pigment to binder will result.

Perhaps the most common problem or failure an investigator will encounter with such coatings is a cohesive splitting of the primer when it is subsequently coated with an epoxy intermediate coat (zinc–epoxy–urethane systems are common in industrial coating situations). Quite often this is because the zinc-rich primer was not adequately cured at the time of topcoating. Consequently, the relatively strong solvents of the epoxy soften and weaken an already cohesively weak film, and the subsequent shrinkage stress of the curing epoxy is sufficient to pull the primer apart. The curing time of inorganic zinc primers can be very humidity dependent, a fact often not recognized on the coating manufacturers' product data sheets.

5.12.3 Polysiloxane Coatings

A relatively new generic type of coating, with commercial development beginning in the 1990s, is polysiloxane coating. These coatings are gaining acceptance for use in a variety of industries, such as new construction, marine, heavy-duty OEM and industrial maintenance.

As discussed in Section 5.12.1, compared to conventional organic resins and polymers, silicone resins have excellent resistance to weathering and ultraviolet degradation, and are also extremely heat resistant. Chemical resistance is also usually very good. These properties have led to the development of the silicone-modified alkyds and pure silicone coatings discussed in Section 5.12.1. However, while the silicone alkyds certainly have advantages over normal alkyds, they still have the basic limitations inherent with alkyd chemistry and are not a panacea for the entire coatings industry. Likewise, essentially pure silicone coatings, such as those used for very high-temperature applications such as heat stacks, have limited use because of their brittleness. Polysiloxane coatings have been developed to circumvent these limitations, and are essentially hybrid coatings which attempt to make use of the desirable properties of both silicone and organic resins.

The nomenclature of silicone resins can sometimes be confusing. Silanes are monomeric silicone compounds with four substituents attached to the silicon atom and are the building blocks for higher-molecular-weight resins. These four groups can be the same or different, and can be inorganic or organic. Inorganic reactive silanes have alkoxysilane groups and can condense via hydrolysis reactions, similar to the inorganic zinc-rich binder curing reactions shown in Figure 5.42. Although the alkoxy side groups are of course organic, these silanes are considered inorganic because when they condense, the reaction product is entirely inorganic. Organic reactive silanes can contain a variety of organic functional groups, such as amine, epoxy and isocyanate, which enable them to undergo curing reactions with other organic resins, similar to the curing reactions of their conventional organic analogues.

A polysiloxane is any resin or polymer that contains repeating silicon–oxygen groups in its backbone or side chains, such as shown in Figure 5.40. Silicones are polysiloxanes which predominantly have two organic groups attached to the silicon atoms (again, as shown in Figure 5.40). By blending organic resins with polysiloxane resins, or by organically modifying the polysiloxane resin itself, hybrid inorganic–organic coatings can be obtained which, ideally, have the beneficial properties of both the inorganic and organic resins. The amount of organic modification varies from one generic type to another, but typically falls in the wide range of about 20–60% based on resin solids.

One of the first steps in realizing the benefits of hybrid inorganic–organic coatings (usually referred to as polysiloxane coatings) was the development in the 1990s of epoxy–siloxane hybrid coatings. These coatings are advertised as having the corrosion resistance of a typical epoxy coating, along with greatly improved weathering resistance. Indeed, one manufacturer of such coatings reports gloss retention of 90% after 18 months of outdoor weathering in Florida, which is dramatically better than that of a conventional epoxy coating.

A typical curing reaction of an epoxy–siloxane hybrid coating is shown in Figure 5.43. It is based on the use of an amine-functional alkoxysilane (aminosilane) with a conventional epoxy resin. Note that this curing reaction is essentially the same as the curing reaction of a conventional amine- or amide-cured epoxy (Figure 5.16) except that in this case, the curing agent is an aminosilane instead of a purely organic amine or polyamide. By the use of this aminosilane, the benefits of silicone resin chemistry are imparted to the hybrid coating.

Figure 5.43 is not the only curing mechanism possible with an epoxy–siloxane hybrid. The alkoxy (O-R) groups attached to the silicon atom are also reactive and under certain conditions (such as the presence of moisture) can react with themselves (self-condense, as shown in Figure 5.42), or react with a low-molecular-weight silanol or alkoxy-silicone intermediate formulated into the coating. These reactive intermediates are usually very low in viscosity, which enable high-solids coatings to be formulated at relatively low viscosity.

Not all polysiloxane hybrids are two-component coatings. Single-component acrylic polysiloxane hybrids exist which are based on pre-reacting acrylic resins that have reactive functional groups (such as hydroxyl functionality) with alkoxy or even isocyanate functional silanes. This results in an alkoxy-functional

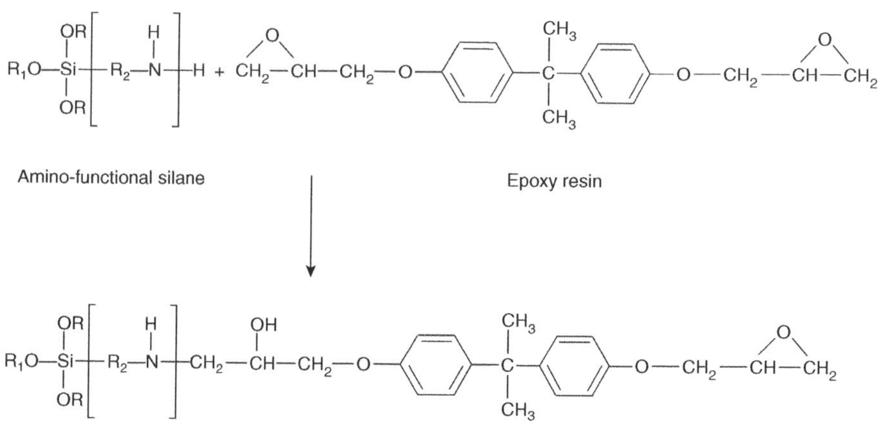

Figure 5.43 Cross-linking reaction of an epoxy–siloxane hybrid coating.

silicon-modified acrylic resin which can subsequently cure via condensation in the presence of moisture.

5.13 POLYUREAS

Polyurea coatings have become increasingly popular over the last decade or so. They share some of the same intermediate materials with polyurethanes and are sometimes confused with them. However, their chemistry and properties can be much different than the conventional polyurethanes.

As previously discussed (Section 5.9), a polyurethane is formed by the reaction between an isocyanate and a hydroxyl-bearing resin (the 'polyol'). In contrast, a polyurea is formed by the reaction between an isocyanate and an amine (—NH)-bearing resin or precursor. The basic reaction is shown in Figure 5.44. The reaction product is a substituted urea, as opposed to a urethane (Figure 5.36). There are also such things as hybrid polyurethane/polyurea coatings. As might be expected, hybrid coatings are made by reacting an isocyanate with a resin blend, where the resin blend is made up of both amine- and hydroxyl-terminated resins. Whereas urethanes can be formulated to range from thin, hard, glossy coatings, such as automotive topcoats all the way to thick, heavy-duty elastomeric liners, polyureas and hybrid polyurea/polyurethane coatings are principally used in the latter category.

One of the big advantages of polyurea coatings compared to polyurethane coatings is their very high solids content. Typically, these coatings are 100% solids, meaning that they are free of any regulations concerning VOC. A second advantage is that they cure extremely fast, even at low temperature. Whereas a conventional polyurethane elastomer may require an overnight cure, most polyurea coatings can be walked on in a few minutes. Furthermore, moisture is always a concern when applying conventional polyurethanes. The isocyanate can react with both the polyol and with the water, resulting in curing problems. However, because the reaction time between an amine and an isocyanate is very fast, polyurea coatings are more tolerant of humidity and moisture during application than are polyurethanes.

Just as polyurethanes come in both aromatic and aliphatic varieties, based on the type of isocyanates used, so too do polyureas. The aliphatic varieties have much better resistance to discolouration (yellowing) than the aromatic types, but are more expensive, particularly for high-build systems. Indeed, aromatic systems

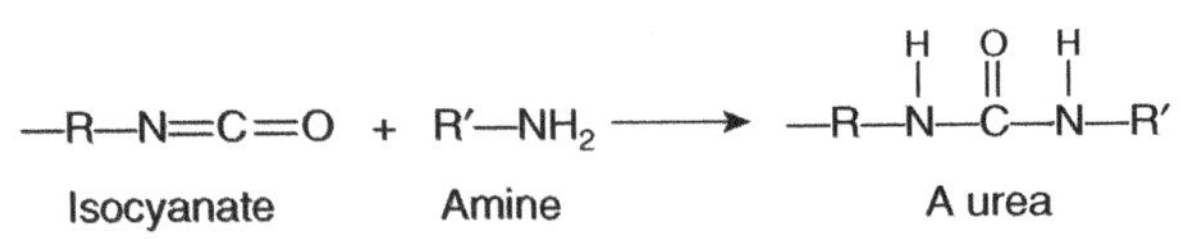

Figure 5.44 The reaction of an isocyanate with an amine to form a substituted urea.

are much more common in the industry. The types of isocyanates used in polyurea technology are very similar to those used in polyurethane technology, such as diphenylmethane diisocyante (MDI) for an aromatic coating and IPDI for an aliphatic coating [8]. It is through the use of di- or trifunctional polyetheramines such as polyoxypropylene amine in the resin blend that the coatings develop the typical toughness and elastomeric nature usually associated with them. These polyether resins have typical molecular weights of 500–5000. They can be used with a chain extender such as diethyltoluene diamine (DETDA) to achieve a variety of properties.

Ironically, some of the advantages associated with polyurea coatings can also work to their disadvantage. Because they cure so rapidly, they also have extremely short gel times, often only a few seconds. This means that they must be applied using very expensive heated plural spray apparatus, where the two individual components are heated to reduce viscosity and are forced into a mixing chamber (the spray gun) using some form of proportional pumping to achieve the specified mix ratio, which is typically 1 : 1 by volume. Some newer isocyanates which reportedly can increase gel times from the normal 2–4 seconds to as high as 7 seconds are available [9].

A second potential disadvantage associated with the extremely short pot life is the fact that the coating reaches a very high viscosity in an extremely short time. While this makes the job go faster, the coating is only liquid for perhaps a second or two, which does not give it much time to flow and wet the substrate. As previously discussed (Section 2.1.2), flow and wetting are very important factors in achieving good adhesion between a coating and the substrate. While polyureas can certainly achieve good adhesion, it is believed that several failures have resulted from their lack of flow and wetting.

5.13.1 Polyaspartic Polyurea Coatings

Since roughly 2000, a new type of polyurea has been developed based on the reaction between polyaspartic ester compounds with aliphatic isocyanates. A polyaspartic ester contains both ester groups and secondary amine groups, as shown in Figure 5.45. It is also possible to use the technology to produce polyurethane/polyurea hybrids.

The commonly available polyaspartic ester products have the structure shown in Figure 5.45. It is the group designated 'X', located between the two nitrogen atoms, which varies and results in a variety of physical and chemical properties of the cured coatings. An examination of the chemical structure of a polyaspartic resin shows that the environment about the secondary amine groups is very crowded. This 'steric hindrance', perhaps combined with electrical effects imparted by the ester portion of the structure, significantly slows down the crosslinking reaction with the isocyanate component of the coating. Indeed, whereas typical polyurea coatings have pot lives (perhaps more appropriately called gel

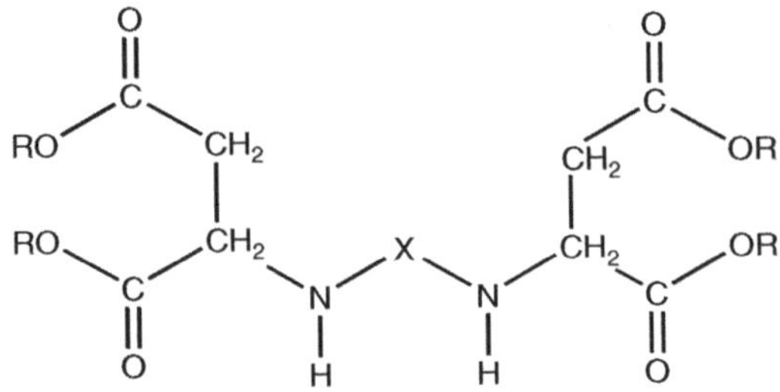

Figure 5.45 The basic structure of a polyaspartic ester.

times) of 5 seconds or less, polyaspartic polyureas (often simply called polyaspartics) can be formulated to have gel times from about 1 to 40 minutes. Not only do the longer gel times provide a longer window of application for the applicator, but the improved flow and wetting of such coatings would be expected to result in improved adhesion compared to polyurea coatings which gel in a second or two. Some of the longer gel-time formulations can also be applied with less expensive conventional spray equipment, as opposed to the use of plural spray equipment.

Because polyaspartic coatings are typically based on aliphatic isocyanates and aliphatic polyaspartic resins, they have good weathering characteristics. While they typically do not cure as fast or build as heavy in one coat application as conventional polyurea coatings, they still cure faster than conventional polyurethane coatings and can go on as thick as 10 mils in one coat.

5.14 POWDER COATINGS

Powder coatings can span a variety of generic coating types. They consist of solventless coatings provided in powder form, where the resins, pigments, curing agents and any additives are mixed, compounded by melting in extruders, cooled and then ground into a homogeneous powder. The virgin powder must have a high enough glass transition temperature (T_g) such that it will not partially fuse during storage, but have a low enough T_g such that it will fuse and level during baking. The coatings are applied by specialized equipment (Section 6.5) either to hot substrates or to room-temperature articles that are then subsequently baked in order to achieve both flow and cure. Powder coatings are used in numerous OEM applications, such as electrical boxes and appliances, and in particular, for buried underground pipelines. The great majority are of the thermoset variety, although some powder coatings are thermoplastic.

Epoxies are one of the most common generic types of powder coating. A curing agent commonly used with epoxy powder coatings is dicyandiamide (Figure 5.21), although modified dicyandiamides are also used. As might be expected,

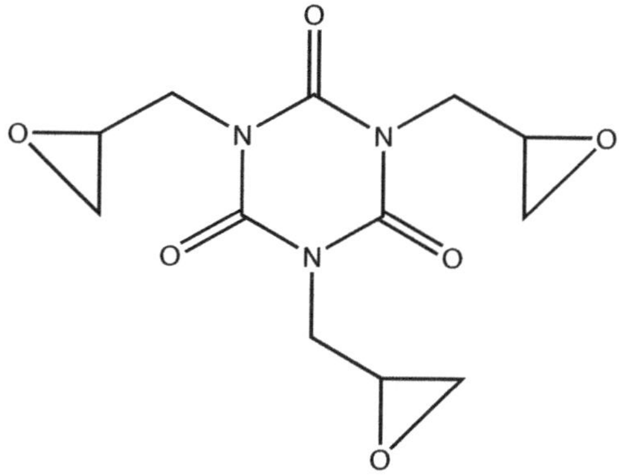

Figure 5.46 Triglycidylisocyanurate (TGIC).

epoxy powder coatings have very good mechanical properties, adhesion and corrosion resistance, but exterior weathering resistance is poor, and they chalk and discolour rapidly. They find considerable use in pipe for underground application, rebar and electrical equipment. For even greater hardness and chemical resistance, novolac epoxy resins or blends of novolac and bis-phenol A epoxy resins are used.

A family of powder coatings referred to as hybrid powder coatings consists of epoxy resins cross-linked with carboxlic acid-terminated polyester resins. These have somewhat better colour and gloss retention upon outdoor weathering than pure epoxy powder coatings. They cross-link primarily by opening of the epoxy (oxirane) ring by the carboxylic acid groups of the polyester.

Polyester powder coatings are also commonly used and have much better outdoor weathering characteristics than both epoxy and hybrid powder coatings. Depending on the type of cross-linking agent, either carboxylic acid or hydroxyl-functional polyesters can be used. A common system consists of carboxylic acid-functional polyesters cross-linked with triglycidylisocyanurate (TGIC, Figure 5.46). Such binders typically consist of 4–10% TGIC and 90–96% polyester resin.

If the polyester resins are hydroxyl functional rather than carboxylic acid functional, they can be cross-linked with blocked aliphatic isocyanates. They have gloss and colour retention equal to or better than TGIC cross-linked polyesters, and excellent mechanical properties and abrasion resistance [10].

Acrylic powder coatings are used for applications such as appliances because of their very good detergent resistance. The acrylic resins can be either hydroxyl functional or carboxylic acid functional, and can be cross-linked with a variety of curing agents. Epoxy-functional acrylics cross-linked with dicarboxylic acids, such as dodecanedioc acid, or with carboxlic acid-functional resins are also available [11, 12].

As might be imagined, the big advantage of powder coatings is the fact that they are solvent free and hence have zero VOC. As VOC is being more and more strictly regulated, this is a big advantage. They can also be formulated to have very good chemical resistance, adhesion and mechanical properties. Flammability and toxicity hazards are reduced, and relatively thick films can often be achieved in a single application. They can also be applied more easily to complicated shapes than liquid-applied coatings. There are no mix ratio problems as there might be with two-component thermoset liquid-applied coatings. Overspray can also be collected and reused.

Powder coatings also have some disadvantages compared to conventional liquid-applied coatings. Very large substrates cannot be coated, because of baking difficulties, nor can heat sensitive substrates be coated. While some powder coatings flow well and provide good appearance, they generally cannot rival liquid-applied coatings for aesthetics. Also, there is a smaller range of raw materials for the formulator to choose from, meaning that powder coatings do not span as wide a gamut of final properties as do liquid-applied coatings.

REFERENCES

1. Z. W. Wickes Jr, S. N. Jones and S. P. Pappas, *Organic Coatings: Science and Technology*, Vol. **I**, John Wiley, New York, 1994, pp. 54, 103, 134.
2. S. G. Croll, *J. Coat. Technol.*, **52**, 65 (1980).
3. J. C. Graham, D. A. Glossky, T. G. Fisher and R. A. Garling, *J. Coat. Technol.*, **60**, 35 (1988).
4. B. L. Burton, presented at the Fall 2001 Epoxy Resin Formulator's Meeting, The Society of the Plastics Industry.
5. I. T. Dinnissen, *J. Prot. Coat. Linings*, December, **22** (10), 40 (2005).
6. C. H. Hare, *Protective Coatings*, Technology Publishing Company, Pittsburgh, PA, USA, 1994, pp. 90, 153.
7. O. M. Considine and G. D. Considine (Eds), *Van Nostrand Reinhold Encylcopedia of Chemistry*, 4th Edn, Van Nostrand Reinhold, New York, 1984, p. 265.
8. W. J. Knight, *J. Prot. Coat. Linings*, September, **18** (9), 48 (2001).
9. C. T. Thurau and M. D. Conner, *J. Prot. Coat. Linings*, February, **24** (2), 13 (2007).
10. Z. W. Wickes, Jr, S. N. Jones and S. P. Pappas, *Organic Coatings: Science and Technology*, Vol. **I**, John Wiley, New York, 1994, p. 489.
11. T. Agawa and E. D. Dumain, Proceedings of the Waterborne High Solids Powder Coating Symposium, New Orleans, LA, p. 342, 1997.
12. C. D. Green, *Paint Coat. Ind.*, September, 45 (1995).

6

Application-Related Problems

While the actual percentage is unknown, there is little doubt that the majority of paint and coating problems are application related. This present chapter will discuss some of the errors and oversights during the application process that can lead to paint failure. Before getting into specifics, a brief introduction to some of the common methods of paint application will be helpful.

The most common methods of paint application are brush and roller, spray, roll coating and powder coating.

6.1 BRUSH AND ROLLER

Many types of high-performance coatings can be applied by brush and roller. However, these techniques are usually restricted to small areas or spot repair. Brush and rollers are mainly used for house-paint or consumer wood refinishing projects.

Brushes come in all shapes and sizes, with either natural (hog hairs) or synthetic filaments. Both natural and synthetic bristles are tapered, with the thicker ends embedded in the base of the brush. The better brushes are chisel cut at the end of the bristles in order to minimize brush marks.

Natural bristles are more expensive than the synthetics and are better for applying solvent-based paints. However, water can soften and swell natural bristles, thus making synthetics the preferred choice for water-based paints.

Rollers can apply paint considerably faster than brushes. Roller covers are made of several fabrics, including acrylic, polyester, rayon, nylon, mohair and sheepskin. The natural fabrics are not recommended for water-borne paints, nor are the synthetics recommended for coatings containing strong solvents.

Failure Analysis of Paints and Coatings, Revised Edition Dwight G. Weldon

In addition to the type of fabrics, rollers also vary regarding their texture, or 'nap'. Shorter naps produce smoother finishes, but they are not suitable for rough or textured surfaces.

6.2 SPRAY APPLICATIONS

Most industrial maintenance and commercial architectural jobs are painted by spray application. Spray painting is much faster than using brushes or rollers and, except for angles or other hard-to-reach places, results in a more consistent application. A large variety of spray equipment are available, including air (conventional), airless, plural spray and electrostatic.

6.2.1 Air (Conventional) Spray

In conventional spraying, an air compressor supplies compressed air to a spray gun, which atomizes and transports the coating to the substrate. Spray guns are equipped with valves to adjust both the volume of air and the spray pattern. For small jobs using low-viscosity paint, the paint is often placed in a cup and is drawn into the gun by suction. For larger jobs or more viscous paint, the paint is held in a pressurized pot and forced into the gun under pressure. For small jobs, the pot is physically attached to the gun. For larger jobs, a bigger pot is used, which is connected to the gun via a paint line.

The requirements for obtaining a good paint job by using conventional spray application are fairly fundamental. The compressor must be able to supply an adequate volume of air at a high enough pressure (usually 55–90 psi) in order to adequately atomize the paint. The air must be clean and dry (requiring oil and water separators), and the spray pattern should be compatible with the dimension of the object being painted. The gun should be kept at right angles to the surface being painted (sweeping arcs should be avoided) and between approximately 5–10 in. from the substrate. Being too close can cause sags and runs, and being too far away can cause a granular or lumpy texture due to excessive evaporation of the solvent before the paint strikes the surface (dry spray). In order to obtain uniform thickness, strokes should be overlapped by approximately 50%.

6.2.2 Airless Spray

In airless spray, a pump is used to force paint under very high hydraulic pressure (approximately 1000–5000 psi) through a small orifice, which results in atomization of the coating into fine droplets. Air compressors are often used but only to operate the hydraulic pump. Electric motors and gasoline engines can also be used. Airless spray is much faster than conventional spray and is commonly used in the painting of large structures.

Airless spray guns have only a single valve to adjust the volume of the paint passing into the gun. The spray pattern is not adjusted by changing the dynamics of the air flow, but rather by changing the tips.

Not only is airless spray faster than conventional spray, but it is also more efficient, with less 'over-spray'. This method can apply coatings that would be far too viscous for conventional spray, such as coal tar epoxies. However, it is difficult to paint small objects with any control of film thickness.

6.2.3 Plural Spray

Most two-component thermoset coatings have a pot life of a few hours, and can therefore be sprayed by normal airless equipment or even by conventional spray if the viscosity is not too high. However, there are many two-component coatings that cross-link almost immediately upon mixing, with pot lives measured in minutes rather than hours. These coatings require the use of special plural spray or proportioning equipment. Usually, airless technology is used for this equipment, sometimes with the added feature of heat. Heating of the individual components prior to mixing them under pressure at the spray gun is sometimes required to reduce their viscosity.

Plural spray equipment requires two paint containers and two hoses, which can be attached to either a twin spray gun head or to a single gun. Both internal-mix and external-mix guns are available. In internal-mix guns, the two components are combined and mixed within the gun. In external-mix guns, the two components meet just outside of the spray tip.

6.2.4 Electrostatic Spray

Electrostatic spray is limited to certain types of coatings applied to conductive substrates and is performed in an assembly line environment. The object to be painted is grounded, and the paint is applied by using modified spray equipment that can provide a high electrostatic charge to the paint particles. Application equipment also includes rotating bells and discs to create certain types of spray patterns. There is almost no over-spray, and hard-to-reach areas are easily coated due to the electrostatic attraction of the paint to the substrate.

6.3 FLOW COATING

Flow coating essentially involves saturating a part suspended from a moving rail by the use of two or more spray heads. The excess paint drains from the part as it moves through the spray booth and is collected and recycled. Compared with actually dipping the part in paint, it has the advantage of reaching corners without being impeded by pockets of entrapped air.

6.4 ROLL COATING

Thin-gauge sheet or coil stock can be coated very rapidly (line speeds up to several hundred feet per minute) and with minimal waste by using either direct roll coating or reverse roll (coil) coating.

In direct roll coating, sheet stock is fed between rollers rotating in the same direction as the moving sheet. If only one side of the sheet is to be coated, one of the rollers is an applicator roller and the other is simply a support roller. If both sides are to be coated simultaneously, then both rollers are application rollers. In either event, the applicator rollers are fed by smaller pickup rollers, which are partially immersed in trays containing the paint. The coated stock is subsequently fed into an oven for baking.

Reverse roll coating, sometimes called coil coating because the stock is unwound and coated directly from the coil, is faster than direct roll coating. As the name implies, the applicator roller actually runs in the opposite direction to the moving stock. Either one or both sides can be coated. After coating, the metal is run through an oven and recoiled for shipment.

6.5 POWDER COATING

Powder coatings can be applied by the use of fluidized beds, electrostatic fluidized beds or electrostatic spray.

Fluidized bed technology consists of a tank or chamber with a porous plate at the bottom. The powder coating is placed in the tank, where it is agitated both by vibration and by air blowing through the openings in the porous bottom. The agitated powder fills the volume of the tank, into which is dipped a heated part. The part being coated is heated to a temperature approximately 100 °F (about 56 °C) higher than the fusion temperature of the powder coating. Parts as large as light poles can be coated by this method. The coating thickness is controlled by the residence time of the part in the tank.

Electrostatic fluidized beds (sometimes called cloud chambers) are extremely similar to fluidized beds, except that the cloud of powder coating is electrically charged. The object to be coated is electrically grounded and not preheated. The coated object is then transported into an oven where the coating is fused and cured. This requires much less powder than a conventional fluidized bed and is more easily automated. However, this technique is restricted to smaller pieces.

Electrostatic spraying of powder coatings is similar in principal to electrostatic spraying of liquid paint, except that some obvious modification to the gun and lines are required. The powder passes a high-voltage electrode at the tip of the spray gun and is therefore charged, and the moving cloud is electrostatically attracted to the grounded article to be coated. The powder adheres extremely well and will 'wrap' around the piece to coat the back and other 'inaccessible' areas. The part is subsequently baked.

6.6 COATING FAILURES RELATED TO APPLICATION PROBLEMS

Improper surface preparation is one of the most common problems leading to coating failures. Coatings must come into intimate physical contact with the substrate in order to achieve good adhesion. This is because the chemical and electrical forces that hold a coating to a surface act over extremely small distances. The roughness of the surface is also an important factor in obtaining good adhesion. Although surface roughness is not measured by using atomic dimensions such as angstroms, it still involves distances of only a few thousandths of an inch.

When it comes to steel, the most common method of surface preparation is blast cleaning. For this technique to be effective, several steps must be taken. The steel must first be solvent cleaned or otherwise degreased if there are any oily contaminants. Blast cleaning an oily surface will only spread the oil around. If rags are used, they should be changed frequently. Oil and water separators should be used on the blast-cleaning lines to prevent the steel from being contaminated as a result of the cleaning procedure.

When new steel is involved, it has been this author's experience that coating failures caused by painting over oily contaminants are few and far between. This may be because most contractors take appropriate precautions, or because many coatings can actually tolerate small amounts of oil, at least on rough surfaces [1].

Painting old steel, however, is another matter. Depending on the environment, considerable amounts of contamination may be present that cannot be removed by the normal procedures employed by the contractor. Two such examples that come to mind are the painting of previously unpainted galvanized steel in a food processing facility and the repainted structural steel in a paper mill.

In the first case, considerable amounts of fatty organic material had built up on the galvanized steel. The latter can be rather difficult to coat in the first place, at least in part, because of its smoothness. Although the contractor made a legitimate attempt at cleaning the steel, the circumstances were such that enough fatty residues remained to cause wholesale peeling of the coating system. In the case of the paper mill, the badly rusted steel was contaminated with large amounts of chloride from the chemicals used in the paper manufacturing process. Although the required 'near-white' blast cleaning was achieved, it could not remove the invisible salt residues present in the pitted areas of the steel, and severe premature rusting of the repainted steel therefore resulted.

The above failure was probably not the fault of the contractor, who simply did what the specification called for. However, on occasion, contractors have been known to omit the blast-cleaning step altogether, or to do it in a haphazard fashion. If loose rust or loose mill scale is painted over, premature failure may be expected.

It is also common to roughen concrete surfaces prior to painting, either by blast cleaning or acid etching. If either of these processes is done incorrectly, a coating failure can result.

It is not uncommon for a coating to peel off a concrete floor as a result of painting over a powdery residue of cementitious material. In some cases, this could be due to efflorescence (water-soluble salts from the cement transported by moisture migration and depositing on the surface). In other cases, it can be a fine residue from the blast-cleaning operation that settled back on the floor and was subsequently painted over.

Roughening of concrete by acid etching is also common. Muriatic acid (hydrochloric acid) is diluted with water and applied to the concrete, where it rapidly reacts with the alkaline matrix to impart a distinct roughening. The technique is fine when carried out correctly but can cause serious problems if not.

Potential problems with acid etching include using acid which is too dilute, not waiting long enough, failure to adequately clean and rinse the floor, or interference from form-release or membrane-curing compounds.

Using acid which is too dilute, or not waiting long enough, results in an etch which is not deep enough to promote good adhesion. Failure to adequately rinse the residues from the concrete will result in a thin layer of powdery material that will interfere with the bond of the coating to the substrate.

Sometimes, form-release agents or membrane-curing compounds are present on the concrete. Membrane-curing compounds can be a variety of materials, including simple hydrocarbons, which are used to trap moisture in the concrete. Cement cures via hydration, and it is desirable to trap a certain amount of water in the concrete to promote curing. However, many of these curing compounds insulate the concrete from the etching action of the muriatic acid. If their presence is known or suspected, steps should be taken to remove them, or the concrete should be blast cleaned rather than acid etched.

Painting over wood that has large amounts of dirt or mildew present can result in peeling of the paint. This type of problem is easily detected with a low-power microscope. What is harder to detect is paint which is peeling due to a thin layer of degraded wood fibres. Fortunately, sometimes at least some of these fibres are attached to the back of the peeling paint.

The problem arises because wood is not stable when exposed to ultraviolet (UV) light from the sun. Sunlight can degrade lignin components of the wood, thus weakening its surface and resulting in premature paint failure. Some deterioration of paint performance can occur after only 2 or 3 weeks of exposure, and sometimes even in cases where the wood was factory primed. Factory primers are not designed to screen out UV light; furthermore, they are usually applied rather thin. A light sanding of unpainted wood prior to painting will usually suffice to remove the thin, degraded surface layer.

Although not exactly a surface preparation problem, painting damp surfaces can readily lead to peeling paint. It is common practice not to apply paint to a substrate that does not have a surface temperature of at least 5 °F (about 3 °C) above the dew point. Although certain types of paint, primarily water-borne ones, can tolerate some moisture, many others cannot. Sometimes primed steel that is topcoated during the cool of the morning will fail, while steel painted with the same coating after lunch will do just fine. Proving that a coating has been applied over moisture can be very difficult, as there are usually no telltale signs.

Another surface-preparation-related failure is exceeding the re-coat window. Certain types of coatings, notably the two-component thermosets, have a window of opportunity during which they can be re-coated. The duration of this window is often temperature related. Beyond a certain point, the coating will simply be too hard (and possibly too smooth) for subsequent coats of paint to adhere to. If it is known that this has occurred, the situation can often be remedied by either applying a strong solvent that will slightly soften the aged coating or subjecting it to a 'sweep blast' to impart some physical 'tooth'. Failure on the part of the applicator to do this can result in peeling of the subsequent coat.

Problems can also occur on the front end of the re-coat window as well. If a high-internal-stress coating is applied to a prime coat that has not had sufficient time to cure and develop cohesive strength, the curing stresses of the topcoat can rip the primer apart. Although this could be an applicator problem, it could also be a material problem if the primer is not curing at the expected rate.

Another common source of applicator error is the 'mis-mixing' of two-component coatings such as epoxies and urethanes. Contractors experienced with such coatings seldom make this error, although sometimes even an experienced contractor may have a new, inexperienced painter on one of his crews. The problem is more prevalent with high-ratio products than with simple one-to-one by volume coatings. It is also more likely to happen if the painters are mixing partial kits, since there is often no real volumetric measuring equipment at a job site.

As a rather extreme example, this author is familiar with more than one occasion where the contractor did not add any of the 'part B'. In one such instance, the contractor's attempt to remedy the situation was to spray apply the 'B' component on top of the coat of sticky part 'A'. Not too surprisingly, it did not work!

A coating that is grossly 'mis-mixed' will not develop the cross-link density of the properly mixed coating. It will be deficient in both chemical and physical properties, even though it may become dry to the touch or even dry hard. A urethane coating mixed with a substantial excess of isocyanate may cure by an entirely different pathway than properly mixed coatings, thus resulting in a 'cured' film with markedly different properties.

Numerous coating failures are the result of the coating being applied either too thickly or too thinly. If a coating is too thin, liquids or aggressive fumes will more readily permeate it, and its protective properties will be compromised. In some cases, shop-applied primers are applied so thinly that the peaks of the blast-cleaning profile are actually at, or even ever so slightly above, the surface of the primer. If the steel is then shipped to a job site and exposed to the elements prior to topcoating, pinpoint rusting can occur.

In the paint industry, the saying that 'if a little is good, more must be better' is usually not true. Not only does excessive paint thickness cost more money, it can often cause its own problems.

For instance, if the interior of a potable water tank is lined with an epoxy coating at two or three times the recommended thickness, additional amounts of solvent will almost certainly be trapped in the thick coating. In an extremely severe case, there may be the possibility of solvents exceeding the threshold limit for drinking water. More common is osmotic blistering of the coating if one of the trapped solvents has any degree of water solubility.

Another problem associated with excessive thickness is stress. All coatings shrink when they cure, while some shrink more than others. This shrinkage is a means of stress relief for the coating, but the problem is that the act of shrinkage imparts stress to the previously applied coating (or to the coating–metal substrate interface). Depending on the relative strengths of the various layers, or the strength of the adhesive forces at the interface, this shrinkage stress can cause cohesive splitting within a previous coat of paint, or adhesive failure at the interface.

Just because one has found a sample of failed coating to be too thick, or even 'mis-mixed', it should not be automatically assumed that this is the cause of the failure. It may be that there are several failing areas that have the correct thickness, or several non-failing areas that have a thick coating. It is important to see if there are trends associated with any of the variables. Indeed, this author was involved in one project where the coating actually performed better when it was 'mis-mixed'.

Application techniques can also cause paint failures. Chapter 2 discussed the concepts of wetting, surface tension and resistance to flow. While these variables are primarily the properties of either the substrate or the coating, the applicator also has some influence on them.

For instance, the surface tension and viscosity of a coating are affected considerably by the type and amount of solvents in the coating. The contractor can greatly change these by improper thinning or by using the wrong thinner. These properties can even be affected by the operator's spray technique. If there is too much air at the gun, or if the gun is too far from the substrate (particularly on hot days), there can be an excessive evaporation of the solvent from the atomized paint before it hits the substrate. While the surface tension of the paint will likely still be low enough for it to wet the substrate, its viscosity, and hence its resistance to flow, may be too high for wetting to occur. Similarly, if a two-component coating is applied beyond its pot life, the viscosity may be too high for flow and wetting to occur, and adhesion failure may result.

Occasionally, the wrong thinner can be used with no consequences. Other times, however, this error may prove to be fatal. For example, if a coating intended for immersion service was mixed with a thinner containing a water-soluble solvent, osmotic blistering could occur. In another example, if a urethane was thinned with an alcohol-containing thinner, the alcohol would react with the isocyanate component, thus effectively stopping the cross-linking reaction.

There is always the chance of an error with any coating that requires baking. Some facilities have very good ovens that are closely monitored, while others do not. Furthermore, some facilities maintain constant, or at least hourly, monitoring and verification of line speed and oven temperatures, while others do not.

A baked finish, such as an acrylic melamine, can be affected by under-baking as well as over-baking. Under-baking generally results in a film that has poor solvent and chemical resistance and which may not weather well. Over-baking can cause a coating to be too brittle for it to withstand post-forming without cracking. Checking specimens from the beginning and end of a coil for solvent resistance and 'bendability' is a good idea, but these are relatively insensitive tests and will provide no information about what may have happened in the middle of the coil.

Another type of failure related to surface preparation is when a peeling failure occurs not in the newly applied paint, but within older layers of existing paint. Such problems are often seen in old homes or other wooden structures having 5 or 10 coats of paint. The reason for this type of failure is usually fairly simple. At some point in the multi-coat system, there is a weak interface. This could be due to any number of reasons, such as cheap paint, painting over contamination, weathering/chalking of one of the older coats and so on. This weak interface may have been under stress for years from the weight and shrinkage stress of the subsequently applied coats of paint. Finally, one too many coats of paint is applied, and the weight and shrinkage stress of this last coat is 'the straw that broke the camel's back'.

Who to blame in a case like this is sometimes not clear. If there was a specification that stated the removal of all old coats of paint, then the contractor is obviously at fault. If the specification simply suggested cleaning of the surface of the existing paint prior to painting, then the contractor is obviously not at fault.

The more typical situation, however, is when the specification says to 'remove all loose and peeling paint' prior to painting. This is a somewhat grey area. One could argue that since the failure occurred at a weak interface between old coats of paint, then by definition, the contractor had not removed all 'loose' paint. One could also argue that there was no visual evidence that the paint was 'loose and peeling', and if 'normal' methods were used (such as putty knives or water blasting) without removing this paint, then the contractor had fulfilled his obligation. Sometimes, determining the cause of a failure is one thing, while assessing blame is quite another.

There are many excellent, ethical contractors highly skilled in the application of paint and coatings. There are also a few who are not. Therefore, when investigating a coating failure, it is important not to automatically rule out a potential cause because 'that just couldn't happen'. After all, coatings fail because something was carried out incorrectly, either with the specification, application or material.

Failures have been known to occur because the contractor used the wrong paint, or painted during freezing temperatures. This author is aware of one floor coating failure that occurred because the contractor intentionally used only half the recommended amount of catalyst. He did so because otherwise the pot life was too short to conveniently mix and apply a full kit. It is important not to assume that such things do not happen, because sometimes they do!

REFERENCES

1. D. G. Weldon, A. Bocham and M. Schleiden, *J. Prot. Coat. Linings*, **4**, 46 (1987).

7

Field Methods

With adequate background information and good samples, many coating failures can be solved in the laboratory. However, with some jobs, the process can be greatly facilitated by a site visit.

The first step in a site visit, or even in a laboratory investigation, is to garner as much background information as possible. What type of structure or substrate is involved? Is it a new paint job or a re-coating project? Was there any surface preparation specified? When was it painted, by whom and where? If done outdoors, what was the time of year? Were there any unusual weather conditions? How much time elapsed between priming and topcoating? What coatings were specified? What did the failure look like and how long did it take to develop?

The above list is certainly not all inclusive, and the answer to one question may suggest a whole line of other questions. It is safe to say, however, that the more information obtained, the better.

Before leaving the laboratory or office for the site visit, one should put together a basic tool kit. A pocket-sized tape recorder can be very useful and more efficient than a notepad. A camera should also be included, as it is usually a good idea to photograph areas from which samples are taken. Some necessities include a sharp knife or scalpel, a utility knife, a permanent marker, ziplock bags for samples, a roll of appropriate adhesive tape for performing the common American Society for Testing and Materials (ASTM) adhesion test (ASTM D3359, Standard Test Method for Measuring Adhesion by Tape Test) and a thickness gauge. There are numerous types of small thickness gauges on the market, and one should select a suitable one for the substrate to be encountered. A small field microscope with magnification of perhaps 10 or 20× is often invaluable. If one expects to encounter liquid-filled blisters, then syringes and small glass vials should be included. A small bottle of a strong solvent, such as methyl ethyl ketone (MEK), and some cotton Q-tips or cotton swabs are useful for assessing the degree of cure. Deionized water, pH paper and a clean cloth are also useful items.

Failure Analysis of Paints and Coatings, Revised Edition Dwight G. Weldon

One of the most important benefits of a site visit is the opportunity to look for patterns. Is the failure always on the sunny side of the structure? If so, degradation due to heat and/or ultraviolet light might be expected. Does the failure only occur on exterior walls near windows or within a few feet of the ground? If so, moisture might play a role.

This author is aware of one instance where an epoxy floor coating was blistering and lifting, and the laboratory concluded that it was because of hydrostatic pressure from ground water. The only problem with this theory was that it was a two-storey building and the failure was occurring on the second floor.

Failures have been known to occur in arcs, matching the trajectory of a painter's spray gun. There was no failure immediately opposite the painter, but where he swung his gun up and about such that there was a much greater distance between the gun and the substrate, the paint was thinner and had not flowed out as well.

It can sometimes be fascinating to see the types of patterns associated with paint failures. This author is aware of the failure of a cylindrical storage vessel coated with an epoxy–phenolic. When put in chemical service, the coating blistered, but only in a pattern of bands approximately 6-in. wide, and running from approximately 'eight o'clock to four o'clock'. The heat for the coating cure was supplied by torpedo heaters placed inside the vessel. The heavy vessel itself sat on steel rungs approximately 6 in. in width, which spanned the 'eight o'clock to four o'clock' positions. Apparently, the extra mass of the steel supports created a heat sink, and the coating did not cure as completely in these locations.

At the very least, the above type of information may provide the laboratory analyst with valuable clues into the mechanism of the failure. In the best case, it may allow for a resolution of the failure without any further testing.

If the failure is one of poor adhesion, then one should not assume that the problem is only confined to those areas where the coating has already fallen off. In such situations, it is very important to perform a series of adhesion tests immediately adjacent to the failed areas and also in those areas that appear to be intact. One may discover that the area of failure is much larger than it appears, and that there are many areas of very poor adhesion where the coating has not actually fallen off (yet).

There is often a great deal of confusion and controversy regarding adhesion testing. It can be argued that adhesion is the single most important property of any coating. Yet, after a century of modern paint technology, the measurement of adhesion is, at best, an imperfect science, subject to error and interpretation.

There are three common ways of assessing a coating's adhesion, namely probing with a sharp object by making a series of knife cuts, using tape to remove the coating (ASTM D3359) or affixing a stub or dolly to the structure with an adhesive and measuring the force required to pull off the coating (ASTM D4541, Test Method for Pull-Off Strength of Coatings Using Portable Adhesion Testers).

Probing or gouging at a coating with a knife or chisel is obviously a fairly subjective process. It is difficult to quantify different levels of adhesion, other than in broad, subjective terms such as good, bad or fair. Different operators can come to different conclusions. Nevertheless, such probing can still be useful if carried out by an experienced investigator.

The 'tape' test (ASTM D3359) is perhaps somewhat more scientific and reproducible than simply probing and picking at the structure, but still has some problems.

Perhaps the single biggest problem with the test is failure to scribe completely through the substrate. This is a very real problem with thick, hard coatings, and often, the test results depend on the physical strength of the investigator and the sharpness of the knife.

Even though the same type of tape may be used, there is no way to validate whether one batch of tape has the same adhesive characteristics as another batch, or whether these properties change with the age of the tape.

If the surface is cold, the tape will not stick as well as if it were dry and warm. If there is dirt, grime, chalk or an exudate on the surface of the paint, the adhesion may appear to be excellent simply because the tape cannot stick to it. If the coating is very thick, even though it may be cut completely through the substrate and be poorly adherent, it is possible that the tape may not remove it. This is because, if the coating is slightly soft, it may 'heal' itself at the cuts, with the simple strength of friction between the cut sides holding it in place.

Pull-off tests that measure tensile strength (ASTM D4541) certainly appear to be more objective than the other methods, but some are notorious for poor reproducibility. It is not uncommon to carry out triplicate tests in a small area and obtain results such as 200, 600 and 700 psi. It is also not uncommon to get relatively high values of tensile adhesion, and yet still be able to pull the coating off with one's fingers.

This author's opinion is that subjective knife probing or the tape tests give more useful information, at least when carried out by experienced investigators, than do the tensile tests.

Regardless of which method one prefers, there is still the problem of making sense of the results. The ASTM test methods are of little help here, for they report either psi or use an arbitrary rating system that runs from a '5' (the best result) to a '0' (the worst result). Is a '5' good and a '1' bad? Is 450 psi good and 350 psi bad?

There are no simple answers to these questions. The real usefulness of adhesion testing is to see if there are any trends. Perhaps, there is a well-defined area of the structure where the ratings are ones and zeros, and the rest are all threes and fours. It is fairly safe to assume, then, that there is something 'wrong' with the area displaying ones and zeros. Obviously, the coating system is capable of performing much better, so there must be some defect or flaw in those areas of lower adhesion.

Another word of caution on adhesion testing: this technique is difficult to use as a predictive tool, especially when there is no actual, spontaneous, existing failure. If the coating is already peeling and adhesion testing shows zeros and ones near the peeling coating and fours and fives in areas removed from the peeling coating, it is safe to assume that the areas of low adhesion are ripe for failure. However, this is very different than doing adhesion testing on a structure that is not failing, and concluding that certain areas are going to fail because they have low results compared with other areas. Depending on the degree of external stress to which the coating system is subjected, it may last its entire expected service life without spontaneously peeling.

When investigating a problem involving poor adhesion or delamination, it is important to take note of the locus of the failure. Is the topcoat peeling from the primer, or is the primer peeling from the substrate, or perhaps breaking cohesively? One should also pay close attention not only to the back of the peeling paint, but to the surface from which it peeled. Is there evidence of contamination on the surface? If it was steel that was supposed to be blast cleaned, is there evidence of a blast profile, or are there large amounts of mill scale? If concrete, is it smooth and hard, or etched? If wood, does it look weathered or mildewed? Often, a piece of the substrate can be submitted to the laboratory where such observations can be carefully made. Often, however, the only chance the investigator will get is at the job site.

In a peeling problem, one will want to take samples. Pieces of coating that have fallen off and are lying on the ground or dangling in the wind should be avoided. Any contamination found on the back of these samples might have occurred after they failed. It is usually not too difficult to find areas where the coating has extremely poor adhesion and where relatively large pieces can be removed. These are the appropriate samples to take.

In addition to sampling the failing topcoat (or intermediate coat), it is also important to sample the coating immediately beneath it. Often the problem is not with the peeling coating, but with the coating it has peeled from.

Although it may seem that anybody who is not blind can tell the front of a paint chip from the back, this is not always as obvious as it seems. In such cases, it is highly recommended that the front or back be identified. With large chips, a pen or marker could be used to physically write on the chip. Another option, especially when the coating flakes off in small pieces, is to take a sample by using tape. The front of this sample will be ruined, but at least it will allow the back to be identified. Once this has been done, it is usually easy to determine the front and back on other loose chips.

One should also take samples of non-failing coating, as well as failing coating. A common protocol in the failure analysis laboratory is to examine failing and non-failing samples to see what is different about them.

Coating thickness can be measured in the laboratory. However, if it is reasonably convenient to do so, they should also be measured in the field. Particularly if the structure is metallic, there are numerous hand-held gauges that will measure coating thickness in a second or two. Numerous measurements can be made, along with adhesion tests, to determine if there is a correlation between thickness and adhesion, or blistering, or corrosion, or cracking – or whatever the failure mode happens to be.

Another common failure mode of coatings, especially those in immersion, is blistering. The investigator will want to know where the blistering is occurring (between primer and steel?) and whether the blister is liquid-filled or not. If it is the former, every effort should be made to obtain a sample of this liquid.

If the blisters are very small, this may be nearly impossible. If they are an eighth of an inch or more in diameter, a sample can probably be obtained.

The trick to obtaining a valid sample of blister liquid is not to contaminate it during sampling. The first step in avoiding contamination is to clean the surface of the paint (the water and clean rag or cotton balls that you packed come in handy here). Once cleaned, a small hole must be made in the blister, and a 1-ml disposable syringe can be used to withdraw the sample. Do not use the needle of the syringe to make the hole. There is the chance that it will either clog, break, or contaminate the blister liquid by carrying material into it from the surface of the paint. Use a pin, knife, or spare syringe needle to make the hole, and then insert the syringe and withdraw the sample.

It may be necessary to use the same syringe on a dozen or so blisters, depending on their size, to get enough liquid for analysis. Most failure analysis laboratories would be thrilled to get 0.5 ml, but some tests can be done with as little as 0.1 ml. The more, however, the better.

Sometimes, thermoset coatings fail because they are not thoroughly cured. In many cases this can be evaluated in the laboratory by using expensive, time-consuming test methods. Sometimes, determinations can be made in the field by rubbing the coating with a Q-tip and solvent (a good reason to bring along that bottle of MEK). Typically, well-cured coatings show little or no colour transfer or softening after 50 double rubs (a double rub is a stroke-up followed by a stroke-down). This test is somewhat subjective, but if carried out consistently by the same operator, it can give reasonably good results. It will not be able to detect slight differences in cure, although gross differences will be obvious.

A simple mistake to make when doing solvent rub tests is to misjudge the amount of paint removed because of its colour. When using a white Q-tip, it is difficult to see the removal of a white paint compared, say, with a red paint. The use of a thin, dark cloth, perhaps wrapped around the Q-tip, can help here.

It is virtually impossible to write a chapter covering all eventualities that might be encountered during a site visit. A keen eye, methodical thinking and experience are the biggest prerequisites to success.

8

Analytical Methods

There are numerous analytical tools that can be used to investigate coating failures. The failure to list a certain method here does not mean that it has no place in failure analysis. There is a bit of art to failure analysis, and the tools used are limited only by their availability and the investigator's ingenuity.

8.1 LIGHT MICROSCOPY

The light microscope may not be the most expensive or glamorous tool in the analytical arsenal, but it is arguably the most important. The need for thorough background information and possibly a site visit has already been discussed. In this author's laboratory, the next step in the failure analysis investigation is to look at the samples carefully.

Looking at the samples begins with exactly that – looking. Are they big or small, green or white, brittle or flexible, soft or hard, dirty or clean? In the chapter on field methods, the importance of looking for a pattern was stressed. If the sample furnished to the laboratory is not a paint chip, but perhaps a section of a painted steel or an entire wooden shutter, a pattern of failure may be evident even on this scale, regardless of whether a 'larger' pattern was evident on the entire structure (such as north versus south, or vertical versus horizontal). One may find that corrosion of post-formed metal roofing occurs only at the bends, suggesting that the coating was too brittle and micro-cracked during forming, or that blistering and peeling of a shutter occurs only on certain distinct, smaller pieces of wood that were joined to make the larger article. Patterns almost always provide valuable clues to the mechanism of the failure.

Once the samples have been examined visually, they should be subjected to close scrutiny by using a stereo microscope. A reasonably good microscope with a magnification range of approximately 10–50× is sufficient, perhaps with a 2× auxiliary lens to reach an upper magnification of 100×. A working distance of 3–4 in. gives one enough room to cut and poke and rub at samples while viewing under magnification. Universal focusing stands, with a counterweighted boom arm assembly, are desirable to facilitate the manipulation of large samples such as pieces of siding or sections of steel beams. A good, bright light source based on fibre optics is also required, not only for viewing but for photography as well. Photographs can sometimes bridge the gap between what the eye sees and what the investigator writes down on paper.

Sometimes, coatings fail for very simple reasons. Paint may peel from wood because it was applied over mildew (Figure 8.1), from steel because it was applied over loose rust scale or other contamination (Figure 8.2) or from concrete because it was applied over laitance. These flaws can be easily observed with a microscope, and perhaps little or no additional work will be required.

A coating system may fail to protect the substrate because it is too thin, therefore allowing water or other agents to readily permeate it. It may fail because it is too thick, thus imparting excess stress to an interface or perhaps retaining too much hydrophilic solvent. These possibilities can be explored by using the microscope.

Figure 8.1 Photomicrograph showing mildew on the back of a paint chip that has peeled from a sample wood.

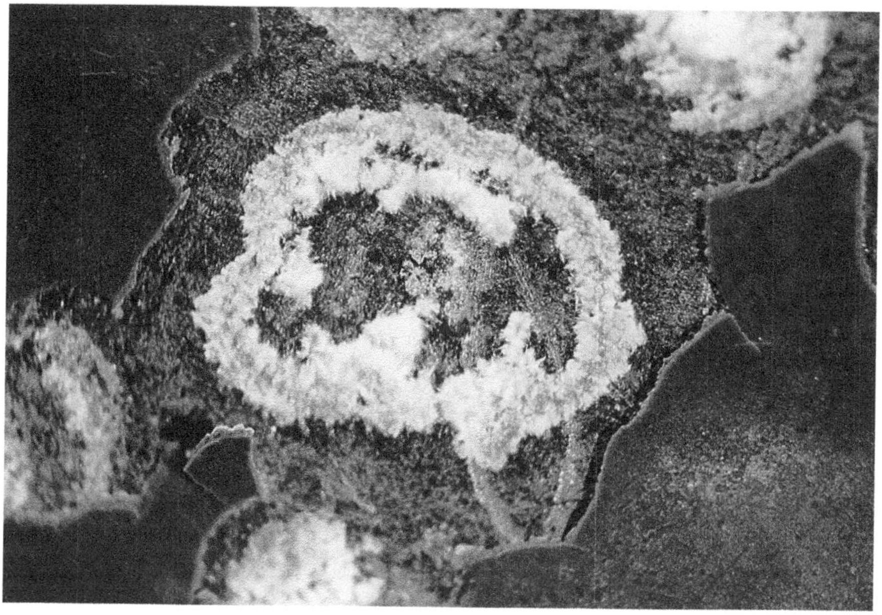

Figure 8.2 Photomicrograph showing heavy deposits of contamination behind poorly adherent paint.

In some cases, the presence of voids within the cross section of the coating can provide clues as to the failure mechanism (Figure 8.3). Voids are due to trapped gases within the coating trying to escape through the viscous film. They could be due to solvent vapours, suggesting that perhaps the wrong thinner was used, or that the coating was applied to too hot of a surface, causing the 'boiling' of the solvent. They could also be caused by excessive amounts of entrapped air, perhaps from overly zealous agitation of the liquid coating or from poor manufacturing or formulating practices.

Another source of voids, at least within two-component urethane coatings, is carbon dioxide. In Section 5.9 concerning urethanes, it was shown that if water is present, it may react with the isocyanate component of the coating to produce carbon dioxide as a by-product. Carbon dioxide is a gas at room temperature and can produce extreme porosity within the coating. The observation of such porosity in a urethane would immediately suggest that certain possibilities should be considered. These might include painting over a damp substrate or application in inclement weather, water in the paint lines, or perhaps the use of a non-urethane-grade thinner, which contained traces of water.

Porosity can sometimes compromise a coating's physical strength. It can also compromise its barrier properties, since its effective thickness is substantially reduced. However, some amount of porosity is normal with certain coatings.

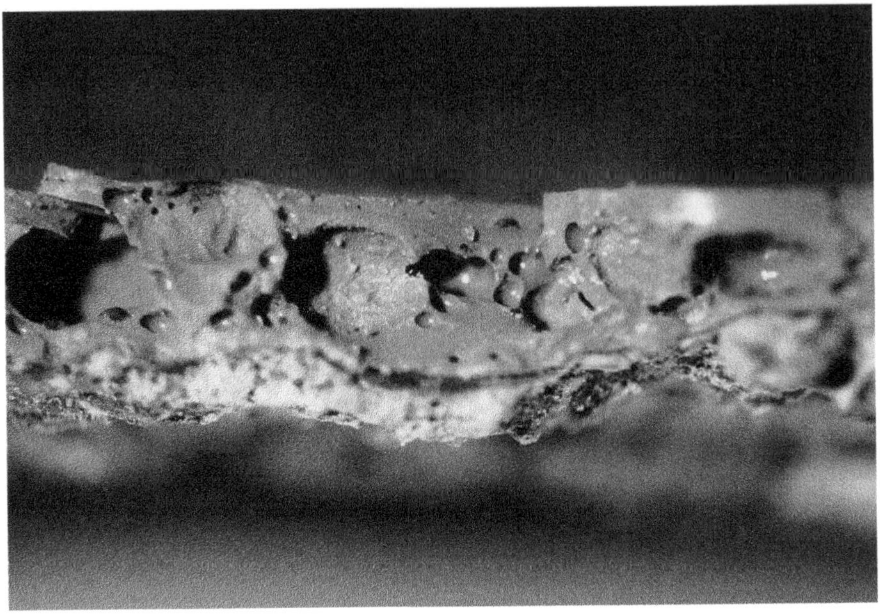

Figure 8.3 Photomicrograph showing voids in a cross section of a coating.

Experience, knowledge of the characteristics of the coating in question, and perhaps an examination of both failing and non-failing samples, will be helpful in arriving at sound conclusions.

One will often see dirt, rust or other contaminants on the back of disbonded coatings. Some questions to consider here are whether the contamination was painted over, whether it formed after painting or whether it got on the surface after the disbonding occurred. The latter scenario can often be eliminated entirely by careful sample collection, such as avoiding pieces of coating that have already fallen off. However, sometimes someone else has taken the sample, and its pedigree may be in question.

For contamination that can be seen with a low-power microscope, it is usually possible to tell whether it was there before or after the coating failed, simply by how loosely attached it is to the coating. If it can be easily blown or brushed off without leaving an imprint on the coating, it is more likely extraneous and should be ignored. If it is embedded in the coating, it was most likely painted over.

Rust will not form on the back of a coating after it has disbonded. The question, then, is whether it was painted over or formed in service. Careful microscopic observations of the cross section of a chip may help to answer this question. If the coating has flowed into and around the rust, perhaps even encapsulating specks of it, then most likely the rust was painted over. If this is not the case, the rust may or may not have been painted over.

If the rust formed after the steel was painted, it is likely that the coating was permeated. In such cases, there will usually be some rust staining, or pinpoints of rust, on the front surface of the coating, or at least within the cross section. If the front and cross section is in impeccable condition, one should suspect that the rust was painted over.

The environment and time frame should also be considered when trying to determine if rust was painted over or formed after the fact. If the coating is only 6 months old, the environment is fairly mild and there are about 10 mil of rust scale under the coating, it is a safe bet to conclude that it was painted over. On the other hand, if there is some visible porosity and/or rust bleed-through to the surface of the topcoat, and if the thickness of the rust layer does not seem abnormally high for the time frame and service environment, it likely occurred after painting.

The importance of flow and wetting has been expressed more than once in this text. Evidence of inadequate flow can sometimes be seen through the microscope by the presence of many small globules on the surface of the coating. These globules, which in extreme cases can impart an almost pebbly texture to the paint surface, are due either to dry-spray or over-spray.

Over-spray, as the name implies, is caused by atomized paint droplets from neighbouring operations settling on an already painted surface. Except for its appearance, it is harmless.

Dry-spray is caused by particles of partially dried paint that have insufficient solvent to properly flow and wet the substrate to which it is being applied. It can be caused by poor application techniques, such as having the spray gun too far away from the substrate or by poor formulation (insufficient solvent, wrong type of solvent or too high a pigment-to-binder ratio). In either case, except for specialty coatings such as hammer finishes, it is never beneficial.

In cases of delamination or blistering, it is extremely important to identify the location, or locus, of failure. Field observations and simple visual observations can sometimes be misleading in this regard. Occasionally, even microscopic observations can lead to the wrong conclusion. If a delamination or blistering problem truly occurred at an interface, then possible mechanisms to consider might include contamination, a blush or exudate, excessive curing of the first coat or simple incompatibility. However, if there is a thin layer of the first coat on the back of the disbonding coat, one might want to consider other possibilities. Perhaps the first coat was inadequately cured at the time of topcoating. Perhaps it could not resist the combined effects of the solvent system and the curing stresses of the topcoat. Perhaps its surface weathered and degraded prior to topcoating. In any event, the location of the failure, coupled with a knowledge of the basic characteristics of the coating system, can greatly narrow the focus of the investigation.

A common mistake, especially with house paints applied to older structures that subsequently peel, is to assume that the newly applied paint was somehow defective. This seems like a logical conclusion, since, after all, the old paint was

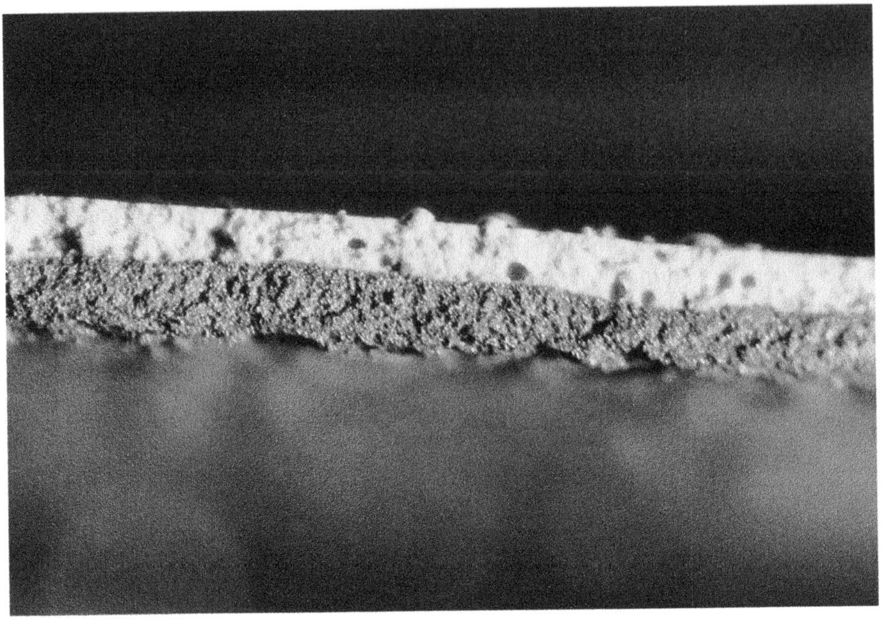

Figure 8.4 Photomicrograph of a cross section a paint chip displaying a primer and a topcoat.

doing fine, and as soon as the new paint was applied, peeling began occurring. This could be true. Alternatively, a microscopic examination may show that there are five coats of paint on the peeling samples, thus indicating that the weak link was a layer of paint perhaps applied 10 or 15 years ago. The extra weight and stress imparted by the new coat of paint was simply a little too much for this weak layer to withstand.

Determining the number of coats by an examination of the cross section is not always as easy as it sounds. If all of the various coats are different colours, then this is a simple task (Figure 8.4). However, sometimes it can be nearly impossible to determine if two coats of the same paint were applied, especially if the painting job was completed in a few days or less. In such a case, there are no colour difference and no thin line of accumulated dirt or debris between the coats. One can sometimes infer the number of coats based on total thickness, but this is not foolproof. Sometimes, differences in temperature, or thinning, may result in a line of voids present in only one of the two coats, or if both coats have voids, the pattern of voids may indicate one or two coats.

It is usually a good idea to measure the coating thickness in the course of a failure investigation. The cracking of an alkyd (Figure 8.5), for instance, can be directly related to excessive thickness. For metal substrates, there are various

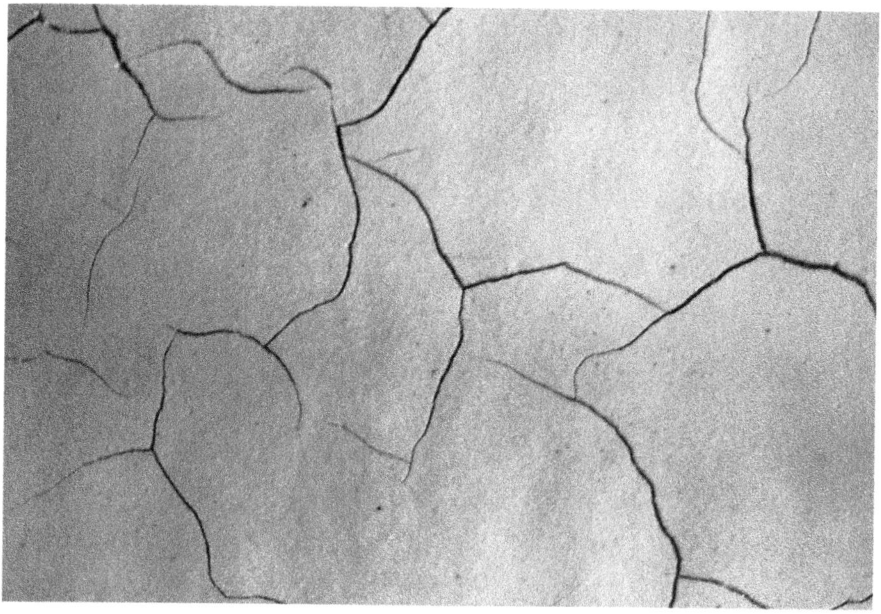

Figure 8.5 Photomicrograph showing the surface cracking of an alkyd resulting from excessive thickness of the coating.

hand-held electronic gauges that can do this, although only the total thickness of all the coats is measured. There are other devices that use special cutting tips with calibrated microscopes that can allow the thickness of individual coats to be measured. The lighting and optics of these devices is rather poor, and adjacent coats of similar colour may be mistaken for a single coat. Furthermore, they are of very limited use on rough, irregular samples or with coatings that shatter easily upon cutting. Additional photomicrographs are shown in Figures 8.6–8.11.

Most of the time, the measurement of coating thickness with a microscope equipped with a calibrated reticle is straightforward. An example where it is not quite so easy is a relatively thin, soft coating, such as acrylic latex, on wood. If the coating has to be forcibly disbonded, a common mistake is to cut a piece off by using a scalpel and then examine the cut edge. The thin soft coating will almost always deform on cutting and smear out. If it smears out only one one-thousandth of an inch (1 mil), and the true thickness is only 2 mil, an error of 50% has occurred.

The correct way to measure the thickness of such a coating, if it cannot be readily flaked off, is to make two closely spaced parallel cuts, perhaps $\frac{1}{16}$ apart. This will usually result in the lifting of a thin ribbon of coating, perhaps with some wood on the back, still attached at both ends to the substrate. If one then takes the tip of the knife, inserts it under the detached ribbon of coating and gently

Figure 8.6 Microscopy is an excellent tool for determining the number and thickness of individual coats of paint.

Figure 8.7 Photomicrograph of salt crystals on the back side of blistered coating.

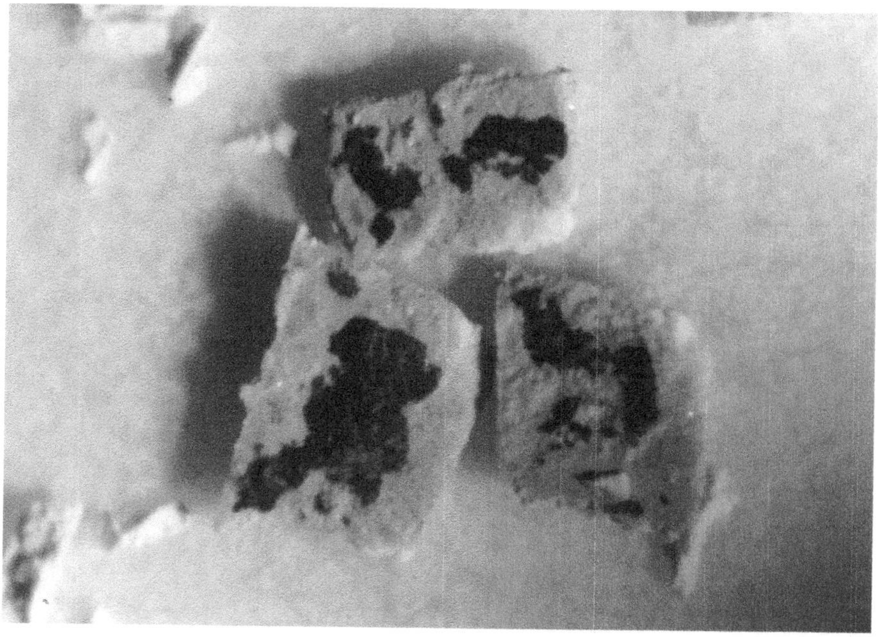

Figure 8.8 Photomicrograph of mill scale on the back side of an epoxy coating. The scale should have been removed by abrasive blast cleaning prior to coating application.

Figure 8.9 Photomicrograph of the back side of a failing inorganic zinc-rich primer. Instead of abrasive blasting, as specified, the steel had been ground (note the grinding marks), which resulted in an adhesion failure.

Figure 8.10 Photomicrograph of the backside of paint which peeled from a plaster wall. The layer of white, powdery plaster on the back of the paint shows that the failure occurred within a weak, upper layer of plaster and, therefore, was a plaster failure rather than a paint failure.

Figure 8.11 Photomicrograph showing extreme porosity of a zinc-rich primer, due to poor application technique. The primer is so porous that there are gaps where the back side of the next coat is visible.

lifts, the coating will usually break in half, and one can then measure the thickness at the broken end without fear of having smeared the coating by cutting.

Another way of measuring the thickness of thin or delicate samples is by mounting a small piece in a clear epoxy resin, and then subjecting the mounted specimens to a series of polishing operations using successively finer and finer abrasives. This can be done manually with an occasional sample or with automated equipment if carried out regularly.

If one is dealing with a thermoset coating such as an epoxy, urethane or inorganic zinc-rich primer, it is usually important to know if the coating cross-linked correctly. This is not always an easy task, even with modern analytical instrumentation. A simple test that can at least reveal gross problems is to determine the coating's resistance to solvent. This is usually done by rubbing with cheesecloth or a cotton Q-tip saturated with methyl ethyl ketone (MEK) and forming a subjective opinion on the amount of colour transfer, or perhaps how many 'double rubs' (a single up and down stroke) are required to wear through to the next coat. This works reasonably well for large samples.

For samples of disbonded paint chips that may be only $\frac{1}{4} \times \frac{1}{4}$ in. or less, solvent rub tests are difficult or even impossible. However, one can often gain some insight into the coating's solvent resistance (and hence degree of cross-linking) by placing a small chip in a watch glass and adding a few drops of MEK while observing microscopically at magnifications of 10× or so. If the chip rapidly swells and curls, it is solvent sensitive. The same test can be carried out on latex coatings, in this case using water, to get a feel for whether or not it is unduly sensitive to moisture. In either case, it is desirable to also perform the test on laboratory-prepared samples, or at least on failing and non-failing samples, to determine if the behaviour is 'normal' for that particular coating.

It is usually important to determine if a failing coating is too brittle or too soft. These are difficult terms to define and equally difficult to measure. A materials scientist would almost immediately think of measuring the coating's modulus from stress–strain curves, or perhaps its flexural strength, as a means of characterizing its brittleness. However, such techniques are typically useless on a real-world sample of failing coating, since seldom does a coating blister or delaminate in the shape of a perfect dog bone of uniform thickness, waiting to be clamped in the jaws of a universal-testing machine.

Likewise, there are several standard American Society for Testing and Materials (ASTM) test methods for measuring hardness. Sometimes, these methods may actually be appropriate for a failing sample submitted to the laboratory, but often only subjective opinions about a failing sample's brittleness or softness can be made.

As long as one does not try to cut too fine a line, subjective 'measurements' of a sample's brittleness (by bending it between one's fingers, for instance) or softness (by observing, microscopically, the effect of indenting it with a knife, spatula, etc.) can be useful. However, the results must be couched in terms of the investigator's experience with similar coatings, or relative to laboratory-prepared controls, or at least failing versus non-failing.

Terms like brittleness and softness are relative. To say that an amine-cured epoxy is too brittle could mean one of two things. It could be 'too brittle' when compared with a properly mixed, applied and cured laboratory sample of that particular coating, or it could mean that this coating, even when properly applied, is 'too brittle' for the particular service in which it is being used. The ramifications of the meaning of the statement are important.

In the first case, if the coating is significantly more brittle than a properly applied laboratory sample, it might mean that it was misapplied in the field, perhaps by using the wrong mix ratio, or that it represents a defective batch of coating. In the second case, the coating is simply not the right choice for that particular job.

It would be impossible to list all of the things which one might see when examining failed samples through a microscope. It is safe to say, however, that the time spent at the microscope will be the most productive if one has background information, knows the significance of each sample (location, failing versus non-failing, etc.) and maintains a curious and open mind.

8.2 INFRARED SPECTROSCOPY

Although it has its limitations, it is hard to overemphasize the importance of infrared spectroscopy as a failure analysis tool. It is often the technique of choice immediately following a microscopic examination.

8.2.1 Theory

A useful definition of spectroscopy is the study of the interaction of light with matter. Light is a form of electromagnetic radiation, which is energy travelling through space at the speed of light, c (where $c = 2.99792458 \times 10^8\,\mathrm{m\,s^{-1}}$).[1]

The classical interpretation of this radiation, as formulated by Christian Huygens in the late seventeenth century, ascribes to it a wave-like nature. Huygens envisioned light travelling through space via a continuous wave-like motion, much like the propagation of ripples or waves caused by tossing a pebble into a quiet pond. This conveniently explained such things as refraction and diffraction. However, certain other phenomena, such as the photoelectric effect observed by Einstein, are more readily explainable by the quantum mechanical interpretation of electromagnetic radiation. In quantum mechanics, electromagnetic radiation travels as discrete packets or 'bundles' of energy, known as photons. The two views of electromagnetic radiation as waves or as photons are actually complementary, rather than being mutually exclusive.

Figure 8.12 shows a simple representation of an electromagnetic wave, which consists of oscillating electronic and magnetic fields moving through space at right angles to one another.

---

[1] This is the exactly defined value of this important physical constant.

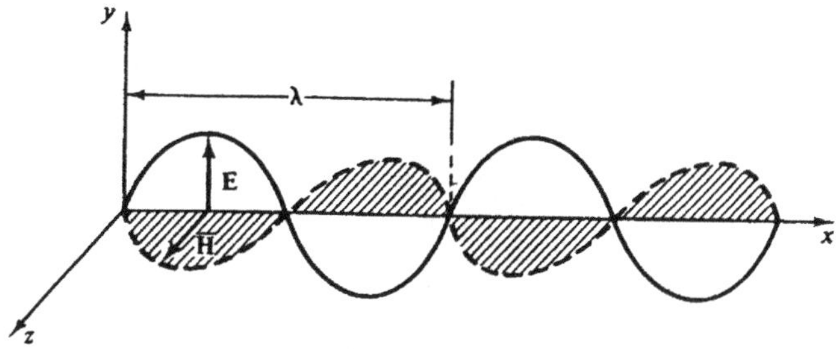

Figure 8.12 A simple representation of an electromagnetic wave. Reproduced from F. Olsen, *Modern Optical Methods Analysis*, McGraw-Hill, © 1975, with permission of The McGraw-Hill Companies.

The wavelength (λ) is defined as the distance between successive maxima or minima (a cycle), while the frequency (ν) is the number of cycles per second.

For electromagnetic radiation propagating through a vacuum, Equation 8.1 describes the relationship between wavelength and frequency, as follows:

$$c = \lambda\nu. \tag{8.1}$$

An additional, extremely useful relationship derived from the quantum mechanical view of electromagnetic radiation is that the energy of the radiation is directly related to its frequency, as follows:

$$E = h\nu. \tag{8.2}$$

In the equation above, h is the Planck constant and has a value of 6.626×10^{-34} J s^{-1}.[2] The higher the frequency (or the shorter the wavelength), then the more energetic the radiation.

The electromagnetic spectrum is depicted in Figure 8.13. This varies from the short-wavelength, high-energy gamma rays to the long wavelength, low-energy radio waves. Since the various regions of the spectrum have different energies, it is not surprising that they interact with matter in different ways. Infrared radiation, which is the subject of this present section, is relatively low-energy radiation that interacts with vibrations occurring within the molecular structure of matter.

Infrared spectroscopy relies upon the fact that molecules are in a constant state of motion. Consider propane, CH_3–CH_2–CH_3. The middle carbon atom of propane is bonded to two carbon atoms and two hydrogen atoms. The angle of each of these bonds (bond angle) is approximately 109°, and the length of the bonds themselves is approximately 1.1 Å for the C–H bonds and 1.5 Å for the C–C bonds. In reality, the molecule is constantly undergoing both bending and stretching vibrations, and the above values are actually time-weighted averages.

[2] This has a value of 6.626×10^{-27} erg s^{-1} in non-SI (cgc) units.

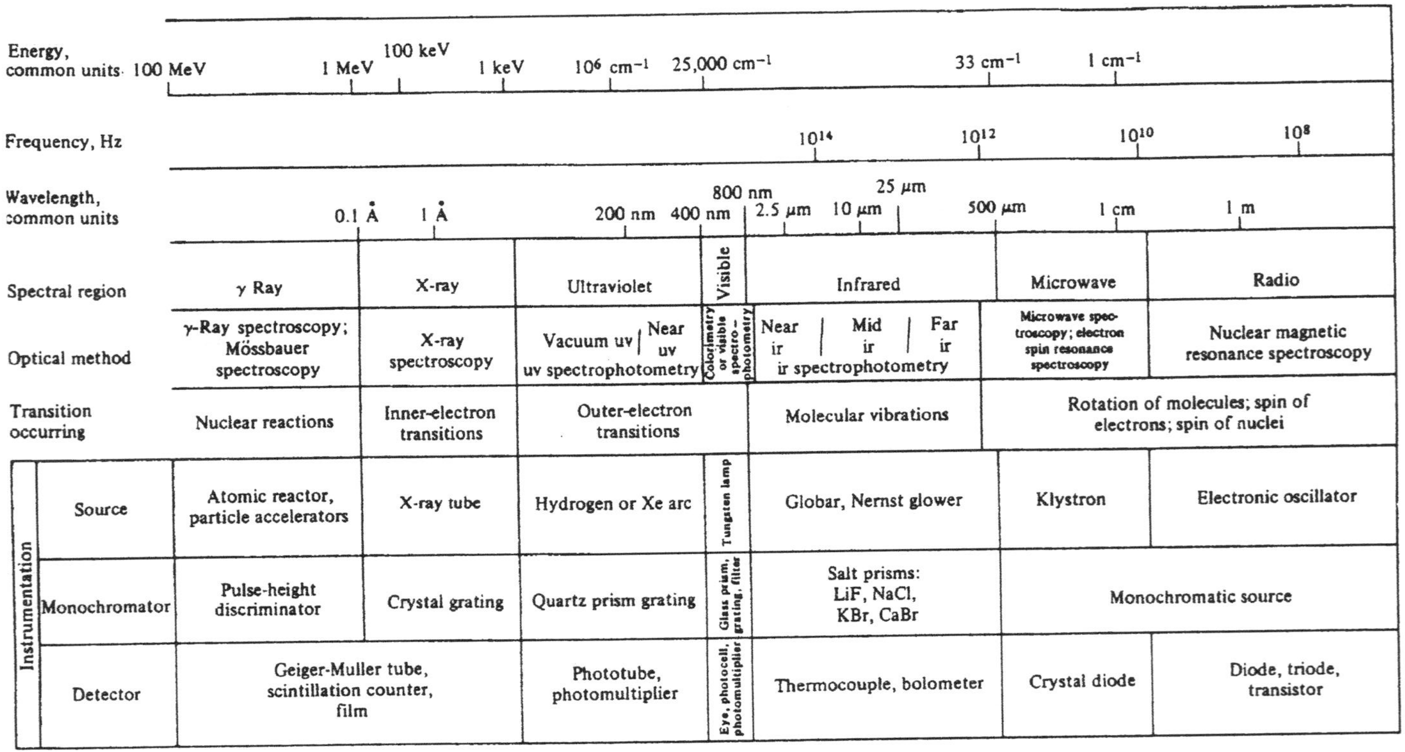

Figure 8.13 The electromagnetic spectrum. Reproduced from F. Olsen, *Modern Optical Methods Analysis*, McGraw-Hill, © 1975, with permission of The McGraw-Hill Companies.

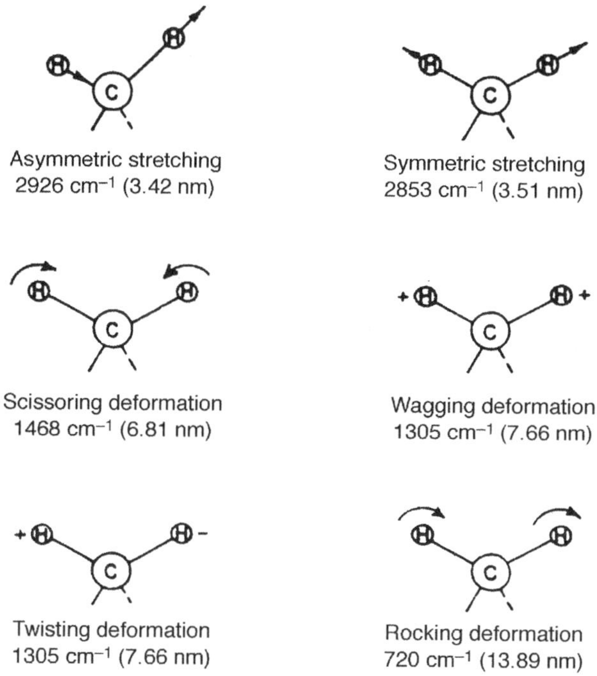

Figure 8.14 Frequencies of various carbon–hydrogen bending and stretching vibrations. Reproduced by permission of The Federation of Societies for Coatings Technology, from D. R. Brezinski (Ed.), *An Infrared Spectroscopy Atlas for the Coatings Industry*, 1991.

Figure 8.14 shows the types of stretching and bending (deformation) vibrations that can occur at a methylene (CH_2) group.

Stretching vibrations involve a change of bond length, while bending vibrations involve a change of bond angle. The vibrations occur at certain frequencies, which are measured in units of hertz or cycles per second. Since, in the infrared region, this would result in such large numbers, it is more common to divide the frequency by c (the speed of light), to obtain what are called 'wave number' units. Wave numbers ($\bar{\nu}$) have the units of reciprocal centimetres (cm^{-1}) and can be visualized as the number of waves (or cycles) that occur per centimetre. This parameter is directly proportional to the frequency.

Luckily for infrared spectroscopists, the various molecular vibrations, which depend primarily on the type of atoms and the type of bonds holding them together, occur within relatively narrow ranges of frequency, regardless of the larger structure of the molecule. The stretching vibrations of an aliphatic C—H bond are going to occur at approximately 2900–2800 cm^{-1}, regardless of whether the C—H group is part of an alkane, a ketone, an acrylic or an epoxy.

As Figure 8.13 shows, the vibrations among the chemical bonds that hold molecules together occur at frequencies within the infrared region of the electromagnetic spectrum. If a molecule is exposed to radiation of a frequency exactly matching one of its molecular vibrations, the molecule can absorb some of this light. The effect is similar to that observed with two perfectly matched tuning forks. If one tuning fork is struck and slowly brought into close proximity with the stationary tuning fork, the latter one will absorb some of this matched 'vibrational energy' and soon begins to vibrate.

This absorption of matched vibrational energy is the basis of infrared spectroscopy. An infrared spectrum is basically obtained by focusing a beam of infrared radiation through a sample and then observing the frequencies at which the sample absorbs this radiation. If two molecules have different structures, their absorption bands will occur at different frequencies or have different relative intensities. Each molecule has its own unique, infrared spectrum, which is sometimes referred to as a 'fingerprint'. Sometimes, as illustrated by the spectra of MEK (Figure 8.15) and toluene (Figure 8.16), the differences are substantial. In other cases, such as the spectra of linseed oil (Figure 8.17) and soybean oil (Figure 8.18), the differences are quite subtle.

Given the structure of a molecule, how would one go about predicting the general appearance of its infrared spectrum? This would be a daunting task were it not for the fact that, in most cases, it has already been done. There are many collections of infrared spectra of both pure compounds and of resins and polymers. Furthermore, the frequencies of the fundamental stretching and bending vibrations of any pair of atoms are only modestly affected by their attachment or proximity to other atoms in the molecule. If a molecule contains a carbonyl

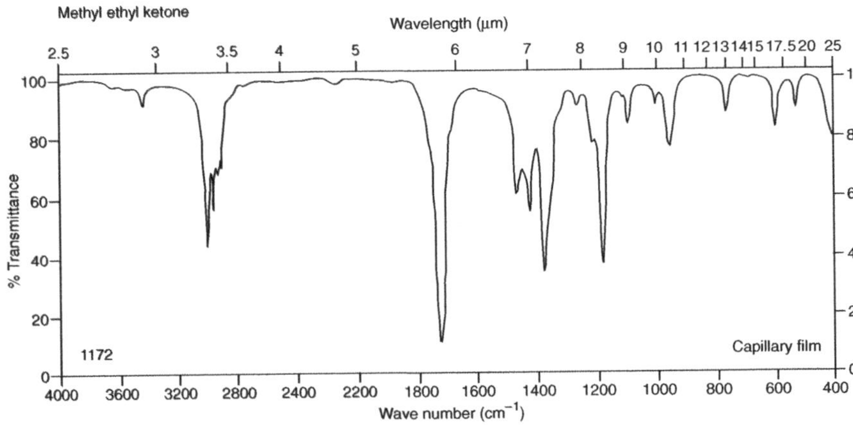

Figure 8.15 Infrared spectrum of methyl ethyl ketone. Reproduced by permission of The Federation of Societies for Coatings Technology, from D. R. Brezinski (Ed.), *An Infrared Spectroscopy Atlas for the Coatings Industry*, 1991.

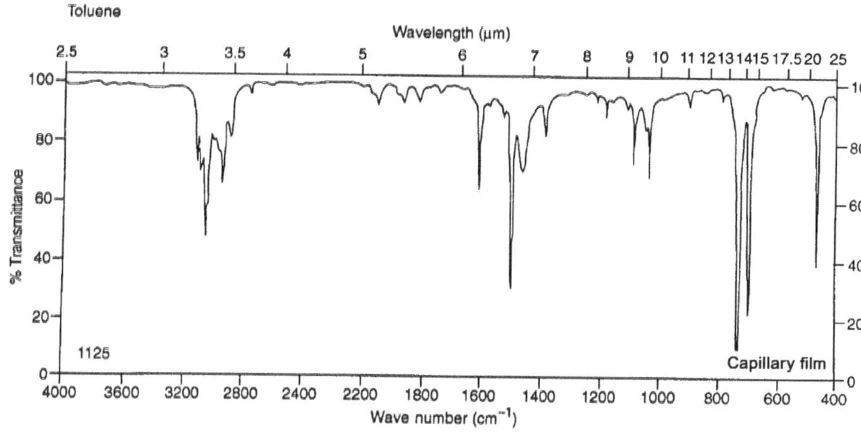

Figure 8.16 Infrared spectrum of toluene. Reproduced by permission of The Federation of Societies for Coatings Technology, from D. R. Brezinski (Ed.), *An Infrared Spectroscopy Atlas for the Coatings Industry*, 1991.

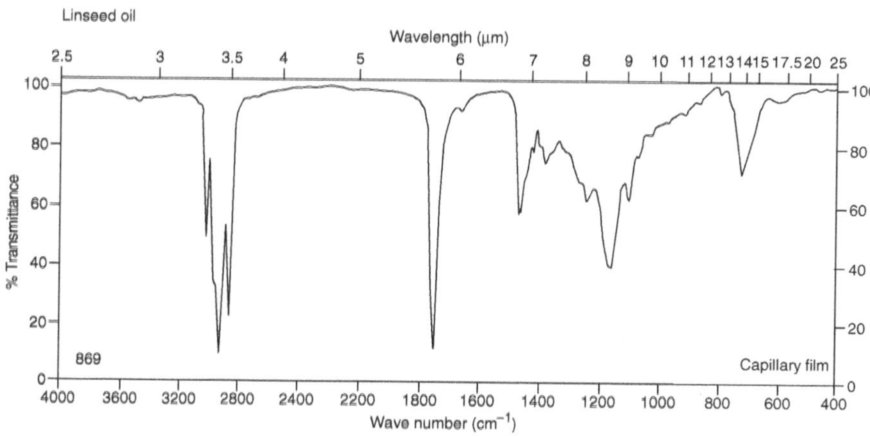

Figure 8.17 Infrared spectrum of linseed oil. Reproduced by permission of The Federation of Societies for Coatings Technology, from D. R. Brezinski (Ed.), *An Infrared Spectroscopy Atlas for the Coatings Industry*, 1991.

carbon, one can expect to see the characteristic C=O stretching vibration somewhere between approximately 1750 and 1650 cm^{-1}. If the molecule contains an aliphatic carbon–hydrogen bond, it will have a stretching vibration in the 3000- to 2800-cm^{-1} region, regardless of the structure of the rest of the molecule. These are just two examples of group frequencies. Group frequencies form the basis of qualitative analysis by infrared spectroscopy. Figure 8.19 lists many of the

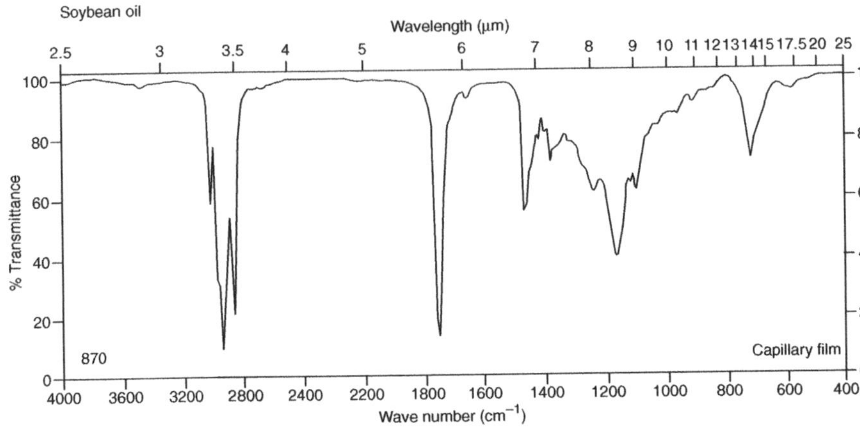

Figure 8.18 Infrared spectrum of soybean oil. Reproduced by permission of The Federation of Societies for Coatings Technology, from D. R. Brezinski (Ed.), *An Infrared Spectroscopy Atlas for the Coatings Industry*, 1991.

common group frequencies encountered in infrared spectroscopy and is known as a correlation chart.

Although correlation charts are the most convenient way of drawing conclusions between spectra and structure, it should be pointed out that the process does not have to be quite so arbitrary. It is possible to actually calculate the approximate location of many molecular vibrations. This is accomplished by considering any two-atom pair to be a simple harmonic oscillator (e.g. two balls connected by a spring). In such a situation, it is intuitively obvious that the frequency of their oscillation would depend on the weight of the balls and the stiffness of the spring.

Mathematically, such a stretching vibration can be described by the following equation:

$$\bar{\nu} = \frac{1}{2\pi c}\sqrt{\frac{k}{\mu}}, \tag{8.3}$$

where k is the force constant (the stiffness of the spring), c is the velocity of light and μ is what is referred to as the reduced mass of the pair of vibrating atoms. For two atoms of masses m_1 and m_2, the reduced mass is expressed as follows:

$$\frac{1}{\mu} = \frac{1}{m_1} + \frac{1}{m_2}. \tag{8.4}$$

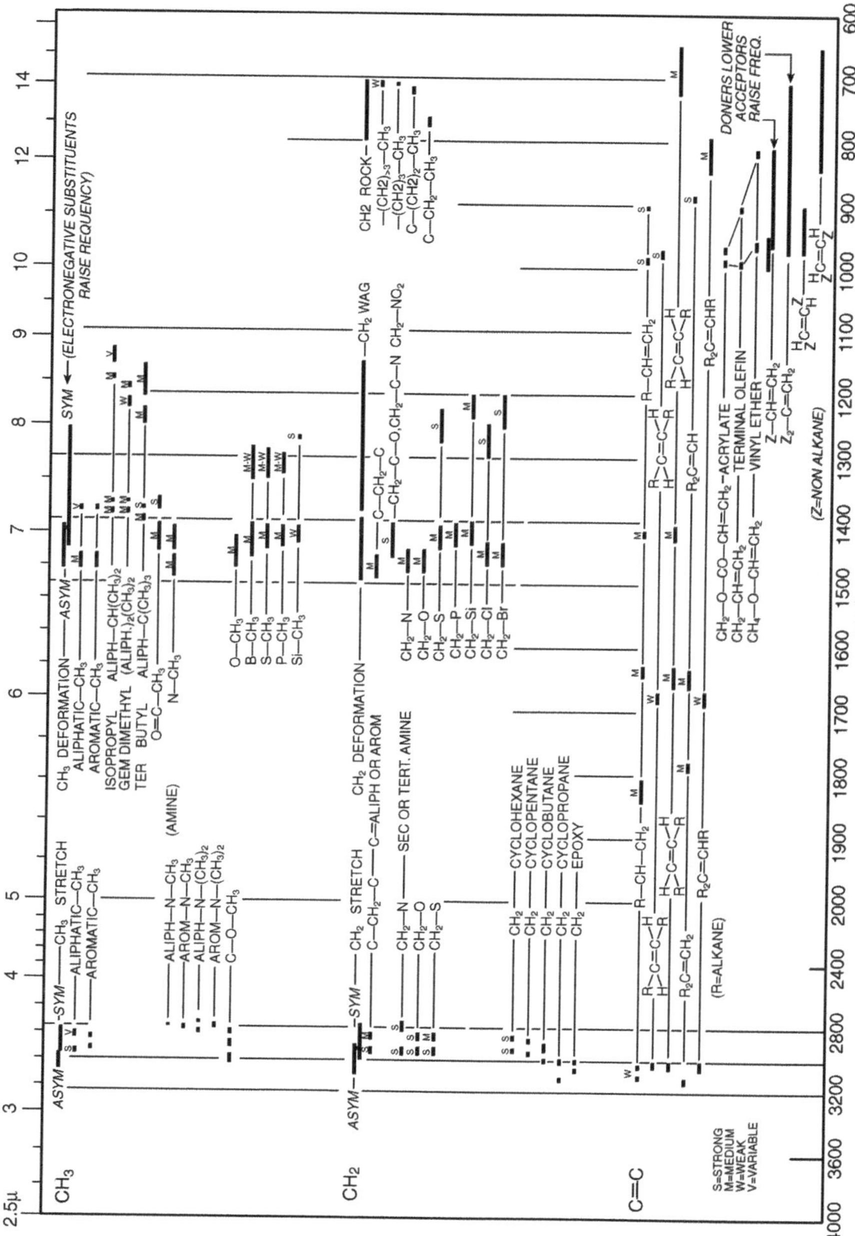

Figure 8.19 Correlation chart of common infrared group frequencies. Reproduced by permission of The Federation of Societies for Coatings Technology, from D. R. Brezinski (Ed.), *An Infrared Spectroscopy Atlas for the Coatings Industry*, 1991.

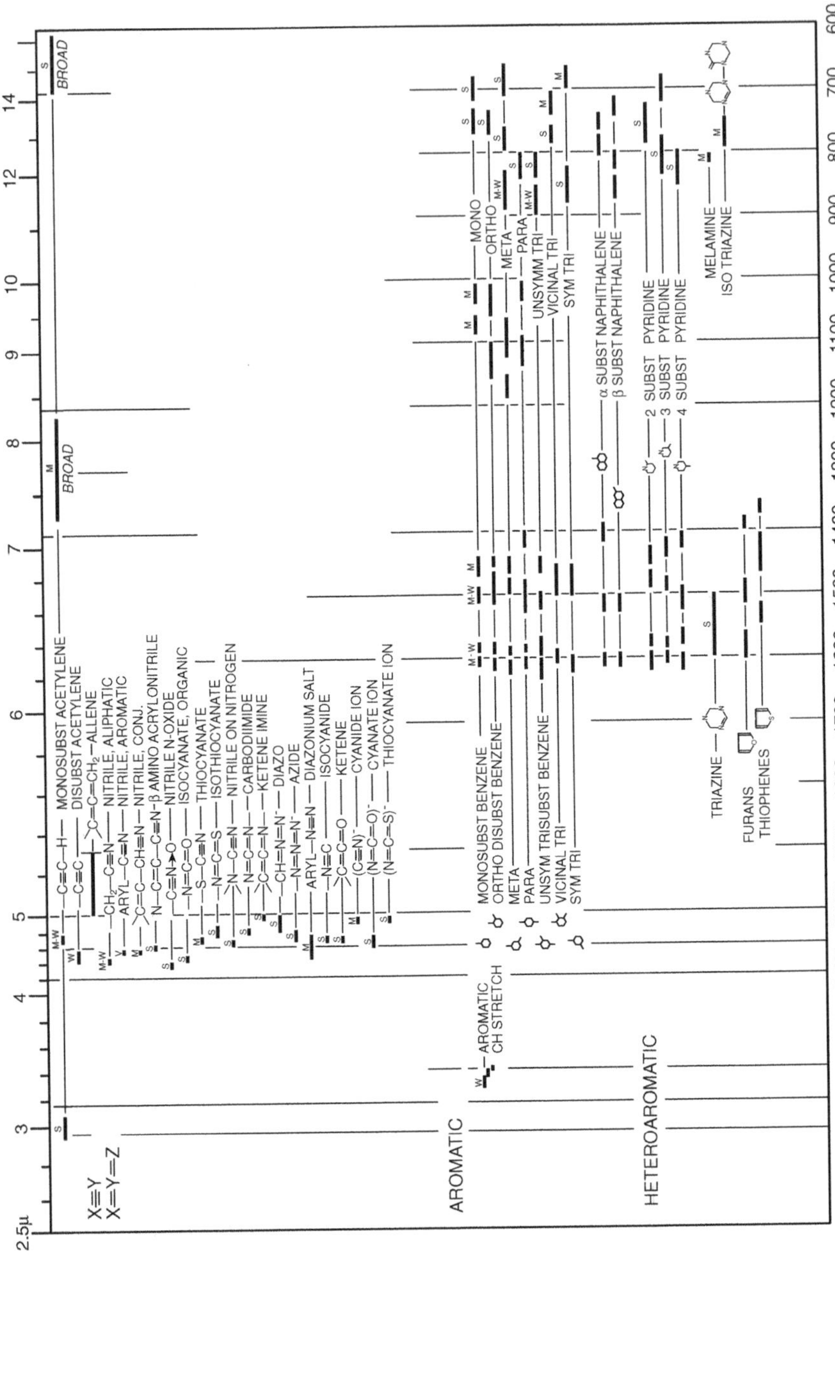

Figure 8.19 *Continued*

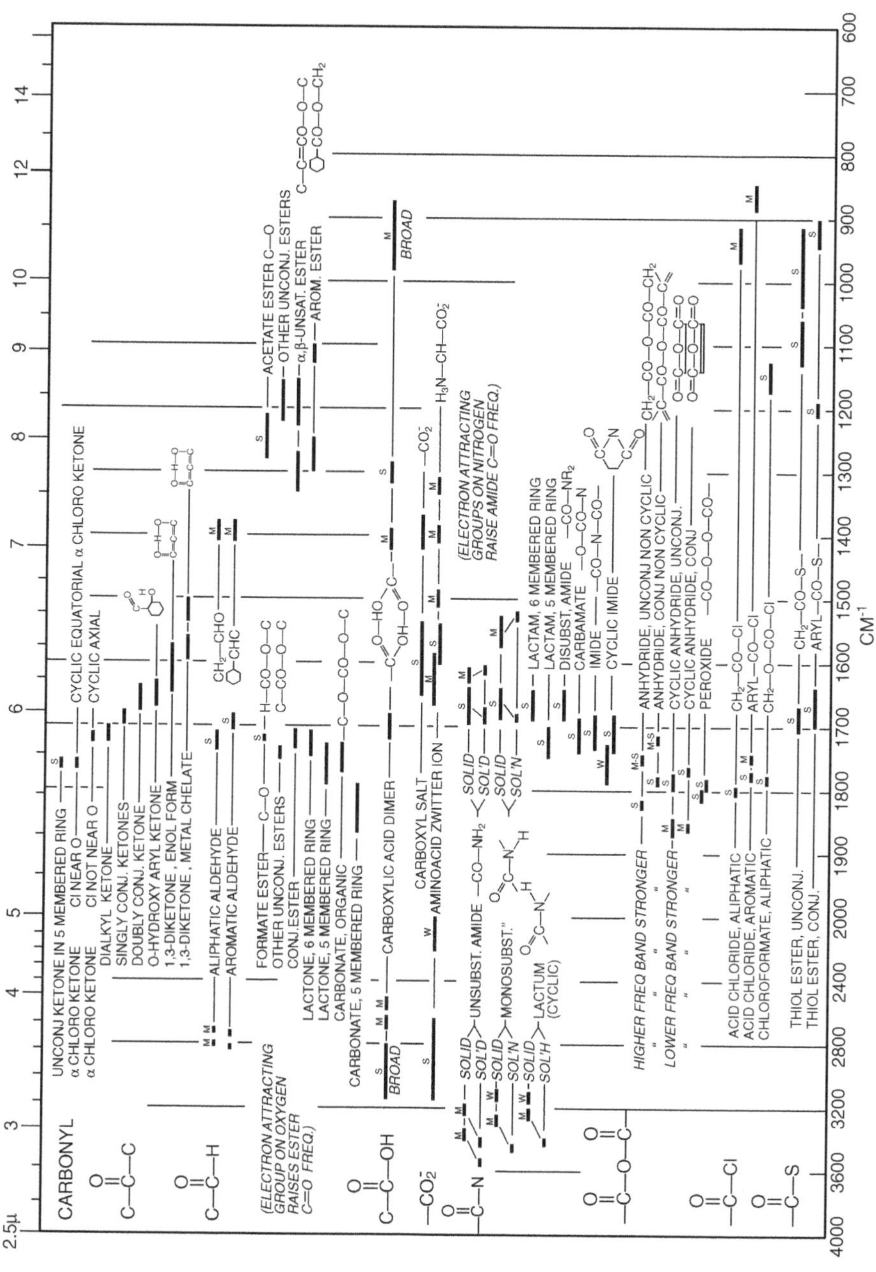

Figure 8.19 *Continued*

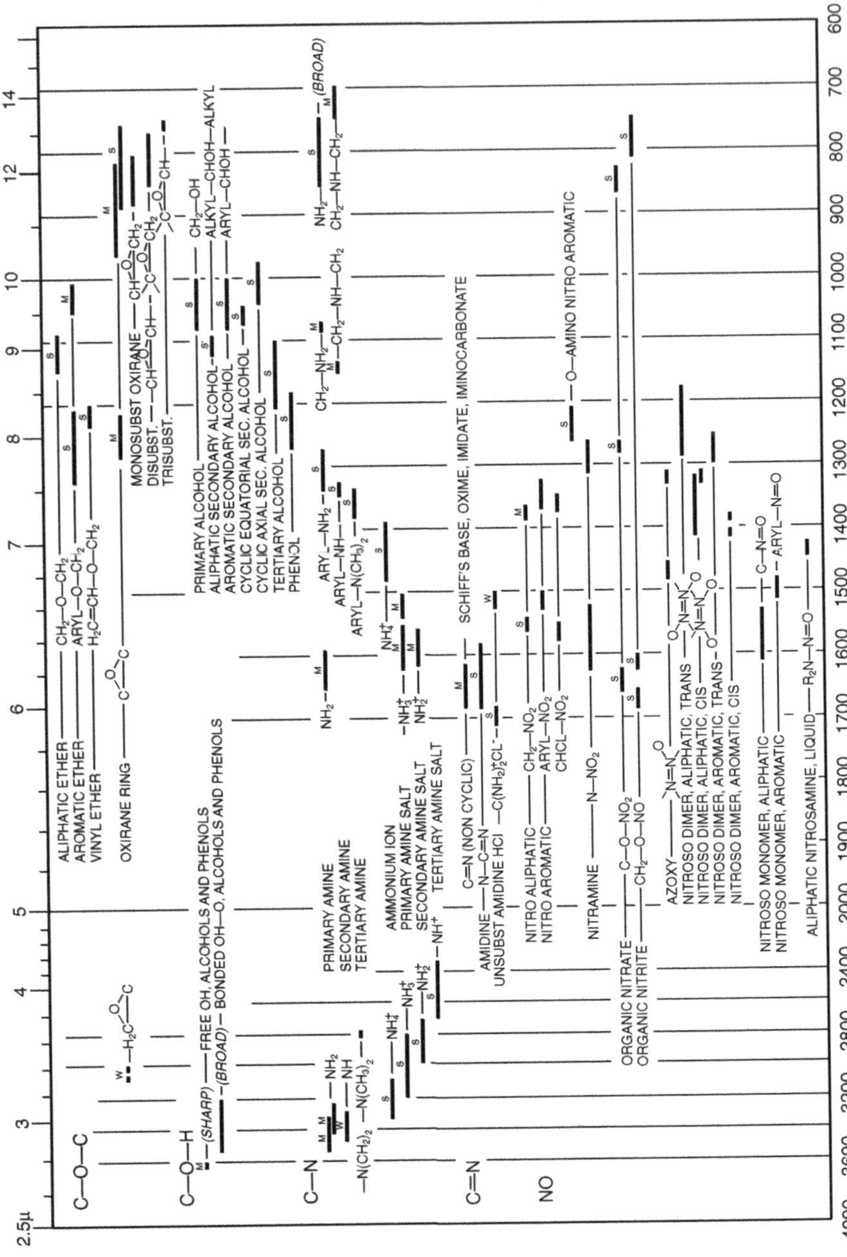

Figure 8.19 *Continued*

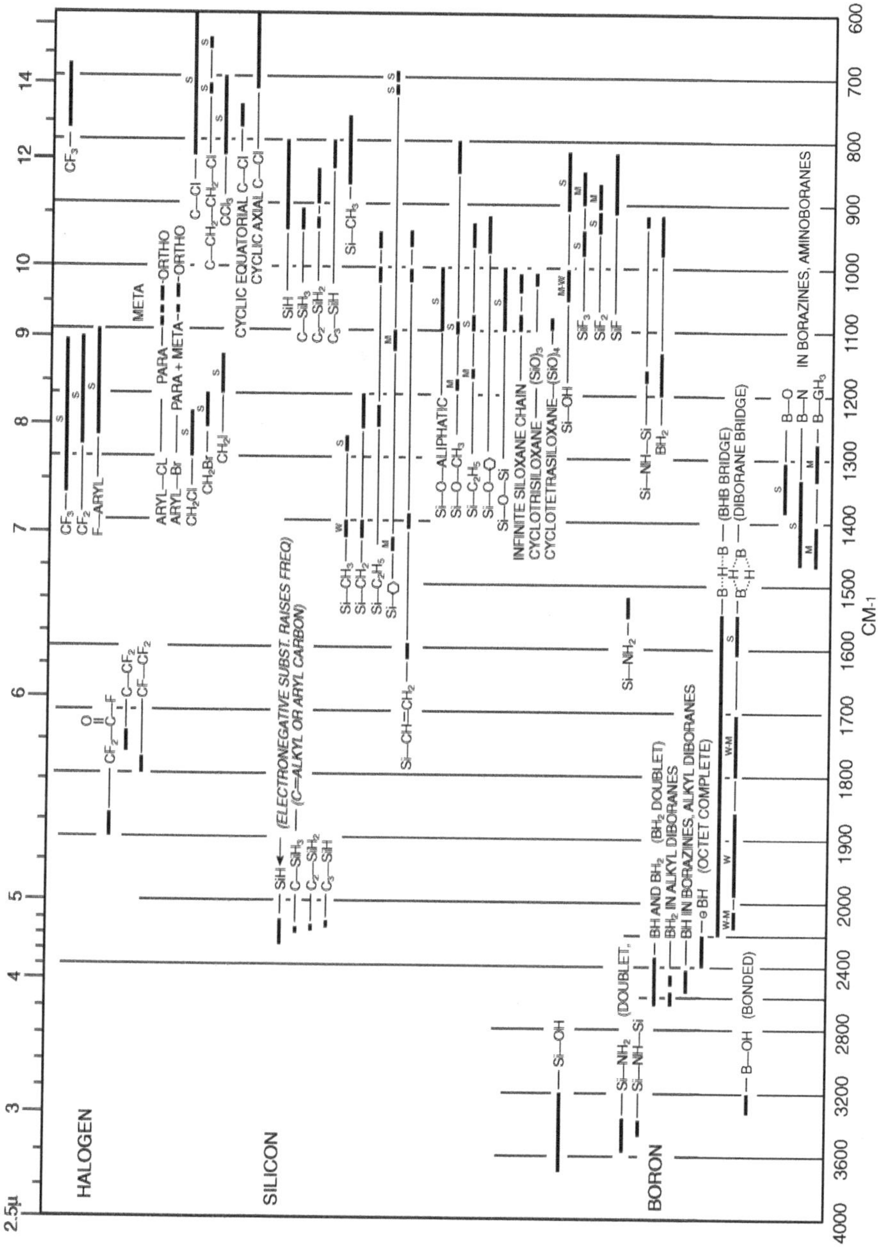

Figure 8.19 *Continued*

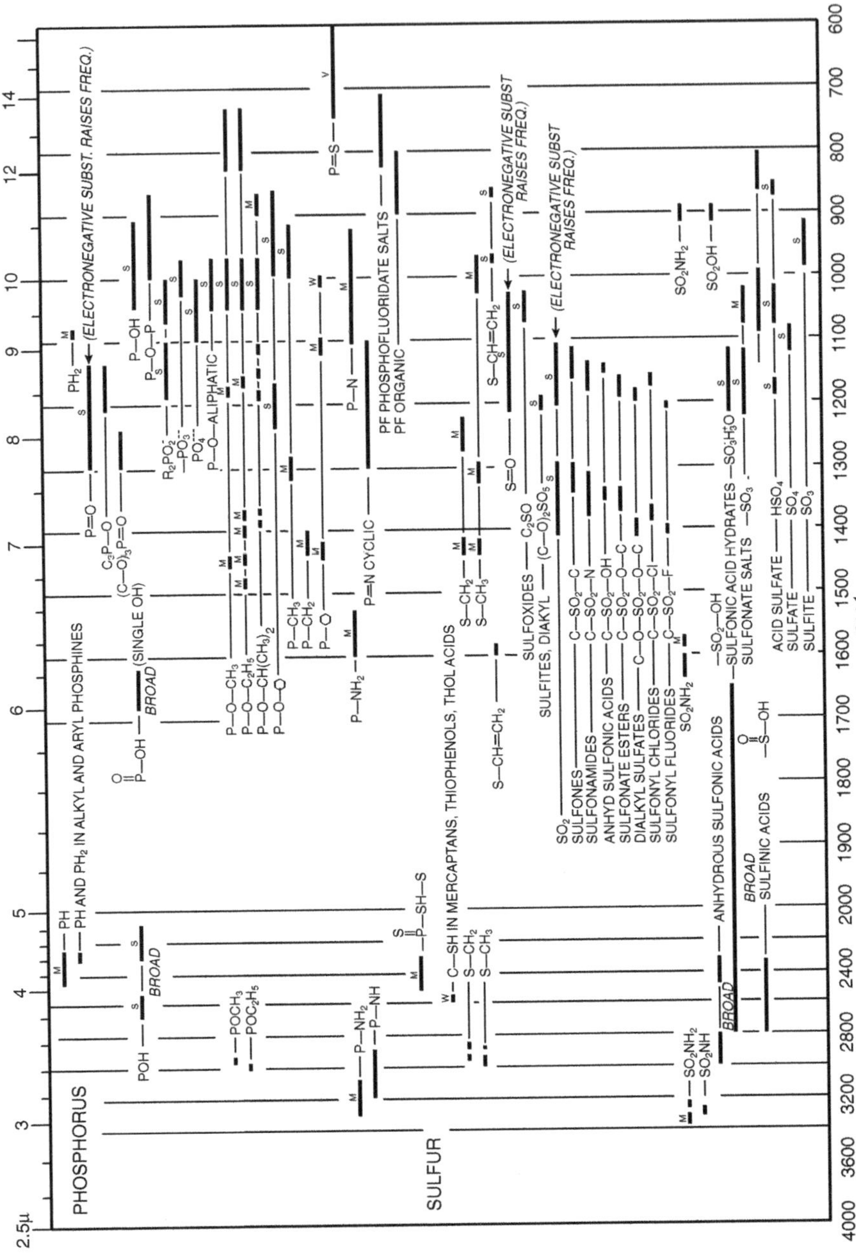

Figure 8.19 *Continued*

Table 8.1 Calculated force constants for various atom pairs. Reproduced from F. Olsen, *Modern Optical Methods of Analysis*, McGraw-Hill, 1975, with permission of The McGraw-Hill Companies.

Bond	Molecule	*k* (dyn/cm)
H–F	HF	9.7×10^6
H–Cl	HCl	4.8
H–Br	HBr	4.1
H–I	HI	3.2
H–O	H_2O	7.8
H–S	H_2S	4.3
H–N	NH_3	6.5
H–C	CH_3X	4.7–5.0
H–C	C_2H_4	5.1
H–C	C_2H_2	5.9
Cl–C		4.5–5.6
C=C		9.5–9.9
H≡C		15–17
N–N		3.5–5.5
N=N		13
N≡N		23
C≡N		16–18
C–O		5.0–5.8
C=O		12–13

Force constants for many atom pairs have been calculated, as shown in Table 8.1. The reduced mass of any atom pair can be calculated based on the atomic weights of the two atoms (in gram per mole), divided by the number of atoms per mole (the Avogadro constant, which has a value of 6.0221367×10^{23}/mol).

By using the above relationships and the data given in Table 8.1, the frequency of the C—H stretching vibration is calculated to be 3057 cm^{-1}. This value is close to the 3000–2800 cm^{-1} frequency range in which most C—H vibrations actually occur. A similar calculation for the carbonyl (C=O) group would show a frequency of 1700 cm^{-1}, again in good agreement with observed values.

While the above discussion gives one an idea of where a particular vibration should occur, it says nothing about its intensity. For a particular molecular vibration to be infrared active, there must be some way to 'couple' the molecular vibration with the vibrating infrared light which is impinging upon it. Whether or not there is a 'coupling mechanism' depends upon whether or not the vibrating molecule can generate its own electric field, and this in turn will depend upon whether or not the particular vibration results in a change of dipole moment.

A dipole, or dipole moment, is basically a partial separation of electric charge, on an atomic or molecular level. It is related to the difference in electronegativity (the tendency of an atom to attract, or hold onto, an electron) of the two atoms and to the geometry of the molecule.

For example, carbon dioxide (CO_2) is a linear molecule, that is, O=C=O. Because oxygen is more electronegative than carbon, there is a small, partial negative charge on the two oxygen atoms and a matching, partial positive charge on the carbon atoms.

If the CO_2 molecule undergoes a 'symmetric' stretching vibration (both carbon–oxygen bonds moving in and out together), because of the symmetry of its geometry, the effective centre of charge does not move. There is no change of dipole moment; hence, no electric field is generated, and the vibration cannot couple, or interact, with the beam of infrared radiation. No band is observed in the infrared spectrum for this vibration.

On the other hand, when the CO_2 molecule undergoes an 'asymmetric' stretching vibration (one carbon–oxygen bond moving out while the other is moving in), the centre of charge moves. There is a change in dipole moment, and this vibration can interact with and absorb infrared radiation.

The strength or intensity of a particular band in the infrared spectrum is primarily dependent upon the magnitude of the change of dipole moment. Bonds with small changes in dipole moment, such as C–C stretching vibrations, are weak, whereas those with large changes in dipole moment, such as C–Cl, are strong.

8.2.2 Instrumentation

There are two basic types of infrared spectrometers to choose from, namely dispersive instruments and Fourier-transform instruments. Most major instrument suppliers now market only Fourier-transform instruments, although it is still possible to purchase a dispersive instrument. Although these latter instruments are bordering on obsolescence, a short discussion of them is helpful both in presenting some basic infrared theory and also in making the advantages of Fourier-transform instruments somewhat clearer.

8.2.2.1 Dispersive instruments

Traditionally, in order to obtain an infrared spectrum, it was necessary to have a source of infrared light that spanned nearly the entire mid-infrared region (approximately 4000–400 cm^{-1}). It was also necessary to have a means of dispersing this continuous source into individual frequencies, a means of selectively focusing the individual frequencies of radiation through the sample, and a means of detecting how much of this radiation was transmitted or absorbed by the sample. Once this was accomplished, one would then have an infrared spectrum similar to those shown in Figures 8.15–8.18.

The source in an infrared spectrometer is a small object that emits infrared radiation when it is heated to a high enough temperature. Common sources are the Nersnt glower (filament), the nichrome wire and the globar.

The Nernst glower is a small rod composed of a mixture of rare-earth oxides which emits broadband infrared radiation when heated to approximately 1900 °C. The nichrome wire is an alloy of nickel, iron, chromium and a small amount of carbon, usually operated at approximately 1100 °C. The globar is a rugged silicon carbide rod operated at approximately 1100 °C, which requires a water- or air-cooled housing.

In a dispersive spectrometer, there is usually a sample beam and a reference beam. The sample beam is focused through the sample. The reference beam travels an essentially parallel path and compensates for variations in source energy with wavelength (the energy output of an infrared source is not constant but varies as a function of wavelength) and also for variations due to water vapour and carbon dioxide in the atmosphere. The infrared spectrum is obtained by taking the ratio of the energy of the sample beam to that of the reference beam.

When the infrared radiation is sent through the sample, it will either be absorbed (attenuated) somewhat by the sample if its frequency matches a suitable molecular vibration, or it will be unaffected. In order to obtain the traditional scan, or spectrum, one must be able to analyse or measure one wavelength at a time. This is the function of the monochromator, which spatially disperses the radiation into its individual wavelengths (frequencies).

Older instruments used prisms to disperse the infrared light into its various frequencies, although diffraction gratings are now more common. A diffraction grating is a small reflective plate containing as many as 300 equally spaced grooves per millimetre. In either case, the infrared light that passes through a prism or reflects off a diffraction grating is physically dispersed into individual wavelengths that span a small, but measurable, distance. In order to scan the spectrum, the prism or grating is rotated slightly such that the different wavelengths of infrared light sequentially fall upon a narrow slit. The slit further limits the width of the radiation that is passed on to the detector. A schematic diagram of a dispersive infrared spectrometer is shown in Figure 8.20.

8.2.2.2 Fourier-transform instruments

Fourier-transform infrared (FTIR) spectrometers still require a source of infrared radiation and a detector, but do not disperse the infrared radiation into its individual wavelengths, and hence do not contain a monochromator. Instead, all of the infrared radiation is simultaneously passed through the sample and subsequently focused on the detector, after having passed through an interferometer. The latter produces what is known as an interferogram, which is converted by a complex mathematical operation (called a Fourier transform) into the traditional infrared spectrum.

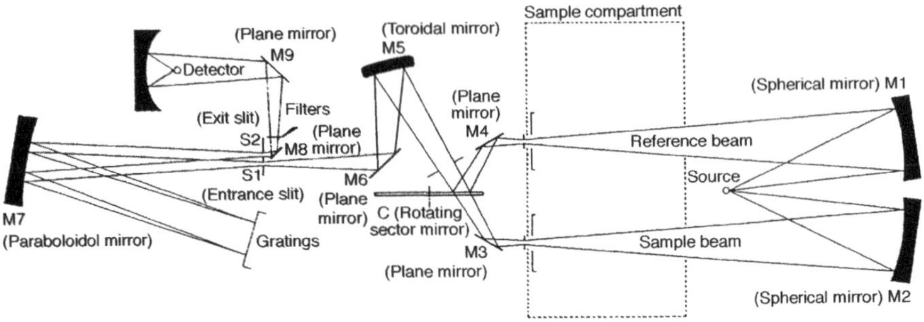

Figure 8.20 Schematic diagram of a typical dispersive infrared spectrometer. Reproduced by permission of The Federation of Societies for Coatings Technology, from D. R. Brezinski (Ed.), *An Infrared Spectroscopy Atlas for the Coatings Industry*, 1991.

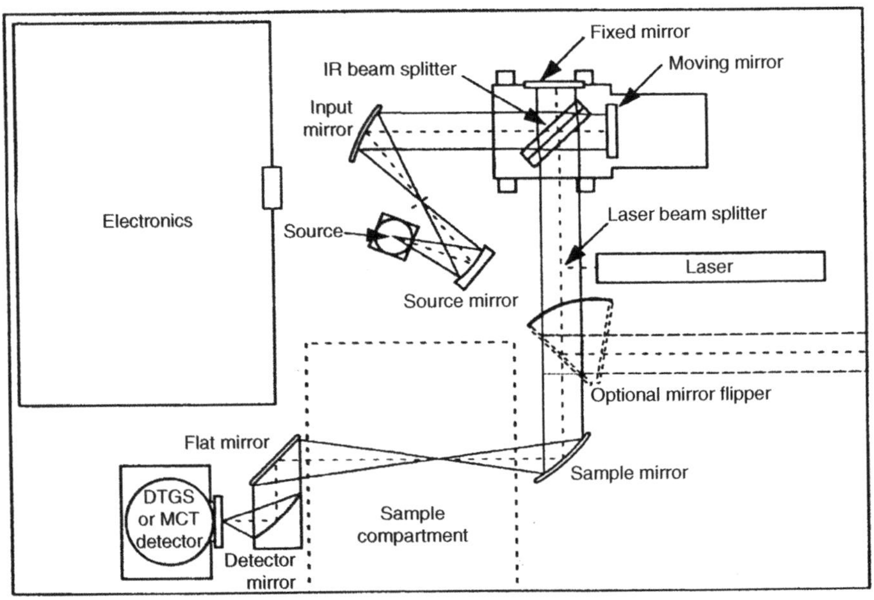

Figure 8.21 Schematic diagram of a typical Fourier-transform infrared (FTIR) spectrometer employing a Michaelson Interferometer. Reproduced by permission of the Nicolet Instrument Corporation.

An interferometer basically consists of a moving mirror, a stationary mirror and a beam splitter, as shown in Figure 8.21.

The incoming beam of radiation is split into two paths. One path has a fixed distance, while the distance of the other path varies by virtue of the moving mirror. The two beams are then subsequently recombined, and their intensity variation is measured by the detector, as a function of their path difference.

An interferogram is based upon the concept of destructive and constructive interferences. Interference of moving wave fronts can be visualized quite easily by dropping two pebbles into a pond. When the ripples, or wave fronts, meet one another, in some cases, they will partially (or totally) interfere with one another and cancel one another out. In other cases, they will 'add' to one another constructively, and the height of the waves will increase. In optics, as with pebbles and ponds, the radiation will interfere constructively if the difference between their path lengths is an even multiple of their wavelength. The interference will be totally destructive if the path length difference is equal to half the wavelength. For intermediate values, the interference will be destructive but not totally so.

It stands to reason that the intensity of the recombined light will be a function not only of the difference in path length caused by the moving mirror, but also by absorption of various frequencies of light by the sample. The Fourier transform is a way of converting the information stored in the interferogram, which measures intensity as a function of path length difference, to a traditional infrared spectrum which measures transmittance (or absorption) as a function of wavelength (or frequency). This concept has existed for a century, but the small, powerful computers required to carry out the mathematics have existed for only a couple of decades.

FTIR spectroscopy instruments have a few, very important advantages over traditional dispersive instruments. One particular advantage of FTIR instruments is speed. An FTIR instrument measures all wavelengths of the spectrum simultaneously. It does not have to measure them one at a time like a dispersive instrument. In fact, it is possible to obtain a spectrum in a few seconds with an FTIR instrument which would have taken perhaps 10 or 15 minutes with a dispersive instrument.

An FTIR instrument is also much more sensitive than a dispersive instrument, since all of the radiation produced by the source is being focused through the sample at once. The detection limits can be improved by one or two orders of magnitude when compared with dispersive instruments.

The fact that the total amount of radiant energy falling on the detector in an FTIR instrument is much greater than in the case of a dispersive instrument leads to other important advantages. There are several types of experiments in infrared spectroscopy where significant energy losses occur. These include attenuated total reflectance (ATR), diffuse reflectance and the use of infrared microscopes. The energy savings due to an FTIR instrument make these experiments much more viable than they might be with a dispersive instrument.

8.2.3 Sample Handling

In the investigation of coating failures, the sample is usually in the form of a solid, although occasionally, liquid samples will be encountered. There are several ways of obtaining spectra of both liquids and solids, as evidenced by the large number of sampling accessories offered by the various instrument manufacturers.

Probably the most common way of obtaining an infrared spectrum of a solid, such as a cured paint sample, is by the potassium bromide (KBr) pellet technique. This technique involves either manually or mechanically grinding a small amount of sample (perhaps 1–2 mg) with high-purity KBr powder (approximately 100 mg). Ideally, the sample will be finely ground to a more or less consistent particle size and will be evenly distributed within the powder. The ground mixture is then placed in a small die and then pressed for 2 or 3 minutes in a hydraulic laboratory press at a pressure of several tons per square inch (hand presses can also be used but are not as convenient).

A perfectly produced pellet (which is rare when working with coatings) will be completely clear and can then be placed in the optical path of the spectrometer. The infrared radiation is then focused through the pellet, and a transmission spectrum is obtained. Note that it is called a transmission spectrum not because the y-axis of the spectrum is in units of transmittance (it could also be plotted in absorbance), but because the detector is measuring radiation that was transmitted through the sample (as opposed to reflecting off of it).

KBr pellets can result in very good spectra, do not require much sample and are usually made in a few minutes. They result in a 'composite' spectrum of all of the materials present in the dried paint, that is, resin, cross-linkers, additives, plasticizers, pigments and so on. Because they result in a spectrum of the bulk material, they should often be used as the first (and sometimes only) step in the analysis of the 'unknown' sample.

Pellet spectra do have some disadvantages. One of these is the hygroscopic nature of potassium bromide. Even if one stores the powder in a vacuum desiccator, some moisture from the atmosphere will be introduced upon handling. The sample itself may have traces of moisture in it. The presence of water will result in a band near $3300\,cm^{-1}$ and a much weaker band near $1620\,cm^{-1}$. Bands due to water do not usually cause a problem so long as one is aware of their source, although they can complicate the analysis of other functional groups that absorb in these regions.

A second disadvantage of pellets is the difficulty in grinding the sample to a uniform particle size. If a small, uniform particle size is not obtained, the infrared radiation cannot interact with the sample in a consistent fashion, and slight variations in spectra can result. These variations are usually of little consequence unless one is trying to do a precise, quantitative work. A much greater problem with the pellet technique is light scattering. Certain types of samples, such as zinc-rich coatings, aluminium flake-filled coatings and coatings containing large quantities of carbon black, are particularly difficult to analyse as pellets. The large quantities of pigment in these samples, particularly aluminium flake, can act as minute mirrors, thus causing severe light scattering. In the worse cases, this can result in drastically sloping baselines and weak signals, since large quantities of radiation never make it to the detector.

With the exceptions noted above, problems related to large and/or inconsistent particle size will in general be minimal for hard, brittle samples, but can be substantial for soft, flexible ones, which grind poorly (or sometimes not at all). In both cases, the technique can sometimes be improved by adding two or three drops of a solvent such as methylene chloride to the mixture during grinding. The solvent has the effect of helping to break apart stubborn samples, thus making grinding easier. The heat generated by the grinding operation, plus the fact that the die is often evacuated during pressing, assures that the volatile solvent is removed from the pellet, without affecting the spectrum.

The fact that pellets result in composite spectra can be either an advantage or a disadvantage. Many times, bands due to inorganic pigments, such as calcium carbonate, silica or titanium dioxide, are very strong and broad, and can obscure certain bands originating from the resin. This can sometimes make generic identification of the resin difficult.

Another way to obtain a transmission spectrum of a solid is by producing a mull. Most mulls are made by dispersing a finely ground sample in a small amount of high-purity mineral oil, such as Nujol. A small amount of this paste is then sandwiched between two sodium chloride (NaCl) or KBr crystals, and a spectrum is obtained.

Mulls have the advantage of no interference from traces of water in the Nujol. They have the obvious disadvantage of interference from the Nujol itself, which will have strong carbon–hydrogen stretching and bending vibrations. The sample, however, must still be ground to a small, consistent particle size.

Another extremely useful means of obtaining a spectrum of a solid (or even a paste or viscous liquid, for that matter) is by ATR. Figure 8.22 illustrates the basic set-up of an ATR experiment.

In an ATR experiment, the sample is firmly pressed against the face of a crystal having a high refractive index (RI). The beam of infrared radiation is focused into this crystal at such an angle (the so-called 'critical angle') that total internal reflection will take place at the back faces of the crystal. This total internal reflection will result in the production of what is known as a 'standing wave' of radiation immediately at the interface between the crystal and the sample of lower RI,

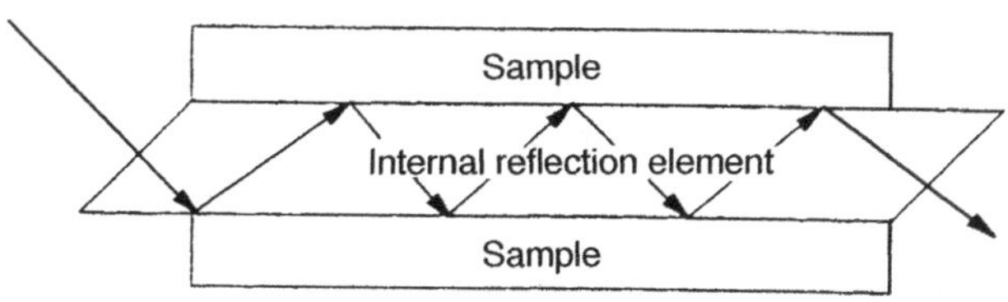

Figure 8.22 Schematic diagram of the basic set-up of a conventional attenuated total reflectance (ATR) experiment. From F. M. Mirabella, *Modern Techniques in Applied Molecular Spectroscopy*, Copyright © 1998. Reprinted by permission of John Wiley & Sons, Inc.

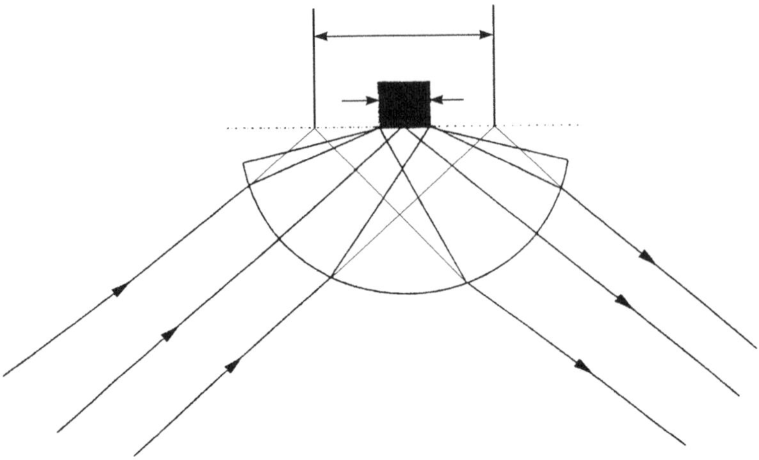

Figure 8.23 Design of a typical hemispherical attenuated total reflectance (ATR) assembly for obtaining surface spectra from samples. From F. M. Mirabella, *Modern Techniques in Applied Molecular Spectroscopy*, Copyright © 1998. Reprinted with permission of John Wiley & Sons, Inc.

which is clamped against it. This standing wave interacts with the first few microns (micrometre) of the sample, such that an infrared spectrum is obtained of its surface.

Obtaining spectra by ATR was a tedious and not totally satisfying task in the era of dispersive instruments, which were not so forgiving of the large energy losses associated with the technique. However, modern FTIR instruments and the ATR accessories developed for them generally make the technique both simple and rapid. It is usually possible to obtain an ATR spectrum from a sample of only 0.5 mm or so in diameter, by using some sort of hemispherical internal reflection element (Figure 8.23). Since the spectrum is not obtained from radiation transmitted through the sample but rather reflected off it, ATR spectra are referred to as reflectance spectra.

There are several advantages to the ATR technique in failure investigations; the most obvious of which is not having to grind the sample. Inconsistent particle size and light scattering are not problems with ATR. The technique is especially adaptable to soft, flexible samples, which can be clamped very intimately against the ATR crystal. Such samples would be difficult if not impossible to run as a pellet or mull. Interference by water in the potassium bromide matrix, or from hydrocarbon bands in mulls, is also not an issue.

Another important advantage of ATR is the fact that it essentially produces a spectrum of a surface (or more accurately, a very thin surface layer), as opposed to a spectrum of the bulk sample. This can be a distinct advantage if one is trying to detect surface contaminants such as oil, grease or exudates suspected to have caused delamination.

ATR does have its limitations. Most collections of reference spectra are presented as transmission spectra, which can sometimes appear substantially different from ATR spectra. In carrying out an investigation, it may be necessary to run one's own 'reference' spectra by ATR rather than by relying solely on published collections.

Secondly, an ATR spectrum can be influenced both by the angle of the incident beam of radiation entering the crystal, and by the degree of contact between the specimen and the crystal. The angle of incidence will affect the depth of penetration of the standing wave into the surface of the sample, and poor contact between the specimen and the crystal can result in significant energy losses, thus contributing to poor sensitivity. Thin, hard, brittle samples, particularly those with wavy or undulating surfaces, will be more difficult to analyse by using ATR than soft, smooth ones.

When evaluating ATR spectra, it is important to note that the depth of penetration of the standing wave into the sample is frequency dependent. The depth of penetration will be greater at the low-frequency end of the spectrum and smaller at the high-frequency end. The carbon–hydrogen stretching bands near 2900 cm^{-1} will appear to be much weaker, while carbonyl bands in the 1700-cm^{-1} region will appear much stronger. This is clearly illustrated in Figure 8.24, which shows the spectra obtained from the same alkyd coating by using (a) the KBr pellet technique and (b) the ATR.

Another aspect of ATR spectroscopy that must be considered is the possibility that the various resins and pigment components in a cured coating may have layered somewhat, or stratified. Particularly in a coating of low viscosity, it would not be surprising that a dense pigment such as barium sulfate or even titanium dioxide would tend to settle to the bottom of the film, while less dense ingredients such as resins or fumed silica would tend to rise. Figure 8.25 demonstrates the differences between the ATR spectra of the front and back of a drawdown of an epoxy coating prepared in the laboratory.

This segregation does not necessarily mean that the coating is somehow defective or unstable. It is possible that such effects, which this author has seen on more than one occasion, are relatively commonplace. However, it can greatly complicate the investigation of a coating failure if the spectrum of a laboratory-applied control sample varies depending on whether the paint was cured vertically, horizontally, or at an angle, and whether the analyst ran the spectrum from the front or the back of the sample.

While most of the samples encountered in a failure analysis are solids, there will be several opportunities to analyse liquid samples. Sometimes these samples will be liquid paint, while at other times, they will consist of liquid contained within an immersed, blistered coating. In some cases, it may consist of dried paint that the investigator has intentionally redissolved. Two of the most useful ways of analysing liquid samples are neat or as cast films.

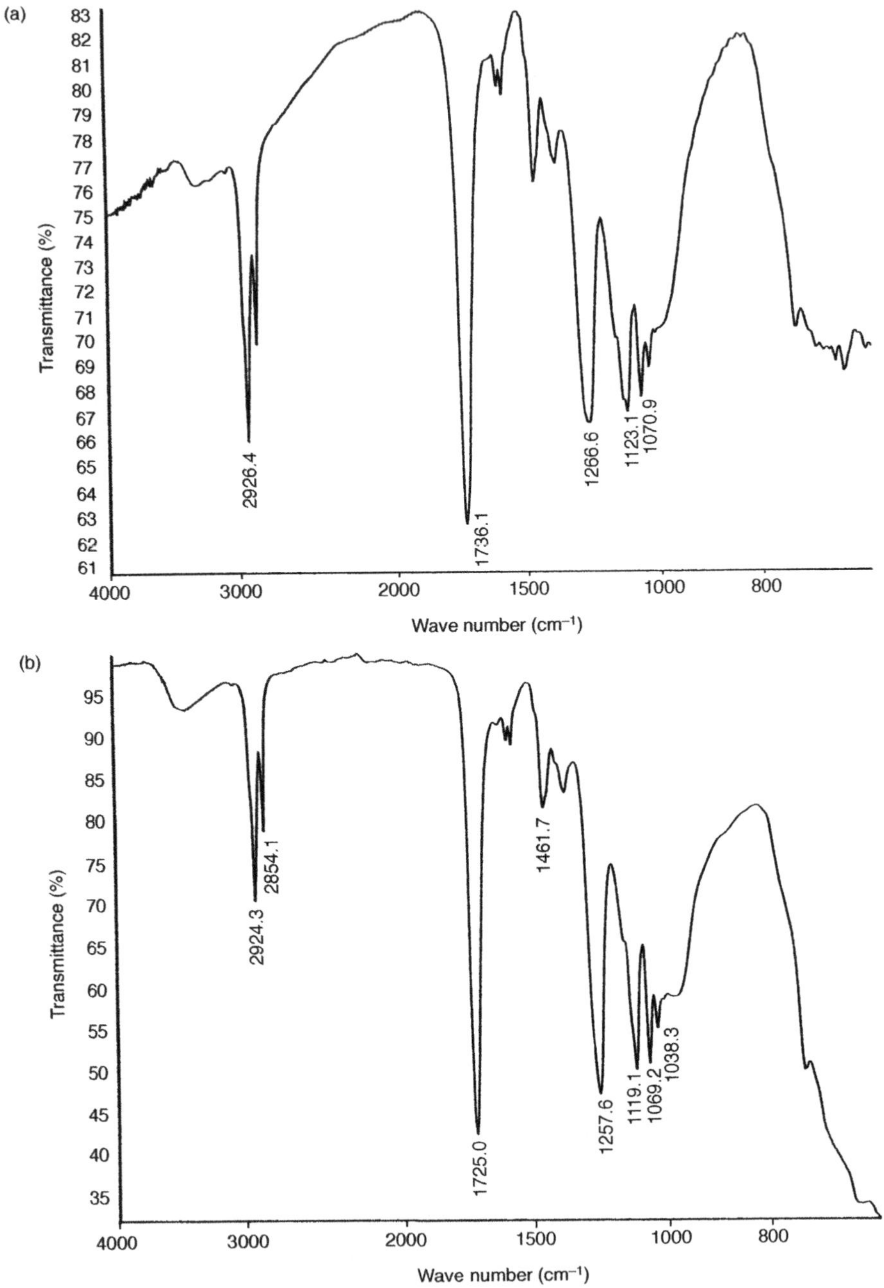

Figure 8.24 Infrared spectra of an alkyd coating obtained by using (a) the KBr pellet method and (b) attenuated total reflectance (ATR).

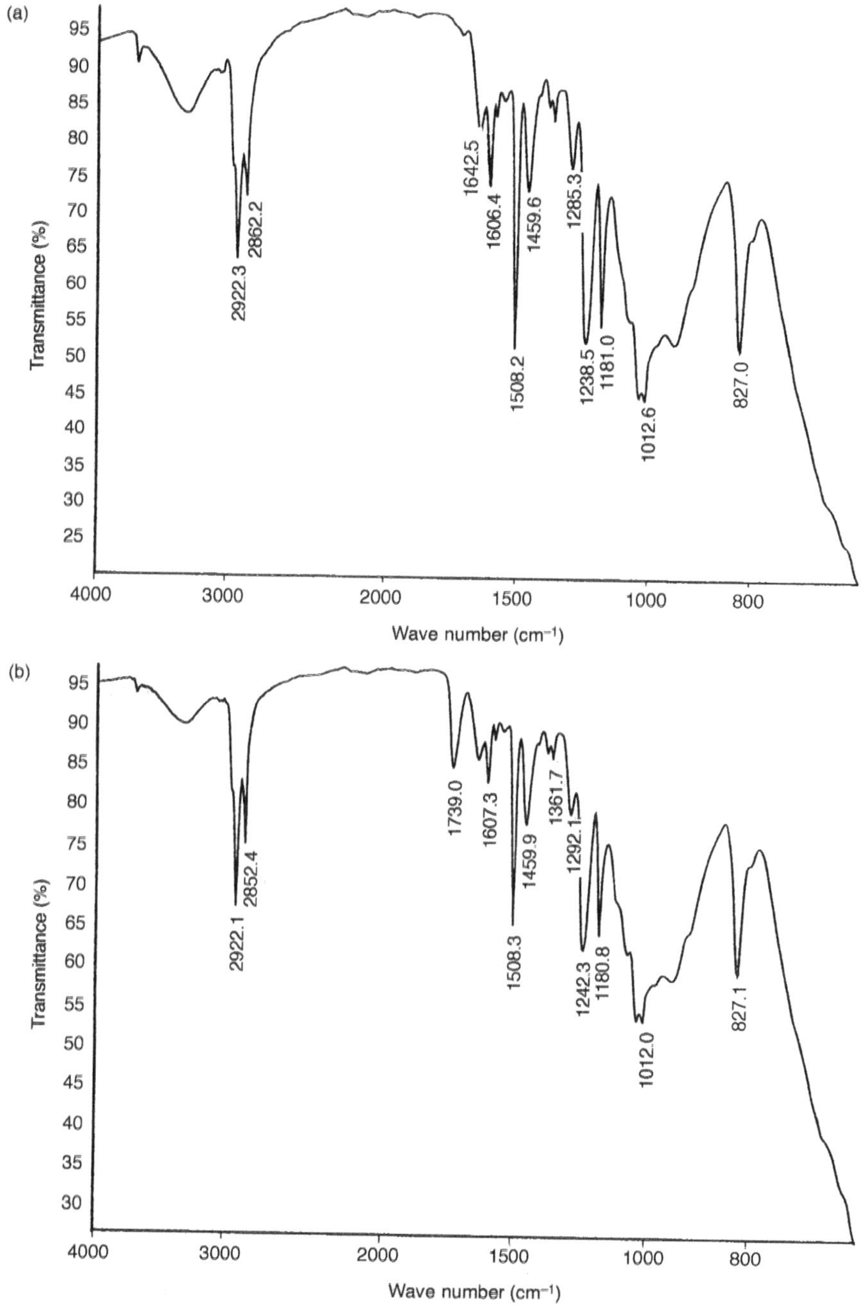

Figure 8.25 Attenuated total reflectance (ATR) spectra obtained from (a) the front and (b) back of an epoxy coating.

Obtaining a 'neat' spectrum simply means obtaining a transmission spectrum of the material 'as is', with no preparation or manipulation. A very easy way to do this is to form a thin capillary film of the liquid paint between two KBr or NaCl plates. This is achieved by placing a drop of sample in the middle of one crystal, placing the other crystal on top, and then lightly squeezing them together.

The technique is fast and simple, and results in very good spectra. It is particularly useful if one is simply carrying out a quality control (QC) 'screening' procedure and trying to conclude if a suspect sample matches a control sample. A spectrum is obtained of the whole paint, including solvents.

In addition to speed, simplicity and good reproducibility, running a neat spectrum avoids the possibility that a volatile component (solvent or otherwise) has evaporated due to sample handling. There are several low-molecular-weight cross-linking agents that could evaporate from the sample if run by other techniques. If the amount or identity of such a cross-linker is important to the investigation, then obtaining neat spectra should be strongly considered.

Obtaining such a spectrum of a water-based coating can be difficult. The common crystal materials used in infrared spectroscopy are quite water-soluble and will dissolve or at least 'fog' when used with aqueous samples. Silver chloride discs, which are insoluble, can be used, but it is usually difficult to get high-quality, reproducible spectra by using this technique.

Depending on what one is trying to learn from the analysis, a potential disadvantage of a neat spectrum is the fact that it contains spectral features from all of the paint ingredients, including resin, pigment and solvent. Sometimes, bands from the pigment or solvent can significantly overlap the bands from the resin component, thus making the identification of the resin difficult.

Another useful and common way of obtaining a spectrum of a liquid paint sample is by casting a film. If one simply wants to get rid of the solvent, a drop of sample can be smeared across a salt plate and allowed to dry, either under a heat lamp or in a conventional oven or vacuum oven. The elimination of the solvent can significantly simplify the spectrum, thus making resin identification simpler. Caution must be used so as not to evaporate a volatile cross-linking agent, which could result in a faulty or misleading conclusion. Water-based samples, however, can still be a problem with this technique.

Another way of obtaining a cast film spectrum and eliminating the contribution from both solvent and pigment is by first centrifuging the sample. If the sample is highly viscous, separation can be aided by first thinning it with a suitable solvent, usually one with a low boiling point, such as methylene chloride or MEK. If carried out effectively, the pigment will compact in the bottom of the centrifuge tube, and the resin and solvent will form a clear supernatant. The latter can be removed with a disposable pipette, and a few drops evaporated onto a salt plate. The solvent is then removed by heat and/or vacuum, and a high-quality transmission spectrum can be obtained of the vehicle, free of interference from solvent and pigment. Again, the possibility of losing a volatile cross-linking agent should be considered.

Obtaining a spectrum of a water-based sample by using the above technique is complicated by several factors. First of all, water is relatively high boiling and difficult to thoroughly evaporate, although this can be alleviated by the use of a vacuum oven. Secondly, water will dissolve many of the common crystal materials. Using silver chloride crystals, or thin, plastic cards, can partly reduce this problem. However, silver chloride crystals are not as easy to work with and greatly reduce energy throughput, and plastic cards will have bands of their own in the carbon–hydrogen region of the spectrum.

A third problem with obtaining a cast film of a water-based coating is the difficulty in centrifuging such samples. The emulsifying agents which are present make this difficult, even under high gravitational forces, to get a clear separation of ingredients. Sometimes a brief, preliminary period of sonication can alleviate this problem.

One way to overcome many of the problems associated with obtaining a good cast film spectrum of a water-based coating is to apply it to a metal panel or glass slide, and simply let it air dry for a few hours or perhaps overnight. This will get rid of nearly all of the water, and, since most water-based samples are either thermoplastic or relatively slowly reacting thermosets, the coating can then be scraped off the glass slide and redissolved in methylene chloride or MEK. The redissolved sample is then centrifuged and handled as though it was a solvent-based sample.

There are many techniques that can be used to obtain infrared spectra of coating samples. Infrared spectroscopy is a versatile technique, limited chiefly by the ingenuity of the investigator. When utilizing infrared spectroscopy in solving a coating failure, the theory of 'apples and oranges' should be kept in mind. A spectrum of the same sample obtained by ATR will look different than one obtained by the pellet technique, which will itself look different to one obtained by the cast film technique. Therefore, except for making certain general conclusions, one should always compare 'apples with apples' and not 'apples with oranges'.

8.2.4 Applications

There are numerous applications of infrared spectroscopy in the field of coatings failure analysis. Although it would be difficult to compile a complete list, some of the more common applications include generic identification, detection of contaminants, identifying mix ratio problems, detecting evidence of coating degradation and monitoring the degree of cure.

8.2.4.1 Generic identification

Sometimes, coatings fail for the simple reason that the wrong coating was used. Infrared spectroscopy is a powerful technique for the elucidation of molecular structure and, therefore, is the technique of choice for determining if, for example, a failing coating really is the specified epoxy or whether an alkyd was used instead.

The location in the infrared spectrum of most of the functional groups commonly encountered in the field of coatings analysis is shown in Figure 8.19. By knowing the basic chemical structure of the resin in question, one can compare the actual spectrum of the sample with the 'theoretical' spectrum suggested by a table of group frequencies. If the characteristic bands of the functional groups known to be in the coating are present in the infrared spectrum, chances are reasonably good that the coating is the correct type. However, if one or more major bands are missing, the coating is almost certainly not what it is supposed to be.

For example, there are many different types of polyesters, but they all have one thing in common, namely the ester group. The carbonyl carbon of the ester group has a very strong, characteristic stretching vibration in the 1700- to 1750-cm^{-1} region. If there is no band in this region, or only a weak one compared with bands from other functional groups, the coating cannot be a polyester.

Negative interpretation is more reliable than positive interpretation. For example, the polyester discussed above has to have a carbonyl band in the 1700- to 1750-cm^{-1} region. If this band is missing, the coating cannot be a polyester. However, aldehydes also have a carbonyl band in this region. Therefore, to conclude that the sample is a polyester, only because it has a band near 1720 cm^{-1}, could be a mistake. One should examine the spectrum for the presence of other, perhaps weaker, bands which are also characteristic of the polyester structure.

In addition to using one's knowledge of the chemical structure of a coating, coupled with the use of correlation charts, there are also excellent collections of infrared spectra of coating materials and polymers [1, 2], as well as inorganic compounds [1, 3]. Figure 8.26 illustrates typical spectra of (a) an orthophthalate alkyd, (b) a vinyl based on poly(vinyl acetate) (PVA) and (c) an epoxy. The three coating types are easily distinguished, even with no knowledge of their chemical structure.

However, failure analysis in general, and infrared spectroscopy in particular, should never be 'conducted in a vacuum'. Without a clear understanding of what the problem is and the significance of the sample one is examining, glaring errors can be made. For example, Figure 8.27 shows the spectrum of an acrylic resin, while Figure 8.28 displays the spectrum of coconut oil. The two spectra are remarkably similar. Although an average spectroscopist could certainly tell the two apart in their pure forms, if one tosses in a few pigments and perhaps an additive or two, it would be relatively difficult to tell them apart.

The collections of reference spectra cited above are extremely useful. It should be kept in mind that these are spectra of *raw* materials and not *formulated* coatings. The introduction of pigments and plasticizers to a resin can greatly alter its infrared spectrum. Many of the inorganic pigments commonly used in coatings, such as calcium carbonate, barium sulfate, titanium dioxide and various silicas

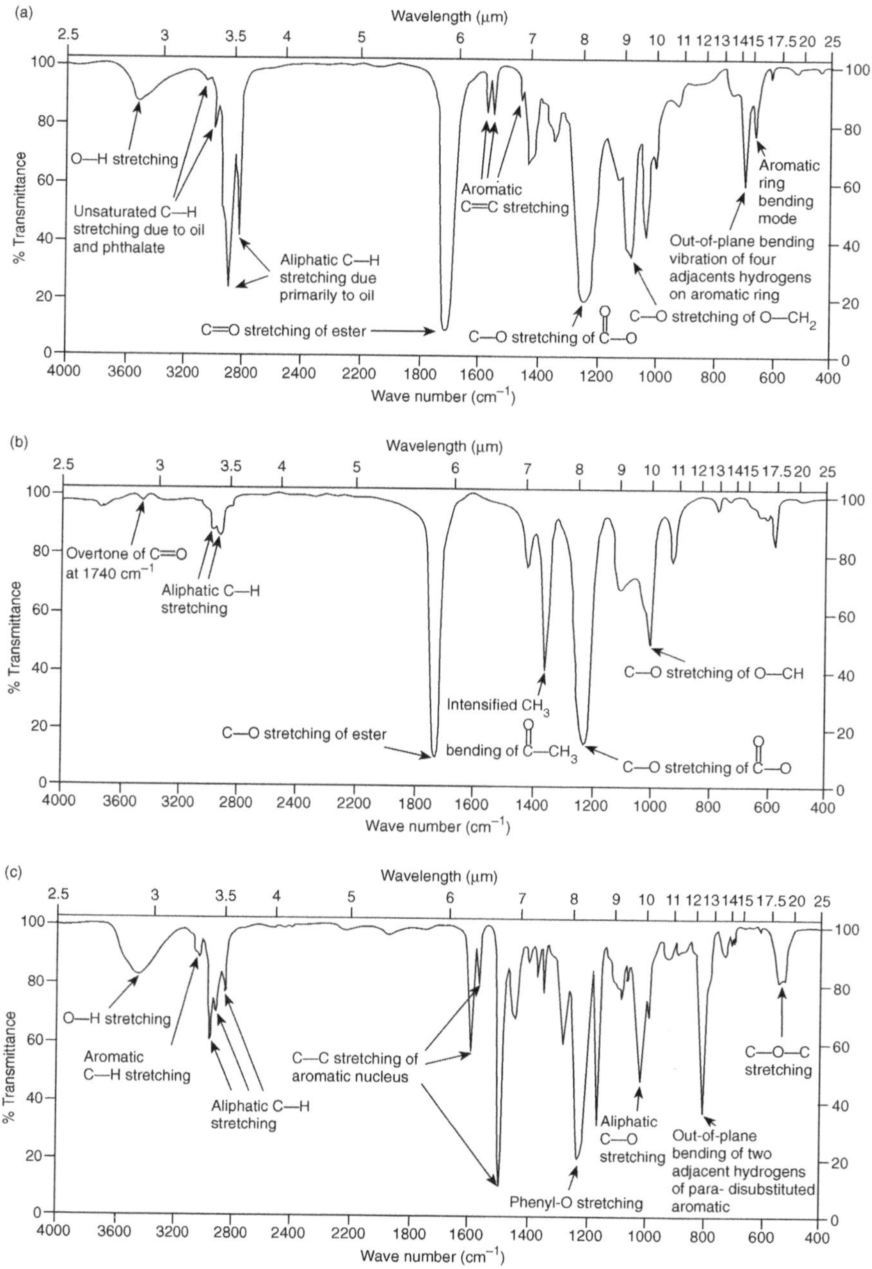

Figure 8.26 Infrared spectra of (a) an alkyd, (b) a vinyl and (c) an epoxy coating. Reproduced by permission of The Federation of Societies for Coatings Technology, from D. R. Brezinski (Ed.), *An Infrared Spectroscopy Atlas for the Coatings Industry*, 1991.

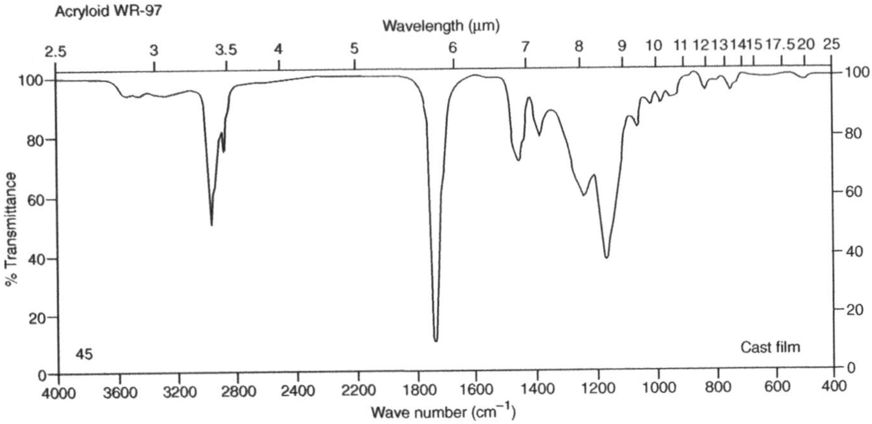

Figure 8.27 Infrared spectrum of an acrylic coating. Reproduced by permission of The Federation of Societies for Coatings Technology, from D. R. Brezinski (Ed.), *An Infrared Spectroscopy Atlas for the Coatings Industry*, 1991.

and silicates, have very strong, broad bands that can easily obscure many of the important, identifying bands of certain resins.

The examples of acrylic resin and coconut oil given above represent one such case where pigment bands could make the analysis problematic. Another example would be trying to determine if a coating containing a large amount of calcium carbonate and silica is an acrylic or a vinyl. Figure 8.27 shows a spectrum of a typical acrylic, while Figure 8.26b depicts the spectrum of a typical vinyl based on PVA. In their 'pure' states, the two resins are easy to tell apart. The vinyl has a very strong methyl bending vibration near 1370 cm^{-1}, versus a relatively weak one for the acrylic. The carbon–oxygen single bond stretching vibration of the vinyl acetate group results in a strong band near 1240 cm^{-1}, versus a weak band at this location in the acrylic. The corresponding C—O stretch for the acrylic occurs much lower, that is, near 1160 cm^{-1}.

Consider the consequences, however, if the two coatings are pigmented with calcium carbonate and silica, the spectra of which are shown in Figures 8.29 and 8.30, respectively.

The broad calcium carbonate band near 1430 cm^{-1} will likely obscure the 1370-cm^{-1} region, and the strong silicate band effectively blocks out the region from 1050 to 1250 cm^{-1}. The two regions most useful for telling an acrylic from a vinyl have been obscured. In flat coatings, the amount of pigment relative to resin will be high. Suddenly, what seemed like a simple analysis is not so simple.

Often, problems such as the above can be solved by dissolving the sample in a suitable solvent, centrifuging to remove pigment and then obtaining a cast film spectrum. It should be kept in mind, however, that some coating types (epoxies, urethanes, aged alkyds, etc.) are not very soluble, and one could then obtain a

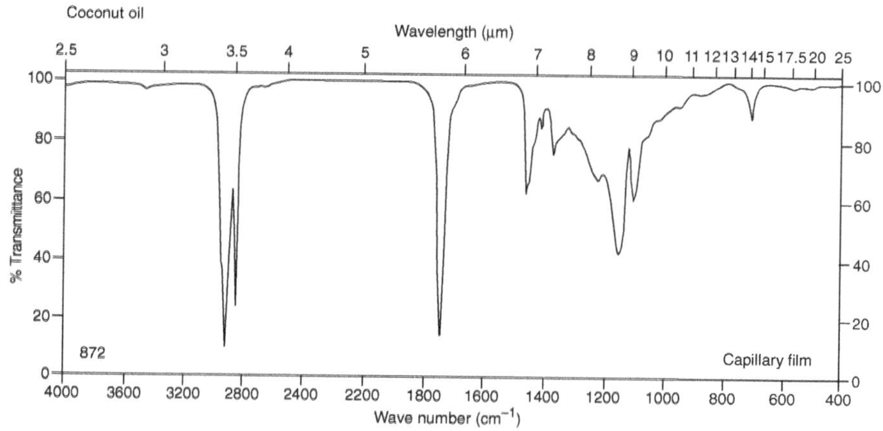

Figure 8.28 Infrared spectrum of coconut oil. Reproduced by permission of The Federation of Societies for Coatings Technology, from D. R. Brezinski (Ed.), *An Infrared Spectroscopy Atlas for the Coatings Industry*, 1991.

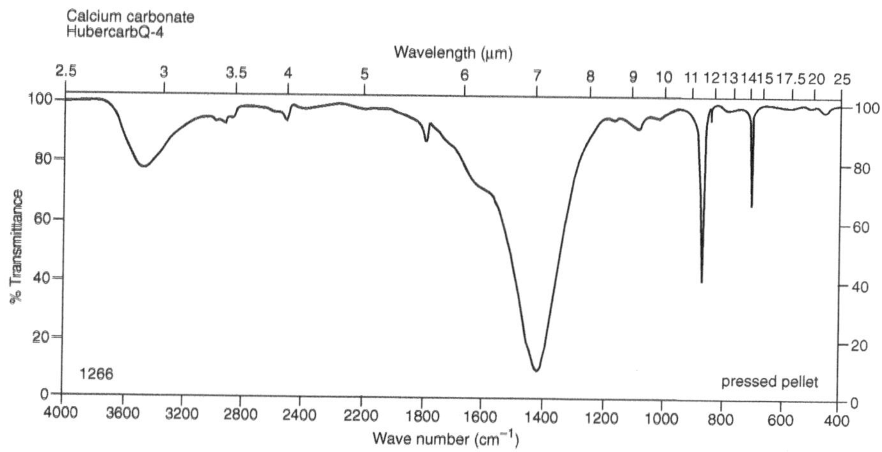

Figure 8.29 Infrared spectrum of calcium carbonate. Reproduced by permission of The Federation of Societies for Coatings Technology, from D. R. Brezinski (Ed.), *An Infrared Spectroscopy Atlas for the Coatings Industry*, 1991.

nice spectrum of an extracted plasticizer while the resin itself remains undissolved at the bottom of the centrifuge tube. As stated before, infrared spectroscopy 'cannot be performed in a vacuum'.

When it comes to determining if the correct, or specified, coating has been used, perhaps the best procedure is to actually obtain a known sample of this coating from the manufacturer and compare its infrared spectrum to that of the

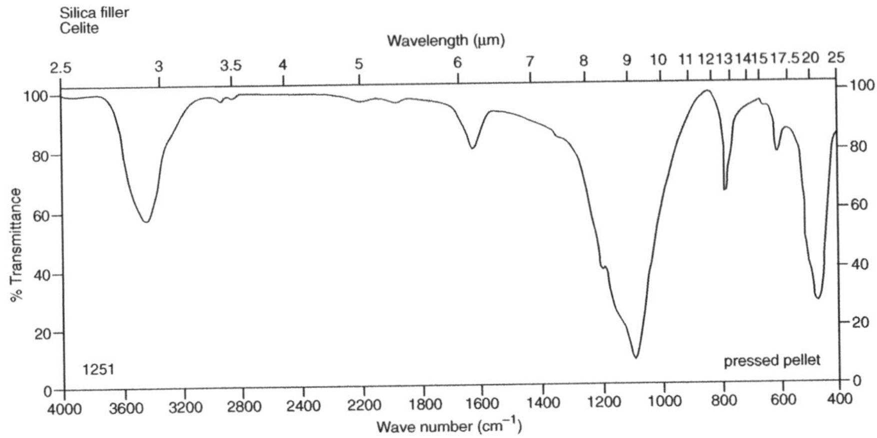

Figure 8.30 Infrared spectrum of silica. Reproduced by permission of The Federation of Societies for Coatings Technology, from D. R. Brezinski (Ed.), *An Infrared Spectroscopy Atlas for the Coatings Industry*, 1991.

unknown. If they match, the sample is almost certainly the specified coating, or at least a coating with an extremely similar composition.

8.2.4.2 Detection of contaminants

A common reason why coatings fail is because they are applied over contamination. Many times, the examination of the back of a disbonded coating with a low-power microscope is sufficient to identify contamination. Contaminants, such as dirt, mildew, concrete laitance, mill scale and rust, are easily seen with a microscope. Sometimes, however, the contaminant is not apparent visually, and spectroscopic methods are required.

The most infamous of contaminants is grease or oil, even though in this author's personal experience they are rarely the actual culprits. There are two effective ways to detect grease or oil. If the coating is a solvent-resistant, thermosetting type, and if the chips are large enough (perhaps 0.5 × 1 in. or larger), they can be angled over a small beaker, and a disposable pipette can be used to repeatedly dribble a millilitre or two of a weak solvent such as hexane over the back of the chip. Hexane is a good solvent for oil but a poor one for paint. The hexane can then be carefully evaporated, either on to a salt plate or on to some pre-ground potassium bromide, and a spectrum is obtained. Figure 8.31 shows a typical spectrum of an oil.

The second alternative, which works for both thermoplastic and thermoset coatings, is to obtain a spectrum of the back surface of the delaminated coating by using ATR. The characteristic aliphatic carbon–hydrogen stretching bands of

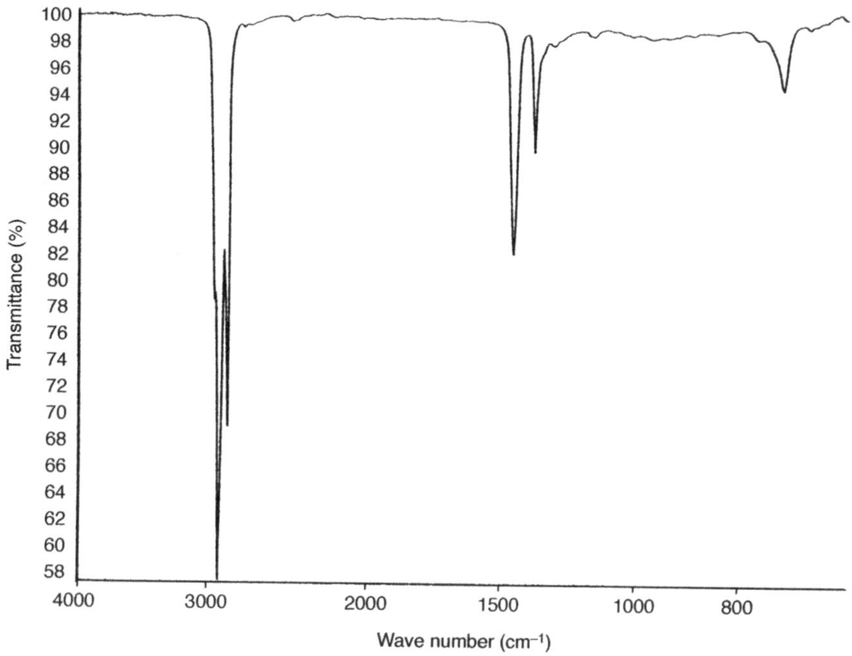

Figure 8.31 Typical infrared spectrum of an oil.

a hydrocarbon contaminant occur nearer 2900 cm^{-1}. The uncontaminated coating will almost certainly have bands in the region. Therefore, it is not the presence of bands in this region that identifies an oily contaminant, but rather their relative intensities.

Occasionally, an agent that might be responsible for inter-coat adhesion failure is not exactly a 'contaminant', but rather a component of the first coat that has exuded to the surface or even the first coat itself which has degraded.

In Section 5.3 concerning epoxies, the concept of an amine blush was discussed. This refers to the exudation, or 'blushing', of some of the low-molecular-weight amine- or amide-curing agents to the surface of the epoxy. Blushing is usually encountered under cool, humid conditions, as many of these curing agents are partially water soluble. The detection of an amine blush by infrared spectroscopy can be performed, but it is not always as straightforward as one would wish. Figure 8.32 shows the ATR spectra of the surfaces of a blushed and non-blushed epoxy. The presence of the curing agent on the blushed epoxy can be seen by the stronger bands near 1650 cm^{-1}.

In this instance, the curing agent was present in its original form. However, as previously discussed (Section 5.3.2), primary and even secondary amines can react with carbon dioxide, and/or carbon dioxide and water, to form carbamates or possibly bicarbonates, and if this happens, the typical infrared bands of the

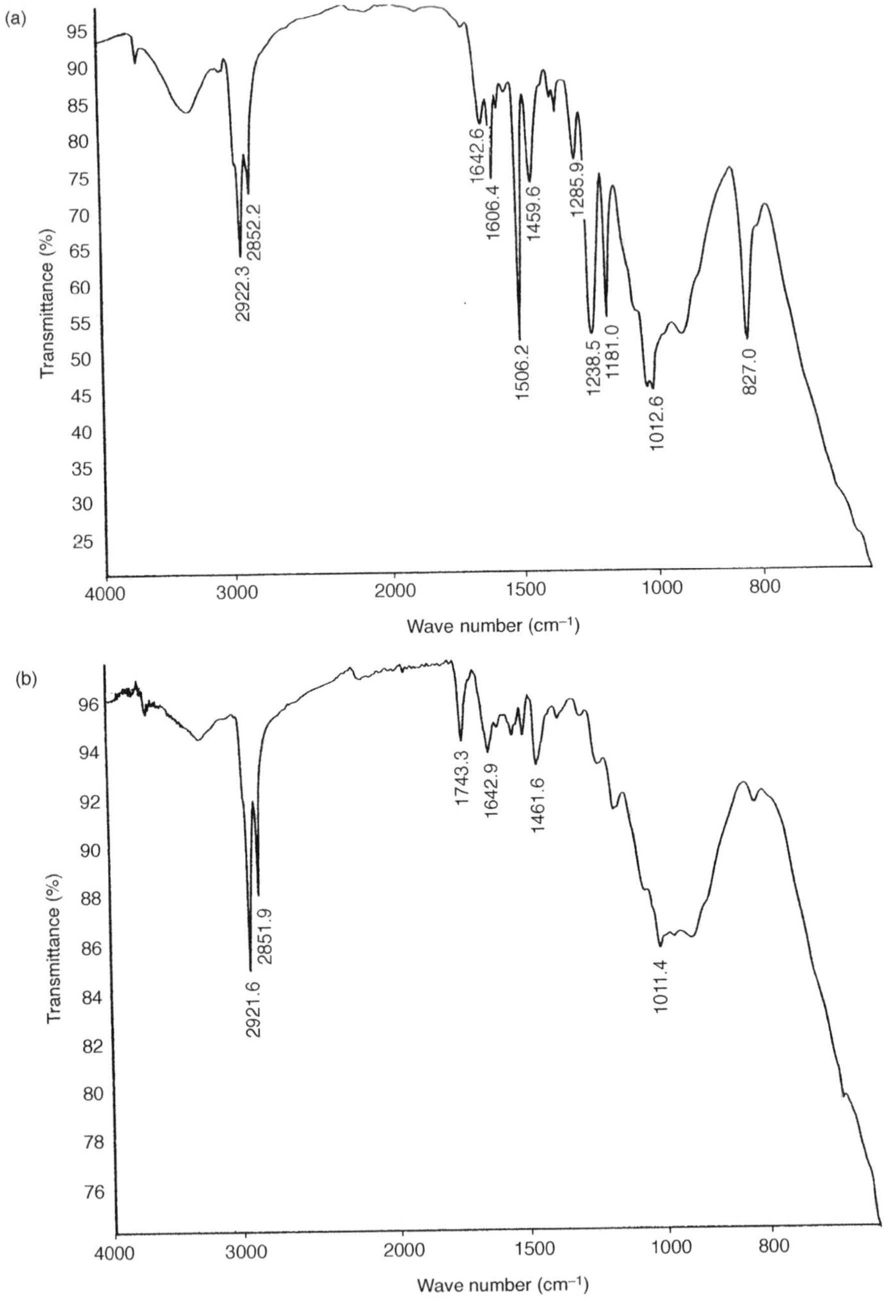

Figure 8.32 Infrared spectra of the (a) non-blushed and (b) blushed surfaces of an epoxy coating.

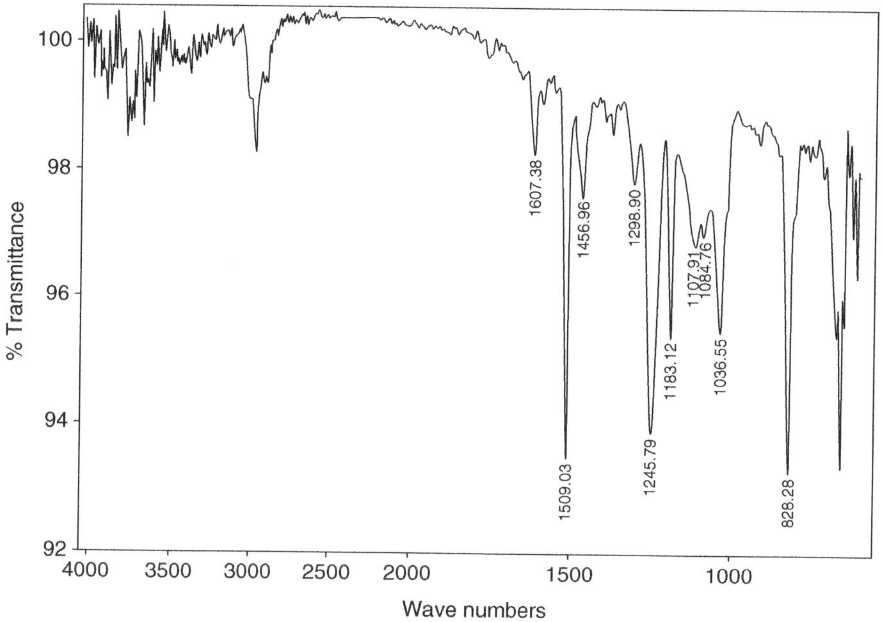

Figure 8.33 Infrared spectrum of the surface of an amine-cured epoxy allowed to cure under laboratory ambient conditions.

amine functional group are lost. Graham *et al.* [4] showed the development of a weak band near 1580 cm^{-1} in ATR spectra of the surface of an amine-cured epoxy exposed to a normal room atmosphere, and also demonstrated that this band was noticeably stronger when the epoxy was cured in a humid environment. This band was attributed to the formation of an amine carbamate. Croll's [5] work also suggested that the reaction of an amine with carbon dioxide in a humid atmosphere would result in the development of a band near 1610–1550 cm^{-1}.

Figure 8.33 is an ATR spectrum obtained in the author's laboratory of an amine-cured epoxy allowed to cure under normal ambient conditions and is typical of such a coating. Figure 8.34 is a spectrum of the same coating which was cured by suspending it over a tray of water. Note the appearance of a moderately strong band near 1565 cm^{-1}, due to the formation of an amine blush which converted to either the amine carbamate or, possibly, the amine bicarbonate.

Based on the above discussion, it is apparent that the infrared spectrum of an amine blush can be variable, depending on such things as the structure of the specific amine-curing agent, the environmental conditions under which the coating cured and perhaps even the length of time before it was top-coated. Furthermore, in failure analysis, one is usually attempting to detect this thin layer of material on the back side of a disbonded topcoat, which might have bands of its own in the same general vicinity as those of the blush.

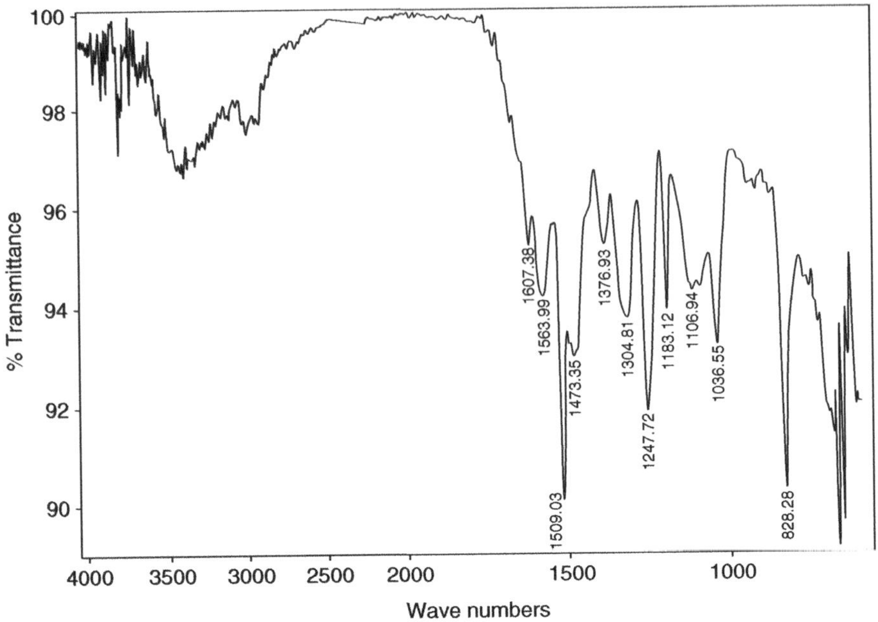

Figure 8.34 Infrared spectrum of the surface of the same amine-cured epoxy shown in Figure 8.33, after the coating had cured for 1 day suspended over a tray of water.

One particular area of failure analysis where it makes sense to look for a contaminant is the disbonding of coatings from concrete floors. Many times this happens because of a thin layer of cement dust on the floor, either from laitance which was not removed or from etching or blasting operations where the residue from the surface preparation was not vacuumed or washed away. These residues can often be seen microscopically and then confirmed by using ATR.

Another reason to run ATR spectra, either on the back of the disbonded coating or on the concrete surface itself, is to look for membrane-curing compounds or form-release agents. There are a wide variety of such materials, including simple mineral oils. Their presence often leads to coating failures, and they can usually be detected by using infrared spectroscopy.

When attempting to detect a contaminant, it is very important to have a control. Simply running an ATR spectrum of the back of a paint chip and announcing that a contaminant was found because the spectrum produced some unexpected feature can be a dangerous practice. Most of the time, the investigator does not have the formula of the paint at hand, so it is difficult to conclude that something 'foreign' was detected. Even if one has the rare advantage of the formula, there are minor or trace components in many raw materials that are not listed on their specification sheets or on the material safety data sheets.

The safest practice is to run both the failing sample and a control. Ideally, the control should be a laboratory-prepared panel of the paint in question. If this is not available, then obtaining spectra of both failing and non-failing field samples is good practice. If this is not possible, then one may have to resort to obtaining spectra from both the front and back of a failing sample, reasoning that the contaminant will only show up on the back of the sample and not the front. This, of course, will not work if the disbonded sample consists of more than one coat of paint. Furthermore, the possibility of layering, or segregation of ingredients, must also be considered.

When all else fails, another alternative is to obtain a spectrum of the back surface of the failed sample, and then use a razor blade or sandpaper to gently remove a thin layer of coating, and subsequently obtain a spectrum of the subsurface. It is reasonable to conclude that materials found on the back of the paint are foreign if they are removed by abrading away a fraction of a mil.

When performing any of the above techniques by using ATR, one must be aware of the possibility that the two spectra could look different, not because of a contaminant, but because of the natural tendency for the various ingredients of some coatings to layer or segregate. This layering effect is likely due to differences in density of the various components of the coating.

If one knows the geometry of how the structure was painted in the field, the laboratory control panel can be allowed to dry in a similar fashion. Failing and non-failing samples may have been taken from similar orientations. Alternatively, perhaps the nature of the spectral difference is so marked that the presence of a contaminant is obvious. At other times, maybe there will be no differences, and a different failure mechanism should be considered.

An example of using ATR to solve an adhesion problem has been presented by Skrovanek [6]. The problem involved poor coating adhesion to a moulded part made of sheet moulding compound (SMC). Skrovanek found that simply washing the parts with soap and water resulted in significantly improved adhesion. The spectra obtained of the surface before and after washing clearly showed that the aliphatic carbon–hydrogen stretching bands near 2900 cm^{-1} decreased after washing, thus leading to the conclusion that a hydrocarbon mould-release agent was responsible for the poor adhesion.

8.2.4.3 Mix ratio problems

Many high-performance thermoset coatings consist of two components that must be mixed together shortly before application. In addition to drying by solvent evaporation, a chemical reaction occurs between the two components, thus resulting in a cured, cross-linked coating. Obviously, there is some optimum mix ratio (hopefully the one specified by the manufacturer) at which the coating reaches full chemical and physical potential.

The types of coatings that could possibly fail due to 'mis-mixing' include epoxies, coal tar epoxies, urethanes and unsaturated polyesters. In the latter case, the second component is usually a small amount of a true catalyst such as peroxide, and infrared spectroscopy is of little use in determining how much peroxide was used. However, infrared spectroscopy can often be used to determine the mix ratio of cured epoxies and sometimes urethanes as well.

The key to using infrared spectroscopy as a tool for determining the mix ratio is being able to find two or more bands whose relative intensities vary in a consistent fashion with mix ratio. It is comforting to know the ingredients, or at least the functional groups, responsible for these bands, although this is not a necessity.

In order to put the technique into practice, one needs to prepare test panels or drawdowns of the coating in question, at different mix ratios. Spectra can then be obtained of these drawdowns and examined for one or more band pairs that vary consistently in relative intensity as the mix ratio varies.

Figure 8.35 shows the spectra of a two-component polyamide epoxy applied at mix ratios of 2:1, 1:1 (the correct ratio) and 1:2 (components A to B, by volume). The epoxy resin has numerous bands associated with it, including those near 1610, 1510 and 830 cm^{-1}. The amide band of the polyamide curing agent has its characteristic carbonyl vibration near 1650 cm^{-1}. As one can see, the relative intensity of the 1650/1610 cm^{-1} band pair varies with mix ratio. The amide band gets stronger relative to the epoxy band as the amount of amide converter is increased.

The spectra shown in Figure 8.35 are presented as % transmittance versus wave numbers, as this is the most common way of plotting infrared spectra. However, it is absorbance, not transmittance, which is directly proportional to concentration. One of the fundamental laws governing the absorption of electromagnetic radiation is Beer's law, as given by the following:

$$A = \varepsilon c l, \tag{8.5}$$

where A is the absorbance, ε is a constant for any given species, known as the molar absorptivity (or extinction coeffcient), c is the concentration and l is the path length or the thickness of the sample. The relationship between absorbance and transmittance is given by the following:

$$A = -\log T. \tag{8.6}$$

Thus, it is possible to manually draw a baseline and calculate the absorbance from the measured transmittance values. It is much easier, however, to let computers do this, and in all modern infrared spectrometers, it is very easy to convert from a transmittance spectrum to an absorbance spectrum.

As Figure 8.35 illustrates, in certain cases, it is possible to determine a sample's mix ratio from its infrared spectrum. Some caution must be exercised in not

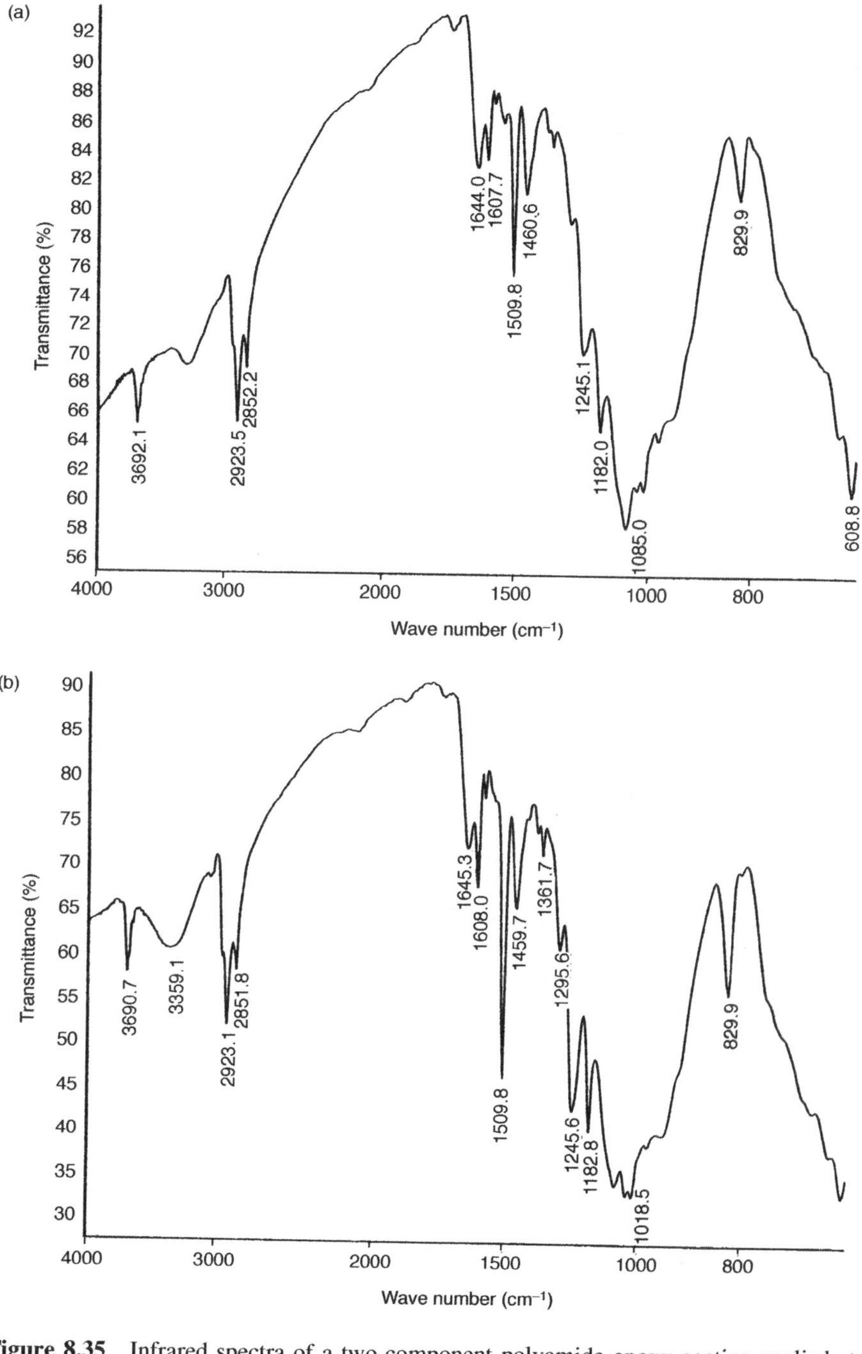

Figure 8.35 Infrared spectra of a two-component polyamide epoxy coating applied at mix ratios of (a) 2 : 1, (b) 1 : 1 and (c) 1 : 2.

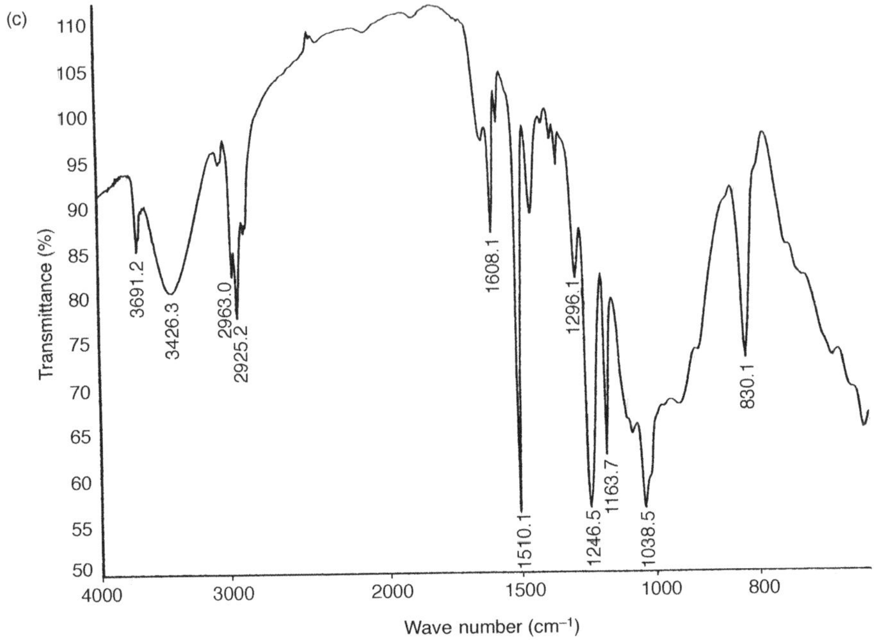

Figure 8.35 *Continued*

forming conclusions that go beyond the accuracy and precision of the method. Beer's law is only valid for dilute solutions, whereas in failure analysis, one is usually dealing with solids dispersed in a solid matrix (such as potassium bromide). Molecules at high concentration are close enough to interact with one another in ways that would not be possible in dilute solution, thus altering their absorption characteristics. Furthermore, when preparing mulls or pellets for infrared analysis, some amount of light scattering will be encountered, which will vary based on the size and uniformity of the ground particles. This is usually very difficult to control.

This author has seen several cases where an investigator has constructed a three-point calibration curve based on single spectra obtained at three different mix ratios, such as 2 : 1, 1 : 1 and 1 : 2, and then listed mix ratios of samples to two or even three significant figures. It is likely that the technique is accurate to two significant figures, if carried out correctly. It is possible, but not likely, that it may be accurate to three. However, without running several replicate samples at each of several mix ratios to establish accuracy and precision, it is best to err on the side of caution and only report approximate mix ratios.

In some cases, it is possible to determine the mix ratio of a two-component urethane by using infrared spectroscopy. However, this can sometimes be a bit more challenging than the epoxy example cited above.

Figure 8.36 shows the spectra of a urethane at mix ratios of 2 : 1, 1 : 1 (the correct ratio) and 1 : 2. This is one of the more straightforward examples, as there is an obvious correlation between mix ratio and the relative intensity of the two bands near 1730 and 1690 cm^{-1}. Many other band pairs also vary with the mix ratio. The 1730-cm^{-1} band is due to the ester carbonyl absorption from the polyol component, while the 1690-cm^{-1} band is the carbonyl absorption of the actual urethane, or carbamate, which is being formed as the coating cures.

Not all urethanes can be analysed so easily. In many cases, there is a significant amount of carbamate functionality already built into the isocyanate component, since these are often urethane pre-polymers with an excess of isocyanate functionality. Furthermore, the intensity of this band will increase upon curing, thus meaning that the slope of the calibration curve can change depending on when it is run. Additionally, isocyanates are very reactive, and variations in temperature or humidity may result in a different reaction pathway than that taken by the laboratory drawdown. The situation is not hopeless, but again, common sense should be used not to 'overextend' the technique (see Section 8.8).

In the case of urethanes, although it is tempting (and often correct) to ratio the ester carbonyl band to the carbamate carbonyl band, it is often fruitful to choose some band other than the carbamate, in particular one that does not vary with degree of cure. The ester band from the polyol is usually a safe one to choose, since it is not the ester group which is reacting with the isocyanate, but rather a hydroxyl group somewhere pendant on the polyester, or acrylic, backbone. If one can find a second band originating from the isocyanate component but not from a functional group actually engaged in the reaction, it is possible to reliably determine the mix ratio of the sample. Sometimes, this 'second band' can be from a pigment. Although the isocyanate component is usually unpigmented and clear, this is not always the case. Some pigment bands are relatively sharp and can easily be used in quantitative analysis.

Determining the mix ratio of a coal tar epoxy by using infrared spectroscopy is difficult. Such coatings usually give poor quality spectra, and a different technique, such as ashing, is usually more successful.

8.2.4.4 Coating degradation

On many occasions, coatings fail because their environment chemically degrades them. Such degradation may result in changes in both the coating's chemical and physical properties. In some cases, infrared spectroscopy is a useful tool for detecting such changes.

One of the most commonly encountered forms of chemical degradation is the saponification of an alkyd coating. As pointed out in previous sections, any ester-containing group is subject to hydrolysis at the carbon–oxygen single bond. When this occurs under alkaline conditions, the reaction is also known as saponification.

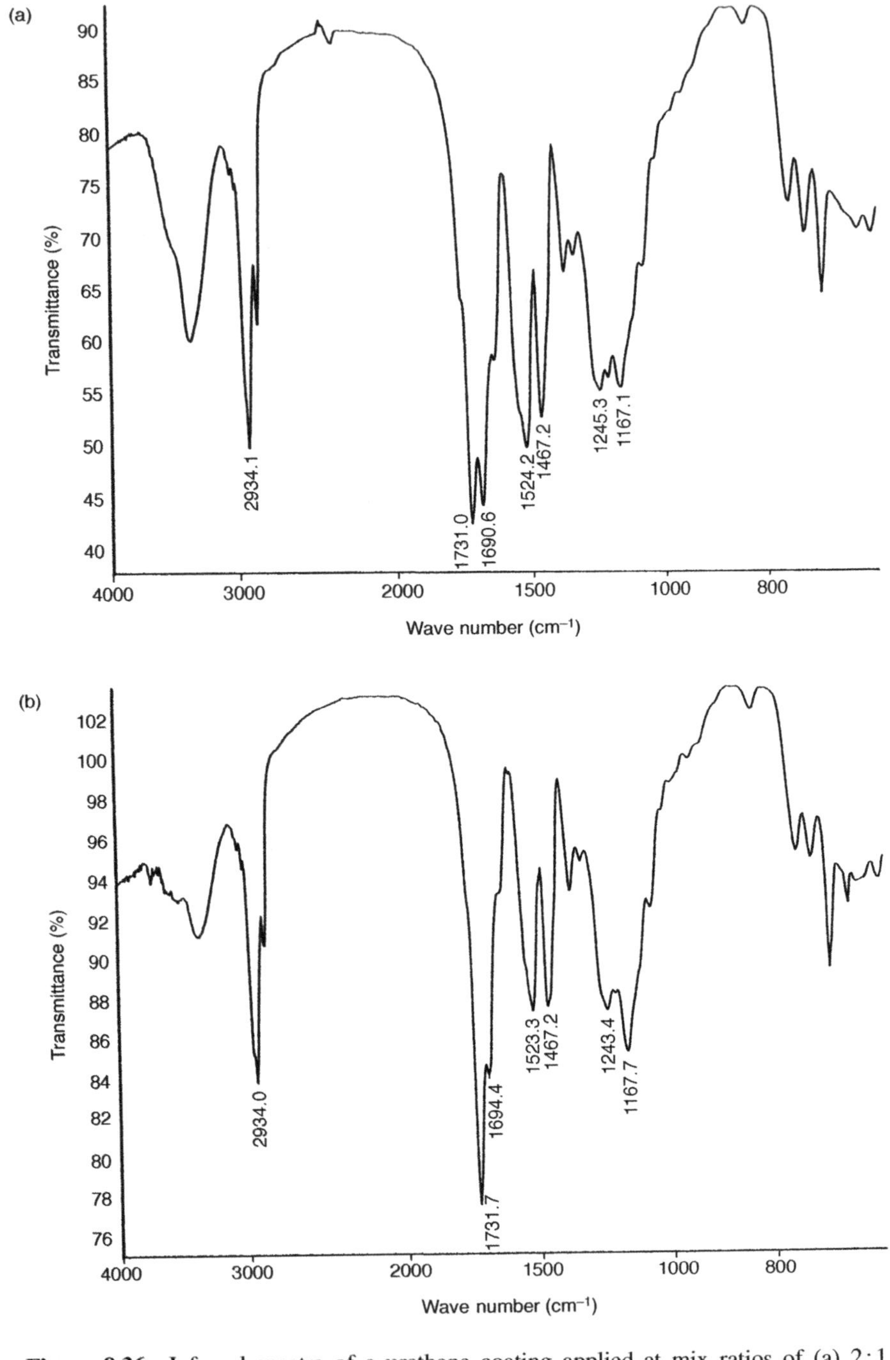

Figure 8.36 Infrared spectra of a urethane coating applied at mix ratios of (a) 2:1, (b) 1:1 and (c) 1:2.

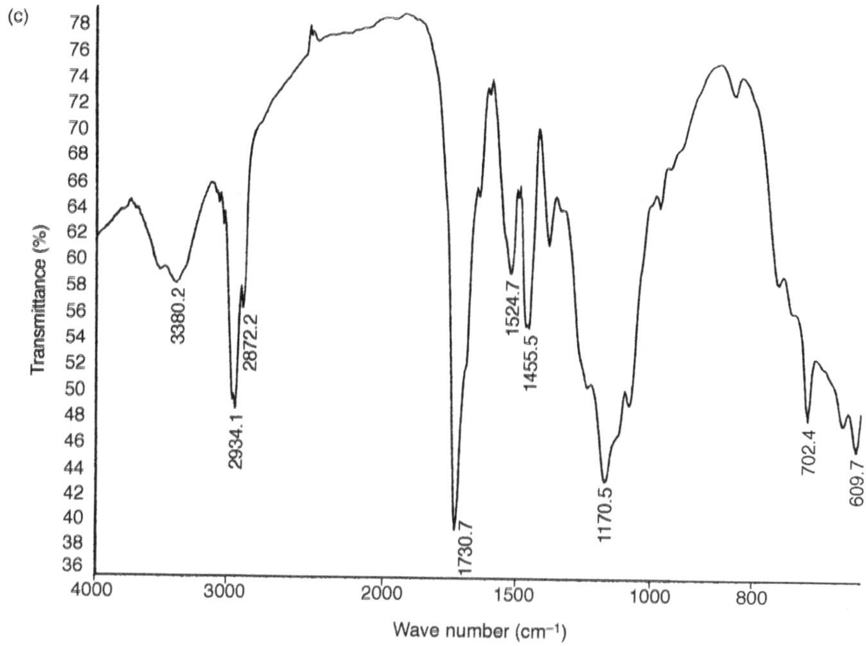

Figure 8.36 *Continued*

Since the reaction occurs under alkaline conditions, the carboxylic acid formed does not exist as the free acid, but rather as the salt of the acid. The cation will depend on the environment. If the base was sodium hydroxide, then sodium will be the cation.

Saponification is easily identified by using infrared spectroscopy. The ester carboyl vibration of an unsaponified alkyd occurs as a strong, sharp band in the 1720- to 1740-cm^{-1} region, as shown in Figure 8.37a. The carbonyl vibration of the carboxylic acid salt formed as a by-product of saponification shows up in the 1560- to 1620-cm^{-1} region, and is often accompanied by an additional, weaker absorption near 1400 cm^{-1}. Figure 8.37b displays the spectrum of a saponified alkyd.

Often, alkyds or polyesters saponify because the surface that they are in contact with is alkaline. Commonly encountered alkaline surfaces include cement and galvanized steel. In severe cases, the alkyd will actually turn into a 'cottage-cheese-like' consistency, and probably any infrared sample preparation technique would detect the gross saponification.

The early stages of saponification, which can still result in failure, can sometimes be missed if the wrong technique is used to obtain a spectrum. In the initial stages, the saponification will only have occurred at the interface between the coating and the alkaline substrate. If a potassium bromide pellet spectrum is

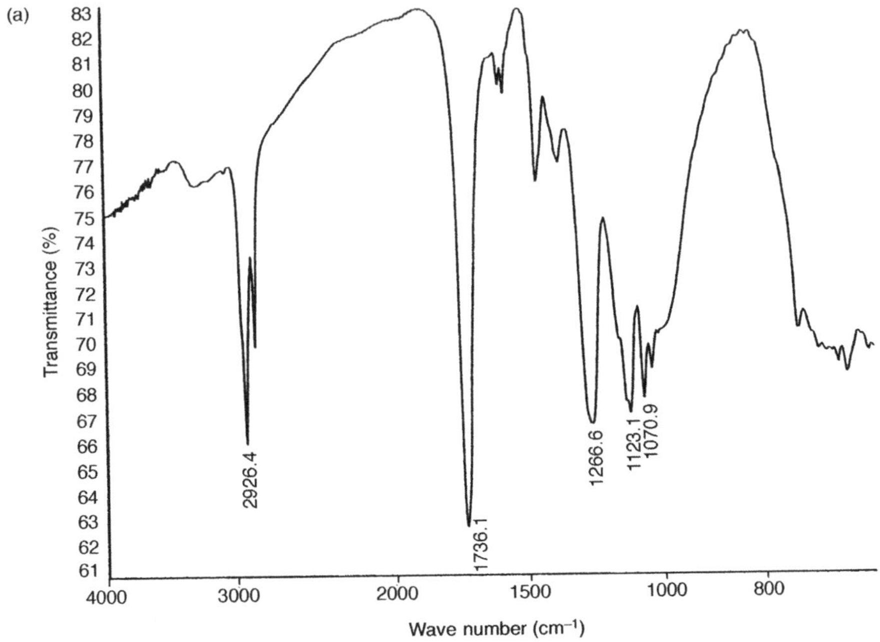

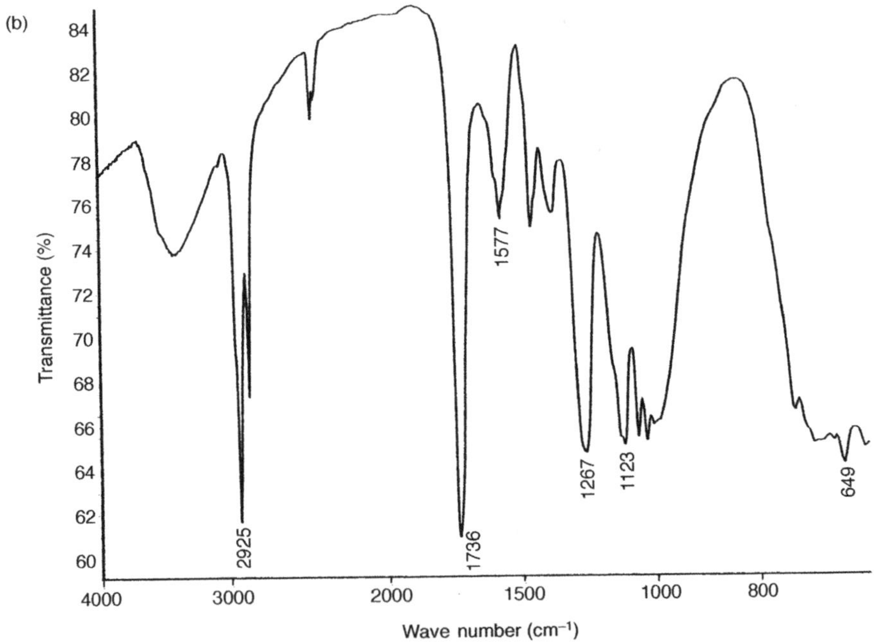

Figure 8.37 Infrared spectra of (a) an unsaponified and (b) a saponified alkyd.

obtained of the disbonded paint chip, the saponified material will make up such a small percentage of the total sample that it may not be evident. An analysis of the back surface of the disbonded coating by ATR will often clearly identify the saponified material.

In Section 5.3, it was pointed out that epoxy coatings do not usually weather well and will degrade by a process known as chalking. The latter is a deterioration of the outer layers of a coating, caused by ultraviolet (UV) light and moisture, resulting in a loose, chalky layer of unbound pigment. In severe cases, the pigment is nearly devoid of any epoxy resin. In other cases, there is still a fairly substantial amount of epoxy resin associated with it, but probably of a lower molecular weight than the bulk resin.

If a chalky layer of degraded epoxy is painted over delamination can result. This author has encountered this situation more than once with urethane topcoats that have peeled from epoxy primers or intermediate coats. It is usually very difficult to see the thin layer of degraded epoxy on the back of the urethane, even with a microscope. However, it is usually easy to detect by using ATR.

Another way for coatings to degrade is by slow evaporation of a semi-volatile ingredient, such as a plasticizer. This will result in embrittlement of the coating, possibly leading to cracking and even delamination. It is often possible to detect this evaporation by comparing spectra of weathered samples with non-weathered ones, or against a laboratory-applied control. The plasticizer will often have one or more bands that appear in a region of the spectrum separate from the resin and pigment bands.

Silicone resins have excellent elevated temperature resistance, but will often eventually degrade. Usually, the degradation is due to charring or oxidation of the organic constituents present on the silicone backbone. The most common silicone resins used in coatings are the methyl silicones. These have a very distinct spectrum, with the strong, broad silicon–oxygen band occurring as a doublet in the 1100- to 1000-cm^{-1} region, and an extremely sharp band near 1263 cm^{-1}, due to the $Si{-}CH_3$ group. As Figure 8.38 shows, the organic portion of a methyl silicone resin seriously degraded in as little as 48 hours at a temperature of 300 °C.

8.2.4.5 Degree of cure

Many coatings fail because, for one reason or another, they did not reach a high enough degree of cure prior to being stressed by their environment. Reasons for this might include a bad formula, a batch of paint deficient in catalyst, 'mis-mixing', low temperature application or, for moisture-cured coatings, low humidity. Sometimes, a lack of cure can be identified simply by rubbing the coating with a cotton Q-tip and a strong solvent and judging whether it softens or dissolves more readily than a properly cured control. In many cases, this can be sufficient. However, there will be other cases where this technique will be too

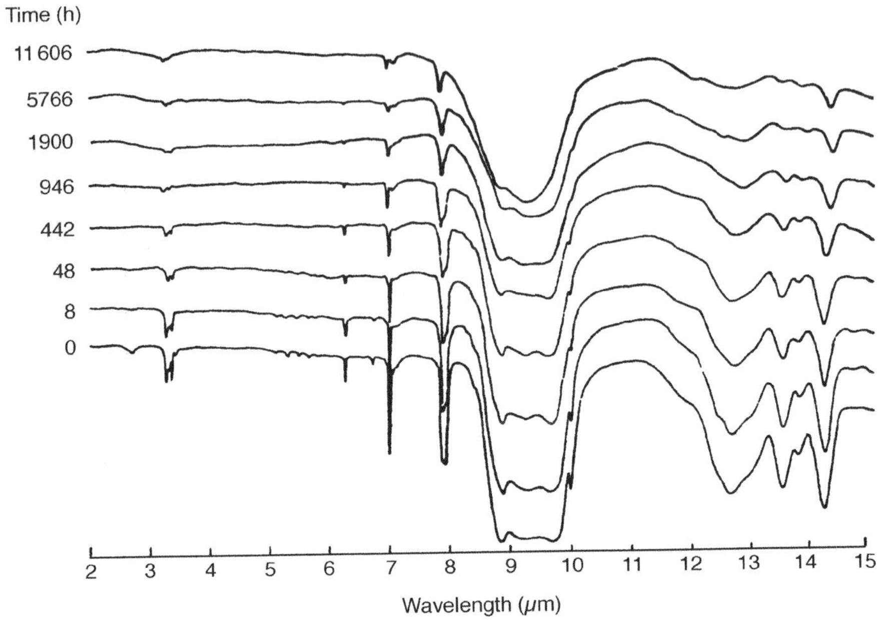

Figure 8.38 Infrared spectra showing the degradation of a silicone resin at 300 °C as a function of time. Reprinted from R. R. Myers and J. S. Long, *Characterization of Coatings: Physical Techniques*, Vol. 2(1), Marcel Dekker, Inc., New York, 1969, by courtesy of Marcel Dekker, Inc.

crude to provide useful information. Depending on the type of coating, infrared spectroscopy may be a useful tool to monitor the degree of cure.

The processes involved in the cure of thermoset coatings involve the breaking and forming of chemical bonds. In many cases, these bonds have characteristic absorption bands in the infrared. It may be possible to track the curing reaction by measuring either the disappearance of a band associated with one of the starting materials or by measuring the appearance of a band associated with the newly formed product.

Urethane coatings, in particular, are amenable to cure studies by using infrared spectroscopy. The isocyanate structure (—N=C=O) has a very strong, distinctive band near 2300 cm^{-1}. Fortunately, this is a region of the infrared spectrum where few other functional groups display any absorption (carbon dioxide being an exception). As discussed earlier in Section 5.9, urethane coatings are formed by the reaction of a polyol with an isocyanate to form the urethane (or carbanate).

Therefore, it is apparent that the intensity of the isocyanate band should decrease as the coating cures. Figure 8.39 shows the spectra obtained for a urethane at different stages of cure, clearly demonstrating the validity of the

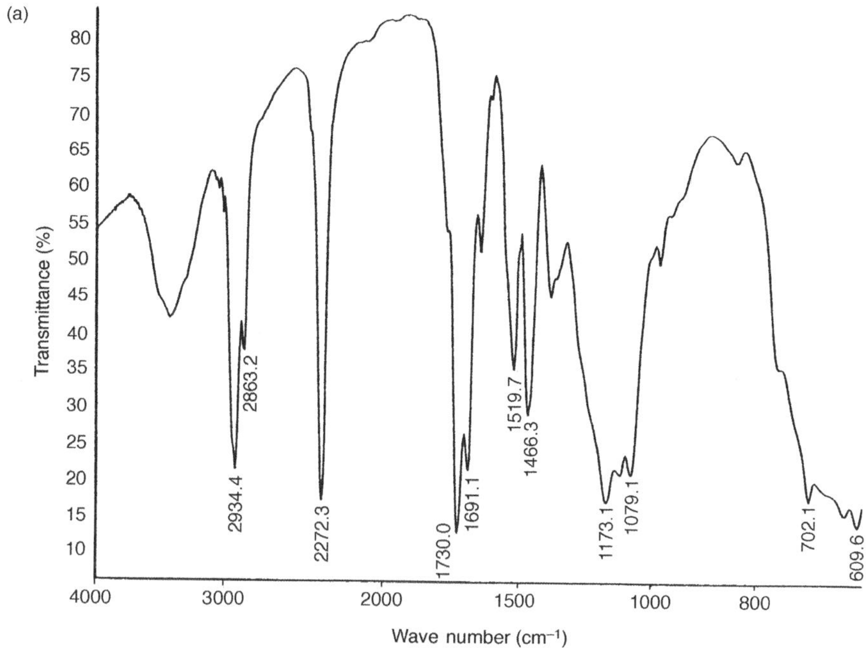

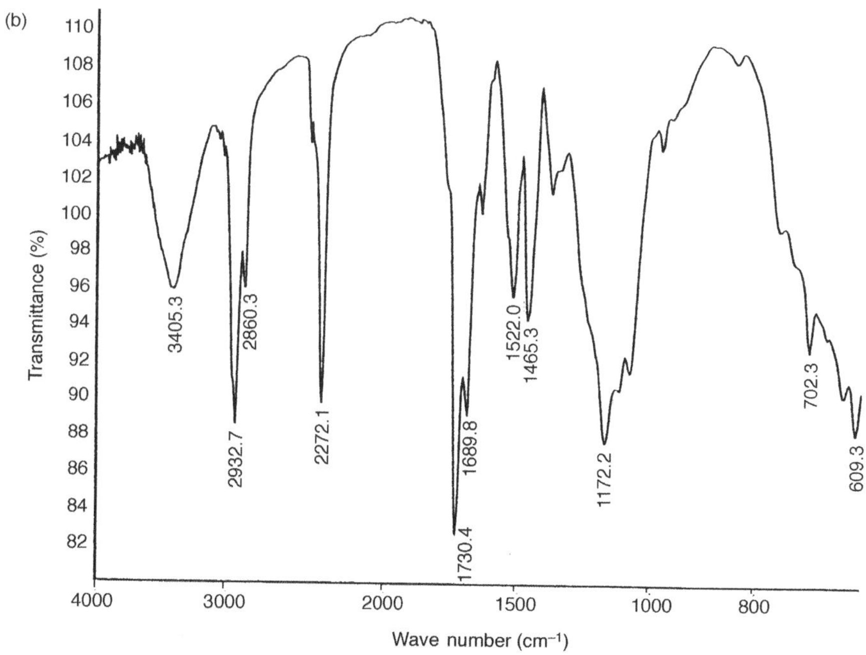

Figure 8.39 Infrared spectra of a urethane obtained after cure times of (a) 2, (b) 18 and (c) 116 hours.

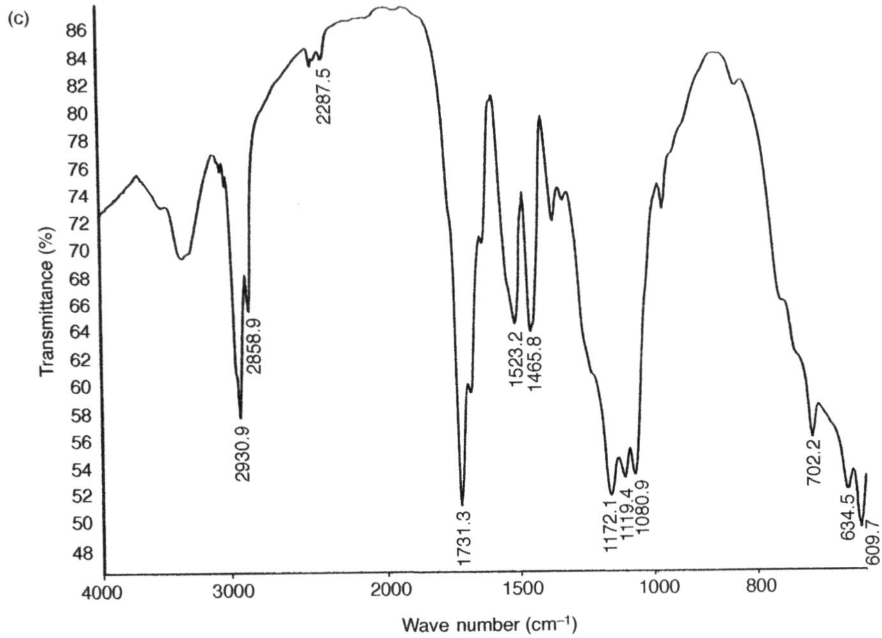

Figure 8.39 *Continued*

technique, as the intensity of the isocyanate band near 2270 cm^{-1} dramatically decreases.

If one suspected that a particular urethane failed because it did not adequately cure, or if one wanted to determine the rate of cure of a suspect batch of urethane with a known control batch, infrared spectroscopy might very well be the preferred technique. However, when investigating coating failures, one also has to consider alternative explanations.

For instance, the above technique will unequivocally tell whether all of the isocyanate has been reacted. However, it says nothing about what it reacted with. As previously pointed out, isocyanates are very reactive and will readily react with numerous functional groups, including water. It is possible that a urethane may have failed because a large amount of the isocyanate reacted with water, rather than with the polyol. If only the isocyanate region near 2200 cm^{-1} is examined, the investigator would be oblivious to the fact that the wrong reaction had taken place.

Fortunately, in the above example, there will be enough clues for the astute investigator to suspect what has happened. If the analyst has looked at the sample first through a microscope, he or she will have noticed the extensive voids present in the coating due to carbon dioxide formation. Carbon dioxide is one of the by-products of the reaction between an isocyanate and water. Secondly, the normal polyol–isocyanate reaction produces a urethane, which has its own characteristic carbonyl vibration near $1690\,cm^{-1}$. By comparing the intensity of this band to that of a properly cured control, one will likely be able to determine that something is amiss.

Thirdly, the reaction with water results in the formation of a disubstituted urea, which has its carbonyl band in the 1640- to $1660\text{-}cm^{-1}$ region. Theoretically, this band should become more intense upon the reaction of the isocyanate with water. In practice, this is not always so noticeable.

There is another possible explanation for why the isocyanate band could be absent – suggesting full cure – and yet the coating is still failing. Perhaps, it was never added. This author has personally witnessed more than one occasion where the contractor simply neglected to mix the isocyanate component with the polyol. Again, a careful examination of the sample will usually indicate what has happened.

Infrared spectroscopy is also useful in evaluating the degree of cure of the alkyl silicate inorganic zinc-rich primers. As discussed earlier in Section 5.12.2, these coatings cure by reaction with atmospheric moisture to produce a highly cross-linked silicate matrix.

It is possible that the metallic zinc dust pigment may also be involved to some degree in the cure. At any rate, the important observation is that the silicate pre-polymer has a certain amount of organic character to it, whereas the cured coating has none. The ethyl alcohol that is cleaved from the pre-polymer during the curing process is volatile and will evaporate from the film. Therefore, if one had a means of tracking the amount of carbon–hydrogen character remaining in the film, one would also have a means of measuring the coating's degree of cure.

Infrared spectroscopy is quite adept at detecting hydrogen atoms bonded to carbon atoms, since this structure results in relatively strong, characteristic bands near $2900\,cm^{-1}$, well removed from any interference due to the strong, broad Si—O band near $1100\,cm^{-1}$. The cure of the coating can be tracked by measuring the disappearance of the aliphalic carbon–hydrogen stretching band [7].

Figure 8.40 shows the spectra of an inorganic zinc-rich coating, cured for 3 and 18 hours at 40% relative humidity, and also for 18 hours at 100% relative humidity. The relative intensities of the carbon–hydrogen bands are seen to decrease with time, compared with the strong silicate band near $1100\,cm^{-1}$, and also to decrease dramatically with higher humidities. By converting to absorbance and measuring the integrated absorbance underneath the band 'envelopes', it is possible to measure a quantity called the 'cure ratio', which has been shown to be proportional to the coating's degree of cure [7].

Theoretically, it should be possible to monitor the degree of cure of a two-component epoxy coating by using infrared spectroscopy. This is because the three-membered epoxy (or oxirane) ring opens as the curing reaction proceeds. This ring has a characteristic band in the infrared near $920\,cm^{-1}$, which gets progressively weaker as the reaction proceeds. This concept was first demonstrated by Dannenberg and Harp [8]. Figure 8.41 shows the decrease in the intensity of

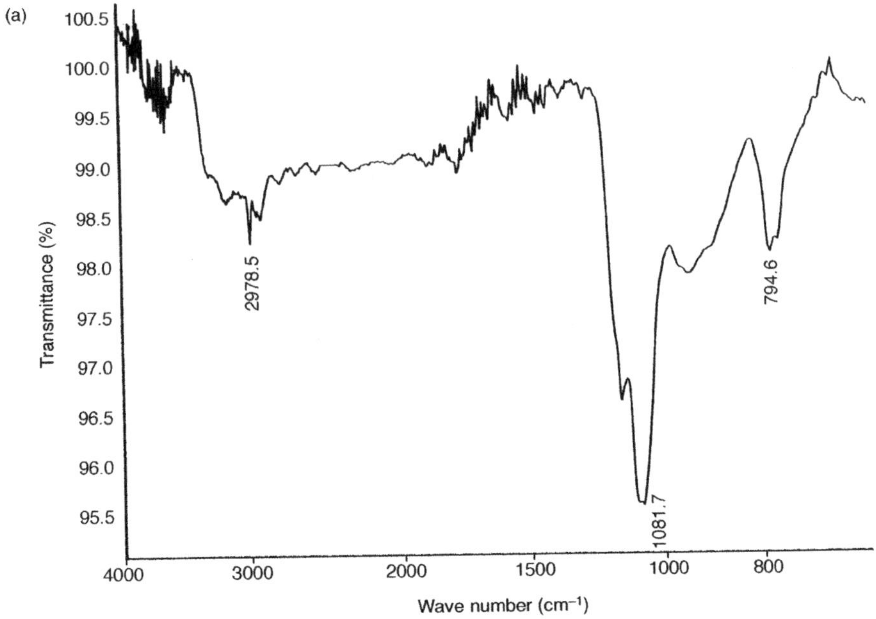

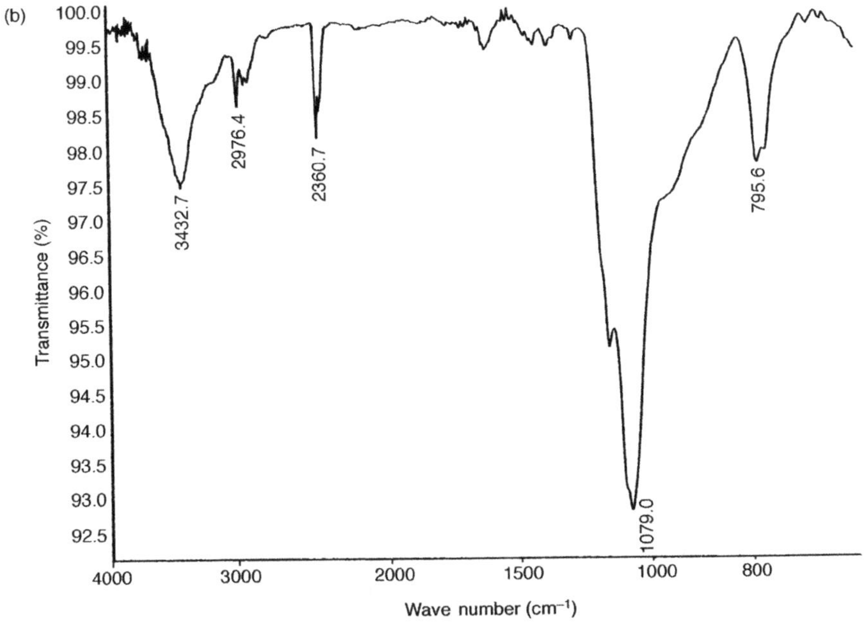

Figure 8.40 Infrared spectra of an inorganic zinc-rich coating obtained after curing for (a) 3 hours at 40% relative humidity, (b) 18 hours at 40% relative humidity and (c) 18 hours at 100% relative humidity.

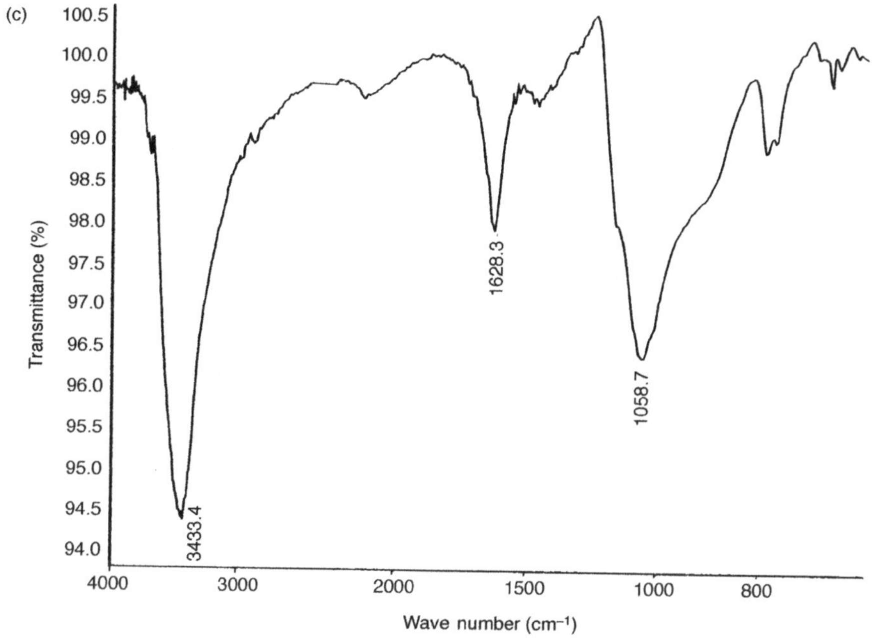

Figure 8.40 *Continued*

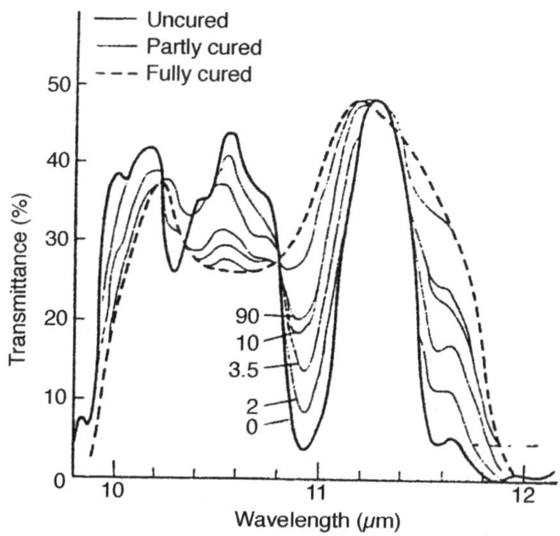

Figure 8.41 Infrared spectra of an Epon epoxy resin, showing the decrease in the intensity of the epoxy (oxirane) band as a function of time of the curing reaction; numbers on curves represent the cure times (hours) at 65 °C. Reprinted with permission from H. Dannenberg and W. R. Harp Jr, *Anal. Chem.*, **28**, 81–89 (1956). Copyright 1956, American Chemical Society.

the oxirane band as a function of time. It has been this author's experience, however, that it is easy to track the disappearance of this band in the early stages of the reaction, but not so easy in the latter stages, which might be critical in achieving full property development.

As with urethanes, the disappearance of a band due to one of the functional groups in the liquid paint, although a very important observation, may not necessarily tell the whole story. In a pigmented coating, the 920-cm^{-1} oxirane band may be partially obscured. Secondly, the wrong curing agent, differing in the number of reactive sites, can result in a complete cure, but perhaps with either a greater or lower cross-link density from what is desired. The cured coating could be more linear than required and hence more permeable and flexible, or more branched and hence more impermeable and brittle. It is possible that a small difference in the degree of cure, perhaps beyond the ability of the infrared spectroscopic method to measure, can result in significant changes in physical properties.

8.3 GAS CHROMATOGRAPHY (GC)

Chromatography can be summarized as the science of separating individual components from a complex mixture. The term is credited to Michael Tswett, a Russian botanist, who in 1906 succeeded in separating individual pigments from plants by using a glass column containing a special packing material. Since the procedure used by Tswett resulted in a series of brightly coloured bands moving down the column, he named the technique 'chromatography', which means 'colour writing'.

Today, there are numerous forms of chromatography, several of which will be discussed in this present text. A very useful form of chromatography in the field of coatings analysis is GC.

8.3.1 Theory of GC

With one exception, all forms of chromatography involve separating an individual component from a mixture through a process of selective partitioning of the various components between a moving, or 'mobile' phase, and a stationary phase. The two phases differ in polarity or some other chemical or physical property, such that the component of interest (the 'analyte') prefers to be in one phase rather than the other. In GC, the mobile phase is a gas. The stationary phase is usually either a fine, powdery solid that has been coated with a small amount of a very high boiling liquid, or an open, tubular column, the inside walls of which have been coated with a non-volatile liquid phase. In the first case, the solid itself is often referred to as the 'support', and the technique is occasionally referred to as gas–liquid chromatography (GLC). The technique is used to analyse volatile or semi-volatile materials, such as solvents.

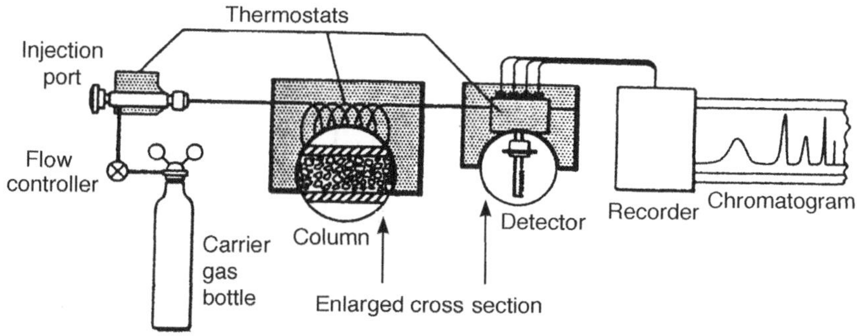

Figure 8.42 Schematic diagram of the layout of a typical gas chromatograph. Reproduced by permission of Varian, Inc., from H. M. McNair and E. J. Boneli, *Basic Gas Chromatography*, 1968.

The basic components of a gas chromatograph, as shown in Figure 8.42, are a source of carrier gas, a flow controller, an injection port, an oven containing one or more columns, a detector and some type of recording device. When a microlitre syringe is used to inject a small amount of sample (such as a paint thinner) into the injection port (which is heated), the solvents are immediately vapourized and the carrier gas (typically helium) flushes these solvent vapours through the column containing the stationary phase. Those solvents that are least attracted to the stationary phase will exit the column first and be detected; those that are more strongly attracted will take longer to move through, or 'elute' from, the column. The time that an individual solvent spends within the column is called its retention time, and this parameter is used to identify it.

This chapter is not intended to present a rigorous chemical and mathematical treatment of chromatography. However, an understanding of some of the fundamental concepts and definitions of chromatography will aid the investigator in understanding and utilizing the technique.

The goal of a chromatographic system is to isolate, or resolve, all of the individual components of a mixture as they pass through the column. The resolution of chromatographic peaks is related to the efficiency of the column and the respective distribution coefficients (or partitioning coefficients) of the analytes in the mobile phase. In other words, a particular mobile phase may be more efficient at separating two analytes because there is a large difference in their partition coefficients (a measure of their relative affinity for the mobile phase versus the stationary phase). In GC, the mobile phase is a gas. Figure 8.43 illustrates the effects of both column efficiency and 'solvent efficiency'.

The efficiency of a column, or its ability to resolve two closely eluting species, is expressed by a somewhat archaic term as the 'number of theoretical plates'. This concept goes back to simple distillation theory. Distillation columns are

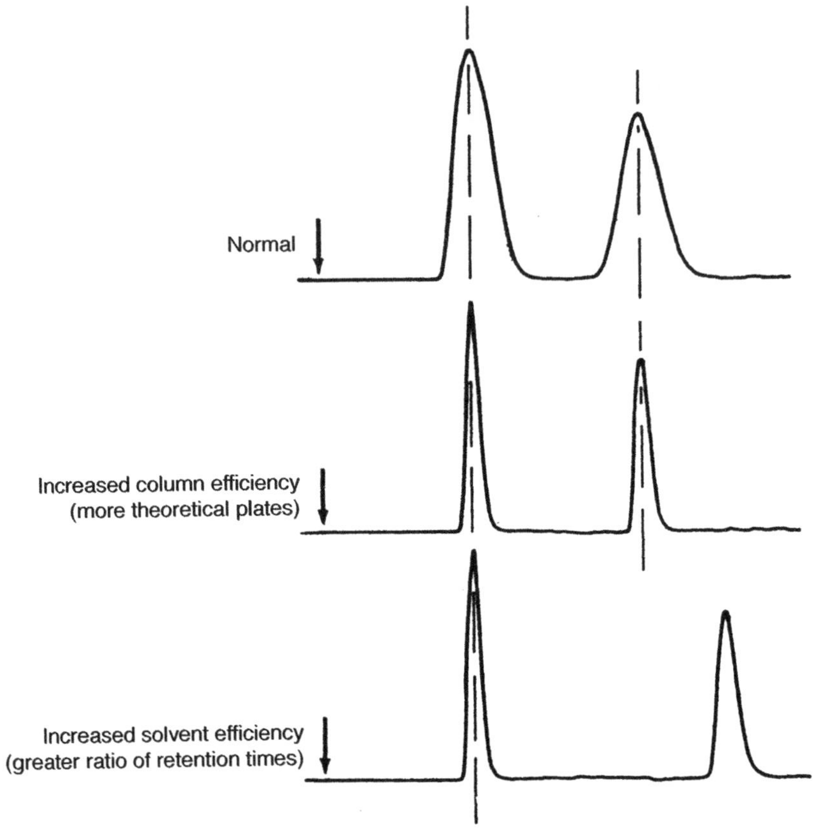

Figure 8.43 Illustration of the effects of column and solvent efficiencies on a gas chromatographic separation. Reproduced by permission of Varian, Inc., from H. M. McNair and E. J. Boneli, *Basic Gas Chromatography*, 1968.

actually composed of discrete, physical plates, and the greater the number of plates, then the greater the ability of the column to separate solvents.

All other things being equal, a chromatographic column containing more 'theoretical plates' will produce sharper, more narrow peaks than one of lower theoretical plates. Sharp, narrow peaks mean that two components which have very similar retention times can be resolved as two closely spaced, distinct peaks, rather than eluting as one broad, overlapping peak.

The number of theoretical plates for a particular column can be easily measured from the printout of the chromatogram, as shown in Figure 8.44. The width of the peak (y) is measured by drawing tangents to the leading and trailing edges. The number of theoretical plates (N) is then calculated as shown in Figure 8.44.

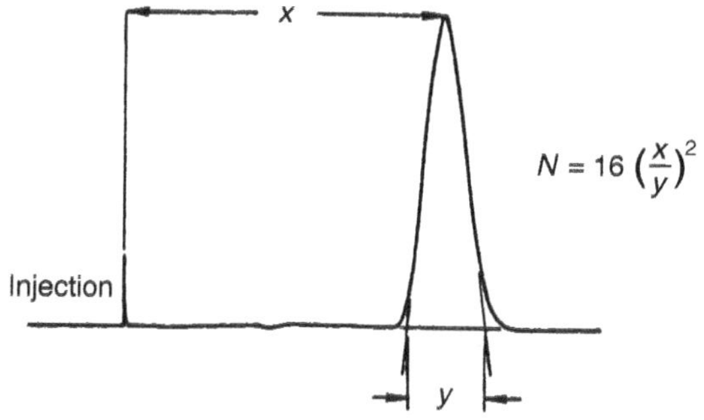

Figure 8.44 Determination of the number of theoretical plates (N) of a chromatographic column. Reproduced by permission of Varian, Inc., from H. M. McNair and E. J. Boneli, *Basic Gas Chromatography*, 1968.

Another term used to express column efficiency is the height equivalent to a theoretical plate (*HETP*). The relationship is a simple one, as follows:

$$HETP = L/N\,, \tag{8.7}$$

where L is the physical length of the column in centimetres.

Ideally, when one injects a sample, such as a paint thinner containing several solvents, into a gas chromatograph, the individual solvents, due to their different affinities for the mobile phase (gas) and stationary phase (column packing), should gradually separate from one another. They should then move through the column as compact 'slugs' of almost infinitely narrow width. This almost never happens, because there are certain processes going on in the column which contribute to the broadening of these bands.

The three major factors contributing to band broadening are as follows:

- eddy diffusion (multi-path effect)
- molecular diffusion
- resistance to mass transfer.

The stationary phase used in packed columns generally consists of a finely ground support (anywhere from typically 80–140 mesh), often coated with a small amount (5–20%) of a high boiling liquid. Because of the huge number of individual support particles contained in the column, there are many different paths that an individual molecule could take to get through the column. These paths will have slightly different lengths, and therefore, the molecules will exit the column at slightly different times. This is the source of band broadening called eddy diffusion or the multiple path effect.

In order to minimize the effects of eddy diffusion, columns should be carefully packed with small particles of uniform size. However, very small particles will cause a greater pressure drop across the column. In other words, the instrument will have to be operated at much higher pressures.

The second factor relating to band broadening is molecular diffusion. This can best be visualized by considering what would happen if the flow of carrier gas through the column could somehow be stopped in mid-analysis. The narrow band, or slug, of solvent would not remain in one place, but rather the molecules would begin to naturally spread out or diffuse into the gas, and a band which was once narrow would begin to broaden. The same effect is happening during normal operation of the instrument, although it is mitigated by the flow of the carrier gas. In order to minimize band broadening due to molecular diffusion, one would want a higher flow rate of carrier gas. Unfortunately, this will also increase the back pressure on the column.

As stated at the beginning of this section, the whole basis for chromatography is the fact that individual analytes will preferentially partition between either the mobile phase or the stationary phase. If two analytes have different partition coefficients, then theoretically, they should be able to be separated from one another. However, the ability, or the rate, at which an individual molecule can move into or out of either the mobile phase or the stationary phase is finite. The process does not happen 'instantaneously' but requires a small amount of time.

The difficulty that a molecule encounters in moving between the two phases is referred to as resistance to mass transfer. It will be affected by such things as the thickness of the liquid phase coating the support material and the viscosities and densities of the two phases. One way of minimizing the contribution to band broadening from resistance to mass transfer is by reducing the velocity of the mobile phase.

Based on the above discussion, the efficiency of the column, as expressed by the *HETP* is given by the following:

$$HETP = A + B/V + CV, \tag{8.8}$$

where A is the eddy diffusion (multi-path effect) term, B is the molecular (longitudinal) diffusion term, C is the resistance to mass transfer term and V is the average linear velocity of the carrier gas.

From the above relationship, which is known as the van Deemter equation, it can be seen that there must be an optimum flow rate that achieves maximum column efficiency. A plot of *HETP* versus V (often referred to as a van Deemter plot) has the form shown in Figure 8.45.

Some other terms that the reader should become familiar with include the partition coefficient, the capacity factor and the resolution.

The partition coefficient, k, is the ratio of the concentration of any particular compound in the stationary phase (C_s) to its concentration in the mobile phase (C_m), as follows:

$$k = C_s/C_m. \tag{8.9}$$

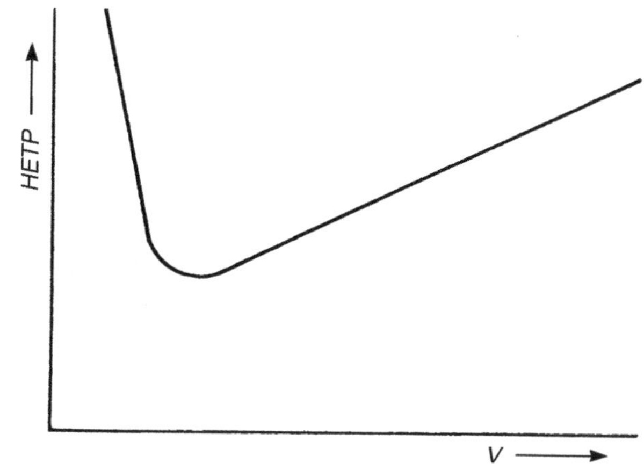

Figure 8.45 A van Deemter plot of height equivalent to a theoretical plate (*HETP*) versus mobile phase rate.

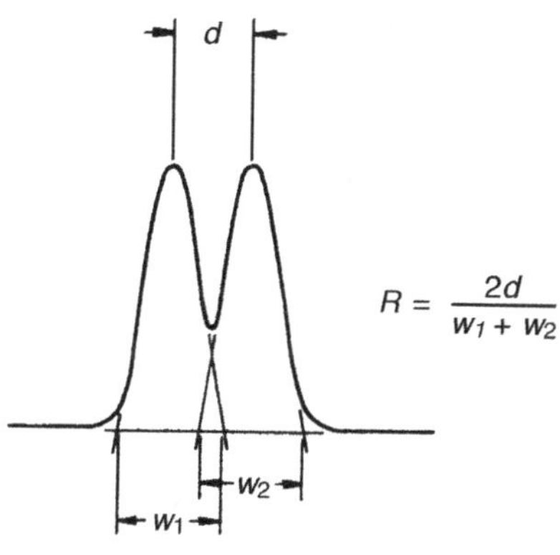

Figure 8.46 Determination of the resolution (*R*) of a gas chromatographic system. Reproduced by permission of Varian, Inc., from H. M. McNair and E. J. Boneli, *Basic Gas Chromatography*, 1968.

The capacity factor, k', is a variation of the partition coefficient. This is the ratio of the total amount (weight) of analyte in the stationary phase (W_s) to the total amount in the mobile phase (W_m), as given by the following:

$$k' = W_s / W_m. \tag{8.10}$$

The resolution (R) is a measure of the efficiency of the total chromatographic system. This parameter is measured from the printout of a chromatogram of two closely eluting peaks, as shown in Figure 8.46. In this figure, d is the distance between the peak maxima (in units of length or time), and w_1 and w_2 are the widths of the two peaks (again, in units of either length or time).

8.3.2 Instrumentation

Although conceptually a gas chromatograph is a very simple instrument (a gas flows through a heated column, carrying the solvents of interest to a detector), it actually consists of a number of sophisticated components. Only the two most important components to the failure investigator will be discussed here, namely the column and the detector.

8.3.2.1 Columns

A plethora of columns are available for GC, varying in material, design and packings.

The traditional type of column, which is still used in many laboratories, is the packed column. Usually constructed of stainless steel or glass, these columns typically range from $\frac{1}{16}$ to $\frac{1}{4}$ in. in outer diameter (od), and usually from 5 to 10 ft in length. They are supplied in a coiled shape so as to be able to fit inside the oven compartment of the chromatograph. Stainless steel columns are more rugged, but glass columns are more inert and will not react with compounds at the high temperatures sometimes used in GC. If steel columns are used, some workers prefer to use a short length of a glass 'liner' at the head of the column, so as to minimize any chance of a reaction occurring at the surface of the hot steel. This is probably not necessary for routine work at relatively high (100 ppm or higher) concentrations.

Packed columns, whether steel or glass, usually contain an inert solid support that has been coated with a thin film of a non-volatile liquid (the stationary phase). Although there are a huge number of packing materials to choose from, most coatings analysis and failure analysis work can be handled adequately by perhaps four or five column types.

The solid support used in packed columns should ideally have the following characteristics:

- mechanical strength;
- large surface area;
- uniform, regularly shaped particles for efficient packing;
- inertness.

Diatomaceous earth is one of the most common supports. It basically consists of extremely small, amorphous, hydrated silica particles. They are very porous and have high surface areas (approximately $1–7\,m^2\,g^{-1}$).

A potential disadvantage of untreated diatomaceous earth is the fact that its surfaces are covered with silanol (Si—OH) and siloxane (Si—O—Si) groups, which can hydrogen bond with certain compounds and cause tailing of peaks. The effects of these surface groups are reduced by the covering of liquid stationary phase. However, sometimes the amount of liquid stationary phase is only 5–10 wt%, and thus experimental conditions are difficult to reproduce accurately.

A common procedure is to mask the surface reactivity of the diatomaceous earth by treating or capping it with various agents. Two common reagents used for this purpose are dimethyldichlorsilane and hexamethyldisilizane.

Many types of liquids are used to coat the support materials used in packed columns, and the choice will depend primarily on the chemical nature of the compounds which one is confronted with. Basically, one wants to choose a stationary phase which is inert and non-volatile, and which is both a good 'solvent' for the sample components, but which is not equally good for all of them. If a stationary phase is a poor solvent for the compounds being analysed, there will be little or no retention in the column, and hence no separation. If the stationary phase has the same, identical solvency for all of the compounds, they will be retained for nearly the same amount of time in the column, hence giving poor separation. It should not be surprising that the choice of a stationary phase is made much easier if one knows the type of components in the sample. Like infrared spectroscopy, GC as it relates to failure analysis should not be 'conducted in a vacuum'.

Liquid stationary phases vary from very non-polar liquids such as long-chain hydrocarbons (squalene) and polydimethylsiloxane types (SE-30 and OV-1), through intermediate polarity liquids such as diphenyldimethyl polysiloxanes (SP-20 and OV-17), to highly polar liquids such as poly(ethylene glycol) (carbowax). The percentage of liquid phase used to coat the solid support is generally in the 5–20% range. Column manufacturers often publish very useful and detailed information on the various stationary phases in their catalogues. If one is performing a standard procedure, such as an ASTM test method, the type of column will be specified.

A variation of the traditional packed column is a column packed with a porous polymer stationary phase, such as Porapak. The latter material is a porous polymer consisting of ethylvinylbenzene cross-linked with divinylbenzene. It can be modified by copolymerization with other materials to vary the polarity. Porapak columns contain only the porous polymer, which has been screened to a certain mesh size. There is no additional liquid coating on the surface of the Poropak. They are especially useful in analysing aqueous samples.

Packed columns are rugged, inexpensive and very useful for analysing relatively simple mixtures. Capillary columns, however, give a much higher resolution and require much smaller quantities of sample.

Capillary columns are long, open tubes of very small diameter (typically 0.01–0.03 in internal diameter (id)), usually made of fused silica. They have no packing in the normal sense of the term, but rather have a very thin film of liquid stationary phase along the inside walls. They are supplied as a coil, often from 100 to 500 ft in length.

Capillary columns have very high efficiencies and low sample capacity. Because there is no particulate solid support inside, even though they are very long, the pressure drop across the column is low. Because of the very small amount of liquid stationary phase, it is easy to overload a capillary column. Since only a very small amount of sample is actually put on the column, they are usually used in instruments equipped with very sensitive detectors.

A capillary column is not much more efficient than a packed column per foot of column. Typical packed columns usually have from 500 to 1000 theoretical plates per foot, compared with about 1000 for a capillary column. However, because of high-pressure drops, packed columns are usually only 10–20 ft in length, for a maximum number of total plates of perhaps 20 000. In contrast, a 300-ft-long capillary column could have 300 000 theoretical plates.

Since capillary columns do not have a solid support coated with stationary phase, there is no component of band broadening associated with the multi-path effect. Therefore, in this case, the van Deemter equation simplifies to the following:

$$HETP = B/V + CV. \tag{8.11}$$

Figures 8.47 and 8.48 illustrate the difference attainable by using a capillary column versus a conventional packed column. The sample consisted of blends of either 16 or 21 different alcohols. Note how much sharper the peaks are on the capillary column, and hence the improved resolution, when compared with the packed column.

8.3.2.2 Detectors

There are numerous types of detectors available for gas chromatographs, some with very specialized purposes. The three that will be discussed here are the thermal conductivity detector (TCD), the flame-ionization detector (FID) and the mass spectrometer.

The TCD is one of the oldest types available for GC. It relies upon the fact that a hot body will lose heat to a flowing stream of gas at a rate which depends upon the composition of the gas stream. Some substances conduct heat more efficiently than others. Therefore, if a stream of pure carrier gas is flowing across a hot filament, it will be able to conduct a certain amount of heat away from the filament. When the composition of the carrier gas changes (such as when a solvent peak elutes from the column), the thermal conductivity of the flowing gas will change, thus resulting in a detectable signal.

In practice, differential thermal conductivity is measured by having two streams of carrier gas, one passing through the analytical column and one passing through

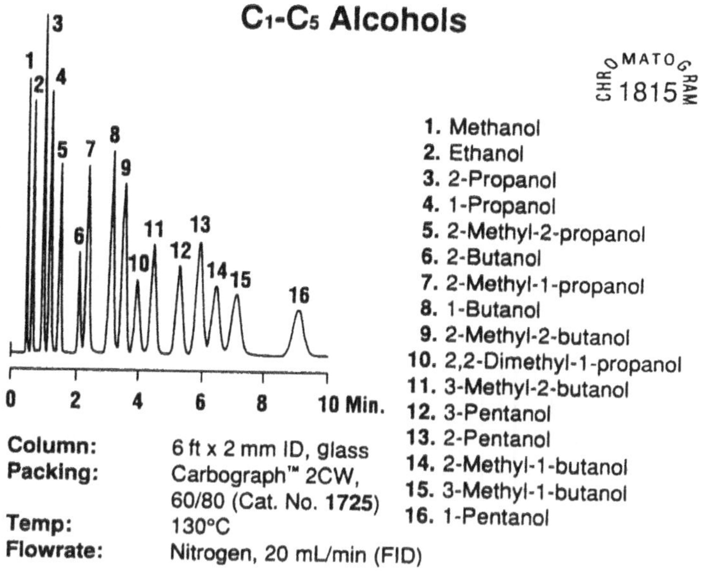

Figure 8.47 Chromatogram of a mixture of various alcohols obtained by using a packed gas chromatography (GC) column. Reproduced with permission of Alltech Associates, Inc. FID = flame-ionization detector.

a reference column, flowing across two different heated filaments. Helium is commonly used as a carrier gas because it is safe and inexpensive and has a thermal conductivity significantly different from most organic compounds.

TCDs are simple, inexpensive, rugged and virtually universal. They can detect any substance that has a thermal conductivity different from that of the carrier gas. They are, however, one of the less sensitive forms of detectors that are available.

In the FID, the carrier gas exiting the column is mixed with hydrogen and burned in air or oxygen. This causes the formation of ionized species, which in turn results in the flow of current across an electrode. The magnitude of the current is proportional to the amount of ionizable solute eluting from the column and the ease with which it ionizes.

All organic compounds ionize to at least some extent in the flame, and hence are detectable by the FID. Some notable compounds that give little or no response include water, oxygen, nitrogen, carbon disulfide, carbon dioxide and ammonia. Carbon disulfide, since it is not detectable by an FID, is a common choice of solvent for many tests.

The FID is much more sensitive than the TCD and has a very wide linear range. However, it is more expensive and less rugged. It will not detect water, which can be either an advantage or a disadvantage.

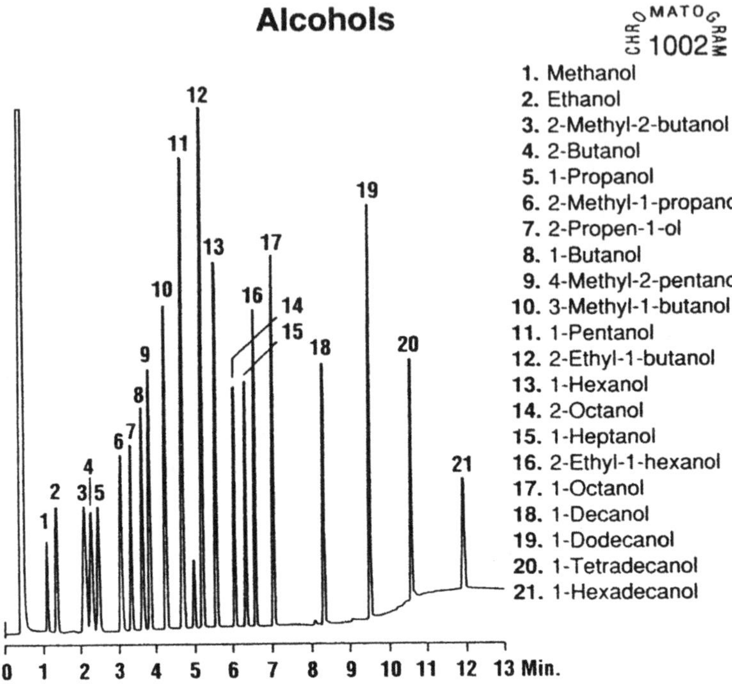

Column:	Econo-Cap™ Carbowax®, 15m x 0.53mm ID x 1.2μm, (Cat. No. **19653**)
Temp:	50°C (2 min) to 220°C at 20°C/min
Carrier gas:	Helium, 8.0 mL/min
Detector:	FID/300°C
Injector:	250°C

Figure 8.48 Chromatogram of a mixture of various alcohols obtained by using a capillary gas chromatography (GC) column. Reproduced with permission of Alltech Associates, Inc. FID = flame-ionization detector.

Both the TCD and the FID are non-specific, and they generate no structural information. Compounds are tentatively identified by their retention times only. For many purposes, this is adequate. If one is doing routine QC work on various batches of the same product, confirming the solvent composition by retention time on a single column is an acceptable practice. Alternatively, if one has good reason to believe that a coating has been adulterated with a specific solvent, if the sample in question has a peak which elutes at the same time as the 'suspected adulterant', chances are good that one has made the correct identification.

However, when truly working with unknowns (which is often the case with failure investigations), making a positive identification based on the retention time on a single column can be risky. It is entirely possible for two or more compounds to elute at nearly the same time on any given column. In such cases, it is recommended that the analysis be performed again on a second column containing a different type of stationary phase. It is very unlikely that two different compounds would have an identical retention time on two different columns.

One way of avoiding the above uncertainty is to use a detector that provides more information than simple retention time. One such detector is the mass spectrometer.

Mass spectrometers basically work by bombarding a molecule with a beam of energetic particles, such as electrons. The molecules are ionized and partially broken up into smaller, charged fragments. These charged fragments (ions) have a particular ratio of mass to charge or *m/e* value. Since in most cases the charge is 1, the *m/e* value is simply the molecular weight of the ion. There are various ways of constructing the detectors, but their basic function is to identify the relative abundance of the various charged fragments of the original molecule and to determine their molecular weights. This plot of the molecular weight of each fragment versus the relative abundances of the fragments is what is termed the 'mass spectrum' of the compound and is extremely useful in identifying it. The upper portion in Figure 8.49 shows a chromatogram of a solvent obtained from pine oil. A half dozen peaks were detected eluting from the GC column between roughly 9–10 minutes. The lower portion of the figure shows the mass spectrum of the peak which eluted at 10.0 minutes, and the computerized library search routine identified this peak as being due to alpha-terpineol, one of the main constituents of pine oil.

Figure 8.50 shows both the chromatogram obtained from a sample of an epoxy thinner, along with the mass spectrum of the peak which eluted at 4.6 minutes. The computerized library search routine identified this peak as methyl isobutyl ketone ($C_6H_{12}O$) or MIBK. MIBK has a molecular weight of 100, and note that there is a peak in the mass spectrum corresponding to a molecular weight of 100. Furthermore, it is not difficult to imagine that as MIBK fragments upon bombardment by the high energy electrons, that a methyl group (CH_3) would break off. The loss of a methyl group would result in a decrease of 15 atomic mass units, producing a fragment with a molecular weight of 85. Note that there is a peak in the mass spectrum of atomic weight 85. It is also not difficult to imagine that the entire isobutyl group (C_4H_9) would break off, leaving a fragment of formula C_2H_3O, with a molecular weight of 43. Note that this is the most intense peak in the mass spectrum of MIBK.

A gas chromatography–mass spectrometry (GC-MS) system basically consists of a gas chromatograph, a mass spectrometer, an interface connecting the two and a data system. The interface module transfers the compounds which were separated in the GC column to the mass spectrometer's ionization source, without allowing them to remix. The simplest interface is simply a direct connection of the end of the capillary column to the sample inlet port of the ionization source. However, since the analyser module of the mass spectrometer must operate at an

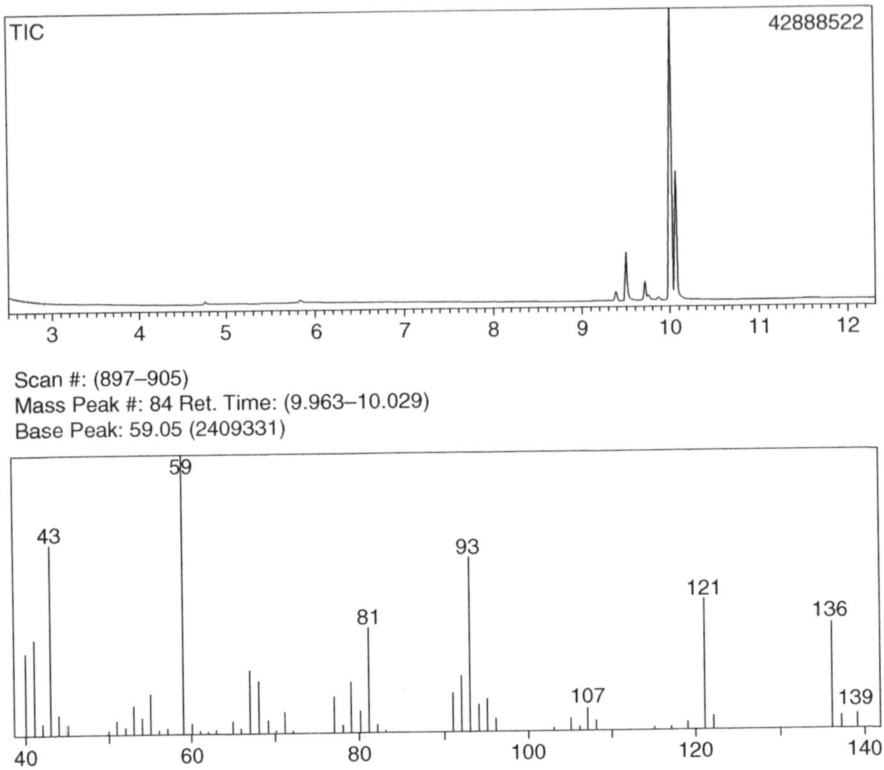

Figure 8.49 Chromatogram of a thinner obtained from pine oil (upper figure). The mass spectrum (lower figure) of the peak which eluted near 10.0 minutes identifies it as alpha-terpineol.

extremely high vacuum, it may not be able to handle the amount of carrier gas which such a simple interface would provide. One way to improve upon this situation would be to add some type of splitter before the ionization source to allow a controlled amount of the sample/carrier gas to be diverted to waste. Yet another way to solve the problem would be to add another vacuum pump to the exhaust coming from the GC column, along with a separator in the line from the GC to the mass spectrometer (MS). This allows carrier gas, which is usually helium and therefore of much lower molecular weight than the sample, to be preferentially removed from the stream, while most of the higher-molecular-weight sample proceeds into the MS source.

The most common type of analyser in GC-MS systems is the quadrupole type. As discussed above, the sample molecules are bombarded in the ionization chamber, typically by high-energy electrons, which result in charged fragments of differing molecular weight, possibly with a small amount of the charged,

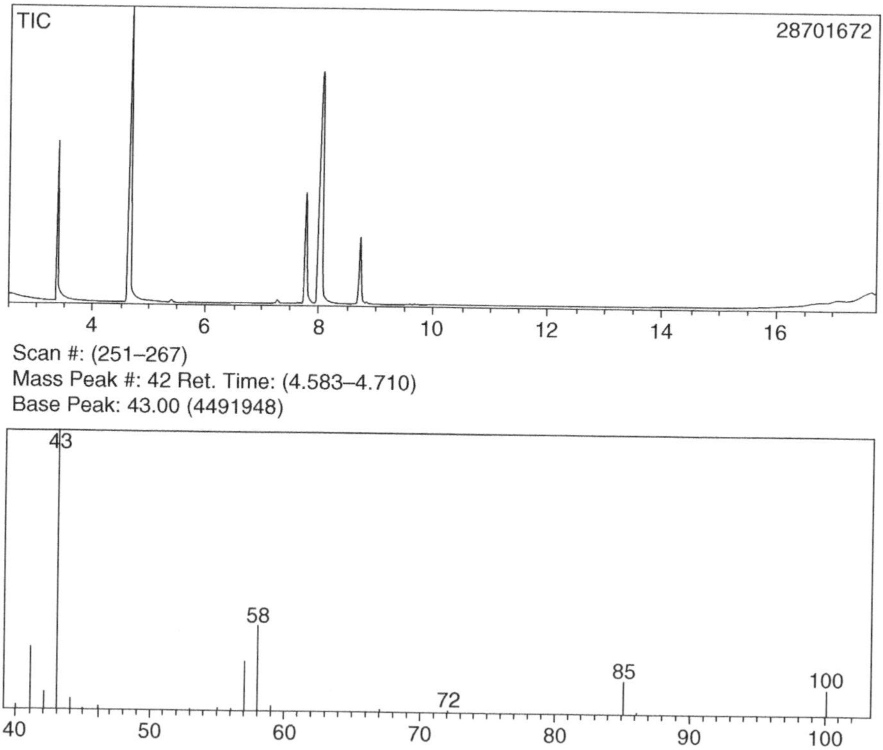

Figure 8.50 Chromatogram obtained from an epoxy thinner (upper figure). The mass spectrum (lower figure) of the peak which eluted near 4.6 minutes identifies it as methyl isobutyl ketone (MIBK).

unfragmented base molecule. These charged fragments are then swept into the analyser, where they are focused via electric 'lenses' and introduced into the quadrupole. The quadrupole, which consists of four electrically charged quartz poles or rods arranged to produce a small 'tunnel', produces a standing magnetic field in which the charged fragments (ions) are aligned. The quadrupole is then swept with a radio frequency signal. As different frequencies are reached, ions of different mass to charge ratio (different molecular weight) are able to escape from the analyser and enter the detector, which results in the production of the mass spectrum.

In MS, the largest peak is called the base peak and is assigned an intensity of 100. The intensities of the other peaks are then expressed relative to the base peak. Not all of the molecules will fragment. There will be a peak that corresponds to the non-fragmented, parent ion, representing the molecular weight of the compound itself. Knowing the molecular weight of the parent ion is the same as knowing the molecular weight of the unknown compound and is obviously important in identifying it. However, the fragmentation pattern is much more important in identifying the compound. If the mass spectrum of an unknown compound is identical to that of a known compound, obtained under similar conditions, then it can be concluded with almost absolute certainty that the two compounds are identical. However, when analysing complex mixtures by GC-MS, there is often the problem of incomplete separation. The MS portion of the instrument can only analyse what is delivered to it by the GC component. Thus, if two solvents have identical or similar retention times, they will exit the GC column together, or as overlapped peaks, and will therefore be analysed together by the MS system. This can result in a mass spectrum which either cannot be identified at all or can be assigned the wrong identity. Some modern software packages have the ability to 'subtract' some or all of the signal from an overlapping peak. This sometimes is sufficient to solve the problem, but the best solution is to improve the chromatographic separation, if this is feasible.

Once the spectrum of a properly separated solvent or compound is obtained, it can be used to identify it. In the review by McMaster and McMaster [9], it was pointed out that the 'determination of the molecular structure of a compound from its molecular weight and fragmentation spectra is a job for a highly trained specialist'. Fortunately, all modern instruments are equipped with mass spectral libraries and software which allow computerized library searching in order to identify compounds. These library search techniques work very well for routine compounds and for many non-routine compounds as well. However, in a very complicated mixture there may be several trace compounds present which cannot be conclusively identified. This could be because they are degradation products which are simply not contained in the computerized library, or because they overlap with other peaks.

The actual process of the interfacing of a mass spectrometer with a gas chromatograph has not proved to be a simple task, and older instruments were extremely expensive. Currently, there are several bench top units in the \$50 000–\$70 000 range that deliver very good performance.

GC-MS is a very powerful tool in routine analysis and in performing failure analysis of paints and coatings. It must be remembered, however, that the first step in the process is still chromatography. An efficient column, compatible with the type of compounds expected to be encountered, must still be used to provide pure compounds, one at a time, to the inlet of the MS detector.

8.3.3 Pyrolysis-GC

A variation of GC that deserves mention here is pyrolysis-GC. Pyrolysis is the breaking down or decomposing of a sample by exposing it to high temperatures. Somewhat akin to MS, this results in many lower-molecular-weight fragments which are characteristic of the original material.

In pyrolysis-GC, the fragments generated by this thermal decomposition are swept into the GC unit for separation and detection. Many of the major peaks exhibited in the chromatograms are identifiable and can provide much information about the sample.

There is no reason to pyrolyse already volatile, low-molecular-weight materials such as solvents. These are usually easily identified by the more convenient, simpler techniques already discussed above. In the field of coatings analysis and failure investigation, the strength of pyrolysis-GC is its ability to analyse non-volatile organic samples, such as resins and polymers.

Upon pyrolysis, some polymers, such as acrylics, generate significant quantities of the original monomers which were used in producing the polymer. These can often be identified by their retention time alone. For other polymers, the 'pryrograms' are more complex but can still serve as 'fingerprints' and possibly be matched up to controls. Some examples of polymer analysis by using pyrolysis-GC have been given by Levy and Wampler [10] and Washall and Wampler [11]. When the fragments generated are isolated by pyrolysis-GC and analysed by an MS detector, the technique is especially powerful for identifying unknowns and providing compositional information on the polymers.

8.3.4 Application of GC

GC has many uses in the area of routine coatings analysis. Table 8.2 presents a compilation of 32 different ASTM methods that use GC in the analysis of coatings and coating-related materials. In the somewhat more specialized field of failure analysis, many of these standard tests are still important if one suspects the cause of the failure is a defective batch of paint, and if the deficiency is such that it can be detected by GC.

The primary applications of GC in failure analysis include the investigation of manufacturing (batching) errors, detecting unauthorized thinners, detecting solvents from a topcoat that may have leached into an older coat and analysing blister liquids for hydrophilic solvents.

There are instances where one may suspect that a coating failure is due to a bad batch of coating. One such scenario would be where a contractor has been painting a very large structure and has used more than one batch of paint. If the failure corresponds to one particular batch of paint, one would be inclined to believe that the problem is more likely with that particular batch, as opposed to application error or environmental influences.

A relatively easy thing to do with GC is to characterize the solvent blend present in a liquid coating sample, at least when one knows which solvents are supposed to be present. If this information is not available, it is still possible to quickly generate important data so long as liquid samples of a 'good' batch and a 'suspect' batch are available. In the latter case, it is simply a matter of looking for differences. If no differences are found, the investigation can be focused

Table 8.2 A selection of American Society for Testing and Materials (ASTM) test methods that are used in the analysis of paints, related coatings and aromatics by using gas chromatography. From J. V. Koleske (Ed.), *The New GARDNER–SWARD Handbook (Paint and Coatings Testing Manual)*, 14th Edn, 1995. Copyright ASTM. Reprinted with permission of the ASTM.

Number	Test method
D 1983	Test Method for Fatty Acid Composition by Gas-Liquid Chromatography of Methyl Esters
D 2245	Method for Identification of Oils and Oil Acids in Solvent-Reducible Paints
D 2306	Methods for C_8 Aromatic Hydrocarbon Analysis by Gas Chromatography
D 2360	Test Method for Trace Impurities in Monocyclic Aromatic Hydrocarbons by Gas Chromatography
D 2455	Test Method for Identification of Carboxylic Acids in Alkyd Resins
D 2456	Test Method for Identification of Polyhydric Alcohols in Alkyd Resins
D 2743	Practices for Uniformity of Traffic Paint Vehicle Solids by Spectroscopy and Gas Chromatography
D 2800	Test Method for Preparation of Methyl Esters from Oils for Determination of Fatty Acid Composition by Gas Chromatography
D 2804	Test Method for Purity of Methyl Ethyl Ketone by Gas Chromatography
D 3008	Test Method for Resin Acids in Rosin by Gas Chromatography
D 3009	Test Method for Composition of Turpentine by Gas Chromatography
D 3054	Test Method for Purity and Benzene Content of Cyclohexane by Gas Chromatography
D 3257	Test Methods for Aromatics in Mineral Spirits by Gas Chromatography
D 3271	Methods for Direct Injection of Solvent-Reducible Paints into a Gas Chromatograph for Solvent Analysis
D 3329	Test Method for Purity of Methyl Isobutyl Ketone by Gas Chromatography
D 3362	Test Method for Purity of Acrylate Esters by Gas Chromatography
D 3432	Test Method for Uureacted Toluene Düsocyanates in Urethane Prepolymers and Coating Solutions by Gas Chromatography
D 3457	Test Method for Preparation of Methyl Esters from Fatty Acids for Determination of Fatty Acid Composition by Gas-Liquid Chromatography
D 3545	Test Method for Alcohol Content and Purity of Acetate Esters by Gas Chromatography
D 3626	Test Method for Tar Acid Composition by Gas-Liquid Chromatography
D 3760	Test Method for Analysis of Isopropyl Benzene (Cumene) by Gas Chromatography
D 3792	Test Method for Water Content of Water-Borne Paints by Direct Injection into a Gas Chromatography
D 3797	Test Method for Analysis of *o*-Xylene by Gas Chromatography
D 3798	Test Method for Analysis of *p*-Xylene by Gas Chromatography
D 3893	Test Method for Purity Methyl Amyl Ketone and Methyl Isoamyl Ketone by Gas Chromatography
D 3960	Practice for Determination Volatile Organic Content (VOC) of Paints and Related Coatings
D 4367	Test Method for Benzene in Hydrocarbon Solvents by Gas Chromatography
D 4457	Test Method for Determination of Dichloromethane and 1,1,1-Trichloroethane in Paints and Coatings by Direct Injection Into a Gas Chromatography
D 4492	Test Method for Analysis of Benzene by Gas Chromatography
D 4534	Test Method for Benzene Content of Cyclic Products by Gas Chromatography
D 4735	Test Method for Determination of Trace Thiopene in Refined Benzene by Gas Chromatography
D 5135	Test Method for Analysis of Styrene by Capiliary Gas Chromatography

elsewhere. If a difference in solvent composition is found, and if solvent-related differences can reasonably be presumed to be responsible for the failure, a more in-depth GC investigation can be conducted.

An example of where a solvent might be responsible for a coating failure would be the use of an alcohol in a two-component urethane. Such urethanes consist of a polyol component (often a hydroxyl-functional polyester or acrylic) and an isocyanate component. When the two are mixed together, the hydroxyl groups react with the isocyanate, thus forming the cross-linked urethane. However, if a hydroxyl-bearing solvent was used (i.e. an alcohol), the hydroxyl group of the solvent will react with the isocyanate, thus making it unavailable for reaction with the polyol. GC might be a useful technique to investigate this type of problem.

If the above scenario actually happened, and if one had liquid samples, it would be relatively easy to detect the alcohol and hence solve the problem. Indeed, most standard GC procedures, such as the ones listed in Table 8.2, are designed for liquid samples. Quite often, however, by the time a coating failure has been recognized, the liquid paint is long gone. Although most manufacturers store liquid retains for 1 or 2 years, these are usually only half-pint or pint samples, and the manufacturers may be reluctant to part with any of them. Sometimes, coating failures are not prominent until a year or more after application, and the retain may have been discarded in the meantime. Can anything, then, be done with only a sample of the dried, failing paint?

Residual amounts of solvents can remain in dried paint chips for long periods of time. One method for detecting them is known as 'headspace analysis'. Many analytical instrument manufacturers offer headspace attachments for their gas chromatographs. In the simple 'homemade' version of this technique, the paint chips are placed in a small vial, which is sealed with a septum cap. The vial is then immersed in a hot oil bath for a few minutes at 100–110 °C. A preheated gas-sampling syringe is then used to pierce the septum and withdraw a quantity of the air (or 'headspace') from the vial. Any solvents that remained in the paint chip will have at least partially evaporated in the oil bath, and can be detected by injecting the headspace sample into the chromatograph. This technique can be carried out with a simple, homemade manual system, or it can be automated. It is strictly qualitative, although one can certainly recognize the difference between major and minor solvents.

Yet another way to analyse paint chips in the laboratory for retained solvents is to extract them with a suitable solvent. In the author's laboratory, this is routinely done by placing a known weight of small chips in an approximately 2.0-ml glass vial and adding a small amount of acetone. Other solvents would probably work equally as well, so long as they do not interfere with the solvents of interest present in the sample. The chips are allowed to extract overnight, and the acetone extract is then analysed in a conventional manner. Figure 8.51 shows a chromatogram obtained from an acetone extraction of paint chips which were nearly 9 months old. The analysis was performed using a GC-MS equipped

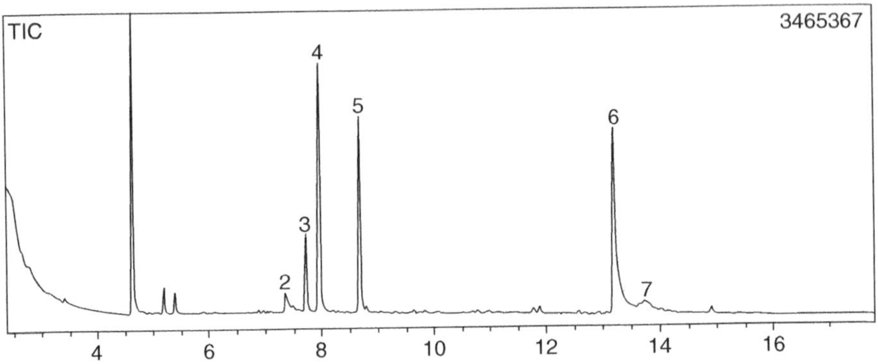

Figure 8.51 Chromatogram obtained from an acetone extract of nine-month-old paint chips, showing that several solvents were still present in the coating.

with a 30 m × 0.25 mm capillary column and shows that several retained solvents were easily detected. The mass spectral library search routine identified some of the solvents as MIBK, xylene and benzyl alcohol.

Would the GC headspace or solvent extraction procedures work in the above example of a urethane that had an alcohol solvent added to it, either at the plant or in the field? The answer is 'maybe'.

If enough alcohol was added such that some still remained after all of the isocyanate had reacted, it should be detectable. However, if a relatively small amount was added, it could still have interfered with the curing mechanism of the urethane, but in the process, it would have all been consumed. GC will not detect it, because there is none of it left to detect.

Sometimes, the use of unauthorized thinner can cause a coating to fail without actually chemically reacting with it. As discussed in previous chapters, the way that a coating flows or wets a substrate can be greatly influenced by its solvent blend. Whether or not voids form in a coating is greatly dependent on the rate of solvent evaporation relative to cure. Excessive voids can cause a coating to be much more permeable than it should be, thus reducing its protective capability. In these cases, GC is a reliable method of detecting unauthorized solvents in dry paint chips.

Another use of GC is in determining whether or not a coating failure is related to solvent from one coat of paint leaching into the coating beneath it. This is difficult to carry out for thin coatings, but if the coating in question is several mils thick, it may be possible to sample and analyse the individual layers. If solvents are detected, additional work will be needed to determine their source. Are they the normal solvents used in the coating, or do they correspond to the solvents used in the adjacent coat?

In Section 2.2.5, the process of osmosis and osmotic blistering was discussed. Briefly, if a water-soluble species is present behind or within a coating film, when the coating is immersed in a liquid (typically water) which has little or none of this species present, an osmotic cell can develop. The osmotic cell consists of a semi-permeable membrane (the paint), separating a solution of high concentration from a solution of low concentration. Nature will not allow this non-equilibrium condition to exist, and therefore more water will pass through the coating in an attempt to dilute the concentrated solution. The process results in liquid-filled blisters.

Sometimes, the water-soluble material responsible for the osmotic blistering is a solvent. Certain glycol ethers are known to cause this type of failure, since not only do they have a certain degree of water solubility, but they also have high boiling points and have difficulty completely evaporating from the coating. In order to investigate such a problem, one needs to collect a sample of blister liquid and analyse it by GC.

Analysing a sample of an aqueous blister liquid by GC can easily be carried out using an FID, since water is invisible to an FID and will not interfere with the detection of small amounts of solvent. Detection limits with an FID are also extremely good. A TCD with packed columns can also be used, but detection limits will not be as good and it may be necessary to judiciously choose two columns with which to analyse the sample. This is because the very strong water peak invariably detected can mask the presence of certain solvents. The chance of an individual solvent being masked by water on two different columns is small. One should also choose stationary phases that do not decompose when exposed to water. This author has found that a methyl silicone phase and porous polymer packing make a good pair when analysing blister liquids with a packed column instrument equipped with a TCD. Capillary column GC-MS also works very well with the appropriate choice of column and is much more sensitive than a TCD. Such samples are routinely tested in the author's laboratory using a 30 m, 0.25 mm 5% diphenyl/95% dimethyl polysiloxane column.

8.4 GEL PERMEATION CHROMATOGRAPHY (GPC)

GPC, also referred to as size exclusion chromatography (SEC), is a subset of a branch of chromatography collectively referred to as high-performance liquid chromatography (HPLC). In this technique, rather than a flowing stream of gas carrying volatilized samples through a heated column (as in GC), the dissolved sample is introduced into a stream of flowing solvent being forced through a column under high pressure by a pump. There are several forms of HPLC. The one that will be discussed in this present section is GPC.

The primary purpose of GPC is to provide molecular-weight information on resins and polymers and to isolate low-molecular-weight additives, such as platicizers or UV stabilizers, from the high-molecular-weight polymers. This is obviously important information in characterizing coatings and investigating failures.

8.4.1 Theory

In the section on GC, it was explained that the fundamental basis for GC was the partitioning of the sample, based on chemical interactions or preferential solubility between a mobile phase and a stationary phase. This is the basic concept underlying all chromatographic techniques *except* GPC. Ideally, the packing material inside a GPC column (the stationary phase) does not interact with the sample, except as a purely physical barrier. The sample does not have a certain solubility in it, or any tendency to adsorb to it. In essence, the stationary phase in a GPC column acts as a selective 'filter', thus separating large molecules from small ones.

A GPC column is usually packed with either a semi-rigid gel, or with small, rigid, inorganic particles. One of the most common packing materials for the analysis of solvent-based coatings is cross-linked polystyrene–divinylbenzene (PS–DVB) copolymer. Several manufacturers have their own trade names for this product, such as Styragel (Waters), Shodex (Showa Denko) and TSK Gel (Toya Soda). PS–DVB packings are popular because they can be obtained in a wide range of pore sizes, and most common polymer types do not interact with them.

The pores in the GPC stationary phase are the key to the technique, as shown in Figure 8.52. As this figure illustrates, the sample, composed in this case of both high- and low-molecular-weight material, is introduced at the top of the column. The sample is then carried through the column by the mobile phase, pumped at a fixed flow rate.

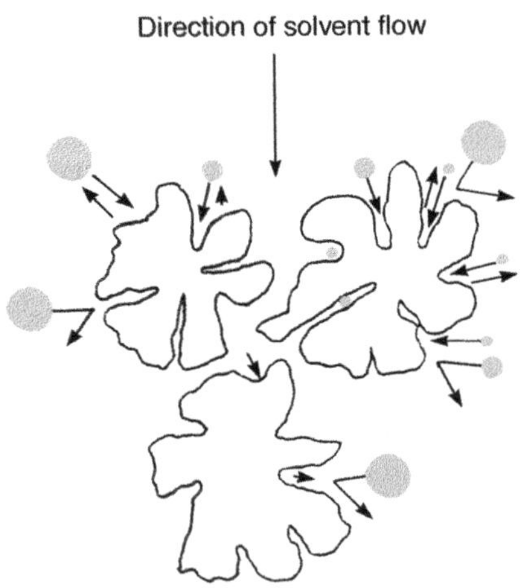

Figure 8.52 Schematic representation of the separation mechanism occurring in a gel permeation chromatography (GPC) column. Reproduced by permission of Varian Inc., from E. Johnson and B. Stevenson, *Basic Liquid Chromatography*, 1978.

When passing through the column, the sample encounters the stationary phase consisting of porous particles or semi-rigid gels containing pores of controlled diameter. The larger polymers are unable to penetrate these pores, and thus take the short way out, travelling around the gel particles in the interstitial space. In contrast, the smaller molecules can fit into some (or all) of the pores, and once they do, they have a long, winding, torturous route before they emerge from the other end of an individual gel particle. They have travelled farther than the large molecules and begin to lag behind them. Once this process has been repeated often enough in the column, the molecules of different molecular weight have been separated, or resolved, from one another, and the larger molecules will elute from the column faster than the smaller ones.

The PS–DVB materials used in GPC columns are formed by suspension polymerization in a mixed organic/aqueous phase system, somewhat similar to the emulsion polymerization process used to form acrylic resins. Micro-beads of polymerized styrene–divinylbenzene, less than a micrometre in diameter, begin to form and precipitate out of solution. The micro-beads then fuse together to form the porous, semi-rigid gel particles. A range of particle sizes can be produced, but those most valuable for GPC have particle diameters of between 3 and 20 μm. Smaller diameter particles will result in more efficient columns (more theoretical plates) and will also cause higher back pressure, which can seriously damage the packing.

The pore sizes of PS–DVB particles are difficult to measure directly, so the material is usually evaluated by testing it with polymers of known molecular weight. Pore sizes are commonly expressed in angstroms, which refers to the size of the calibration standard (usually polystyrene) which is just large enough to be excluded from the pores.

One big advantage of GPC over other forms of chromatography is that the components of the sample will elute within a relatively narrow range of time. This is because the only process that prevents a molecule from getting through the column is how many pores it fits into.

Consider a polymer with a very high molecular weight, which is too big to fit into any of the pores in the packing material. Ideally, there is no partitioning between the mobile phase and stationary phase and no tendency to be adsorbed onto the stationary phase. Thus, the only thing determining how long it takes for this molecule to get out of the column is the rate at which the mobile phase is being pumped, and the void volume, or interstitial volume, within the column. This volume is often referred to as V_0.

Consider again the case of a molecule that is small enough to fit into all of the pores. The largest volume in which this molecule could possibly travel would be the void volume plus the total volume of all of the pores in all of the particles (V_p). This is a finite volume, known as the total permeation volume, which is often referred to as V_t.

Barring any adsorptive effects, all components of a given sample must elute from the column at some point between V_0 and V_t. This is usually between

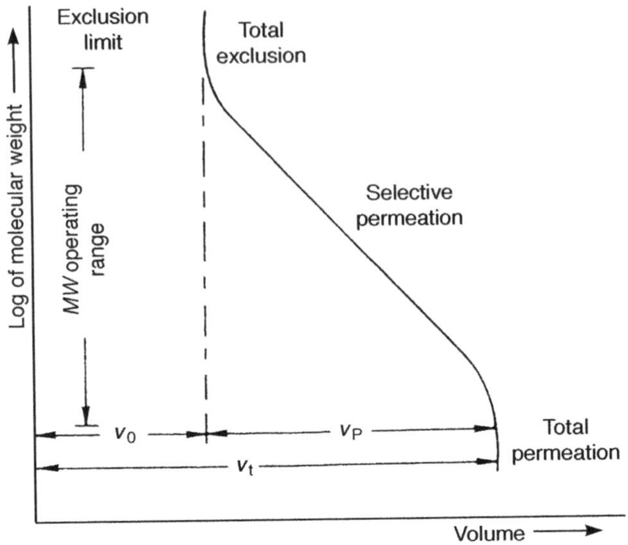

Figure 8.53 Relationship between the logarithm of the molecular weight and retention volume (or time) in a gel permeation chromatography (GPC) column. Reproduced by permission of Varian Inc., from E. Johnson and B. Stevenson, *Basic Liquid Chromatography*, 1978.

approximately 10–60 minutes, depending on whether one is working with a single column or two or more columns connected in series. .This is in contrast to other forms of chromatography, where an unknown component might be on the column for hours due to a strong affinity for the stationary phase.

The relationship between retention time and molecular weight in GPC is a logarithmic one. A typical plot of the logarithm of the molecular weight versus the retention time (or volume) is shown in Figure 8.53.

The molecular-weight region where there is a straight-line relationship between the logarithm of the molecular weight and the retention time is the useful range, or resolving range, of the column. Molecules that are above a certain maximum molecular weight (MW_{max}) all elute at V_0, and those that are below a certain minimum molecular weight (MW_{min}) all elute at V_t.

The above plot also illustrates that any one column's packing can only separate molecules spanning a certain range of molecular weight. Therefore, especially when one is working with an unknown sample, it is common to use two or three columns connected in series.

Although much of the above discussion has used the term 'molecular weight', it is apparent that a GPC column actually separates molecules based on their size in solution, which corresponds to their radius of gyration, or hydrodynamic volume. There is, of course, a relationship between the molecular weight of a molecule and its effective size, but this is different for different types of polymers. How then does one calibrate a GPC column such that useful information about a sample's molecular weight can be obtained?

Two of the most common means of calibrating a GPC column are the direct (or peak position) method and the universal calibration method.

In the direct calibration method, the column is calibrated by injecting known, narrow-molecular-weight distribution standards of the same polymeric type as the sample. By plotting the log *MW* of the standards versus their elution times or volumes, a straight-line calibration curve can be obtained. Although it might seem that the next step is simply to inject the sample and assign it the molecular weight corresponding to the retention time of the peak maximum, this is not quite so simple. Most real-world polymers have a broad range of molecular weights and, therefore, result in a broad peak that may or may not be Gaussian in shape. To carry out the calculation for an average molecular weight by hand is tedious. Luckily, this is not necessary in our modern computer age, and many software packages are available that make this task much easier.

One major catch in the direct calibration method is that the calibration standards must be of the same type of polymer as the sample. In some cases, this is possible, although it is rarely so in the coatings industry. The errors associated by calibrating with a different type of polymer than the one being analysed can easily be a factor of 2 or 3.

Despite this error factor, it is common practice to determine the molecular weight of a sample polymer, regardless of type, by performing a direct calibration. Polystyrene standards are often used because they are readily available over a wide range of molecular weights. The technique is fine so long as one realizes that only differences in molecular weight are being measured. This is usually sufficient in failure analysis work, where a failing or suspect sample is being compared with a non-failing or 'good' sample.

The second common way of calibrating a GPC column is known as the universal calibration method. This method is based on the fact that the intrinsic viscosity of a dilute solution of polymer is directly proportional to that polymer's volume in solution (its hydrodynamic volume). A plot of the logarithm of the product of intrinsic viscosity and molecular weight versus the retention time (or retention volume) should produce a straight-line calibration curve that is generally valid for all types of polymers.

In practice, a universal calibration curve is prepared much like a direct calibration curve. In fact, the first step is to determine the retention times of a series of standards of narrow, known molecular weight (polystyrene is typically used). The sample is then injected, and its 'apparent' molecular weight is calculated in the same fashion as if the direct calibration technique was being employed. The data are then mathematically manipulated by using what are termed the Mark–Houwink constants. The latter relate the molecular weight of the polymer to its intrinsic viscosity. A detailed discussion of this method has been described by Snyder and Kirkland [12].

Although the universal calibration method extends the accuracy of GPC data to a number of polymeric types, it still has limited use in the analysis of coatings, where the Mark–Houwink constants are not readily available for many of the specialized polymers and natural resins used in such coatings. The technique is also complicated if more than one type of resin is used in the formulation.

A third way of calibrating a GPC column is not really calibration in the usual sense of the word, since external standards and a calibration curve are not involved. In this method, a detector is used; the output of which is molecular-weight sensitive. Such detectors can provide a direct means of absolute molecular-weight determination without the need to resort to external standards. The most popular of such detectors rely on light scattering phenomena.

8.4.2 Instrumentation

As in GC, the high-pressure liquid chromatographs used in GPC consist of several precision components melded together, and a detailed discussion of each component is beyond the scope of this present text. Two components that will be briefly mentioned, however, are pumps and detectors.

The pump must be capable of delivering a precise flow of mobile phase, which is often tetrahydrofuran (THF), at rates ranging from perhaps 0.5 to 1.5 $ml\,min^{-1}$ and pressures up to 1000–2000 psi, in a relatively pulse-free fashion. The semi-rigid gels cannot withstand pressures much higher than this, although the rigid supports can. It is very important that the flow rate be reproducible or else errors in calibration will result. A reproducibility of less than 1% is required.

Most high-performance pumps are of the dual-piston reciprocating variety. These use small chambers with reciprocating pistons to force the solvent through the columns. The two pistons work 180° out of phase, in order to provide a more or less constant flow of solvent.

The two most commonly used detectors in GPC are UV and RI detectors.

UV detectors rely on measuring the absorption of UV light by the sample as it passes from the column through the detector. If there is a functional group in the sample that absorbs strongly in the UV region of the electromagnetic spectrum, such as conjugated double bonds or aromatic rings, these detectors are very sensitive. UV detectors are reliable, easy to operate and relatively insensitive to changes in flow rates and temperature. If the sample to be characterized has a UV-absorbing chromophore, a UV detector is probably the detector of choice.

On many occasions, the polymer being analysed does not have an appreciable absorbance in the UV region or the composition of the sample is simply unknown. In such cases, it is best to use a 'universal' detector, such as an RI detector. Such detectors function by monitoring the difference in RI between the pure mobile phase and the mobile phase coming off of the column. When a solute having an RI different from that of the pure mobile phase elutes from the column, the detector senses this change in RI, and a signal is generated, thus resulting in a peak.

RI detectors are much less sensitive than UV detectors to compounds having UV chromophores, but potentially more sensitive to other types of samples, and are therefore more versatile. They are reliable and relatively easy to operate but sensitive to modest variation in flow rates and temperature and usually require a long warm-up time.

8.4.3 Applications

The generic type or chemical composition of a polymer largely determines its chemical resistance properties. However, its molecular weight largely determines its physical properties, partly because longer polymer chains tend to entangle to a greater extent than short ones. Long, entangled polymer chains enhance the physical and performance properties of the coating, unless the molecular weight becomes so high that the coating becomes brittle. Highly cross-linked coatings are much more impermeable than low-molecular-weight ones, and therefore exhibit superior corrosion resistance. The ability to measure molecular weight is obviously important in characterizing coatings.

The primary applications of GPC in failure analysis include the examination of suspect liquid coating samples (or raw materials) to see if they conform to specification and the examination of failing paint chips for signs of chemical degradation.

8.4.3.1 Investigation of liquid samples for manufacturing or raw material errors

If there is reason to believe that a coating failure is due to a defective batch of paint, there may be several reasons why GPC would be an appropriate technique to employ. Although infrared spectroscopy would probably be the first step in comparing a suspect batch to a known ‘good’ batch, it has two major weaknesses. Firstly, it will not provide any molecular-weight information on the sample. Secondly, it will also have difficulties identifying any constituents of the coating that are present at concentrations below about 3–5%. GPC can help fill in these gaps.

As already pointed out, the molecular weight of a resin has a great deal to do with the performance of the coating. This is a particularly critical factor for thermoplastic coatings, where the molecular weight of the polymer does not increase upon curing. What you have in the can will be what you get in the dried film. There is no way to ‘make up’ for starting with a resin which is either too low or too high in molecular weight.

Obtaining a molecular-weight profile of a resin in a liquid sample is relatively straightforward, so long as one has a good or non-failing sample to compare it to. In such cases, a simple direct calibration using polystyrene standards is adequate to obtain relative differences in molecular weight between good and bad samples. One should not, however, analyse a suspect sample in this fashion and compare the results to those obtained from other sources, such as manufacturers' literature values or even other analytical laboratories. The discrepancies associated with other analytical methods, even if the experimental differences appear subtle, can be much too large to allow for confident conclusions.

Kuo and Provder [13] have provided several examples of how GPC can be used to investigate problems related to coating materials. One such example involved two batches of blocked isocyanate cross-linking agent for use in a powder coating. Blocked isocyanates are stable at room temperature but 'unblock' and react at the elevated temperatures used in baking these products, thus resulting in curing of the coating.

In this example, one batch of blocked isocyanate cross-linkers was acceptable, while another batch was much too reactive. When the two batches were analysed by GPC (Figure 8.54), it was found that the bad batch had a noticeable excess of a component that eluted near 40 ml. This component corresponded to a species having free (unblocked) isocyanate, which accounted for the greater reactivity of the batch.

Kuo and Provder [13] also analysed two batches of isocyanate cross-linkers, one of which was too tacky. As Figure 8.55 shows, the tacky sample had considerably more of a component eluting near 50 ml than did the control sample.

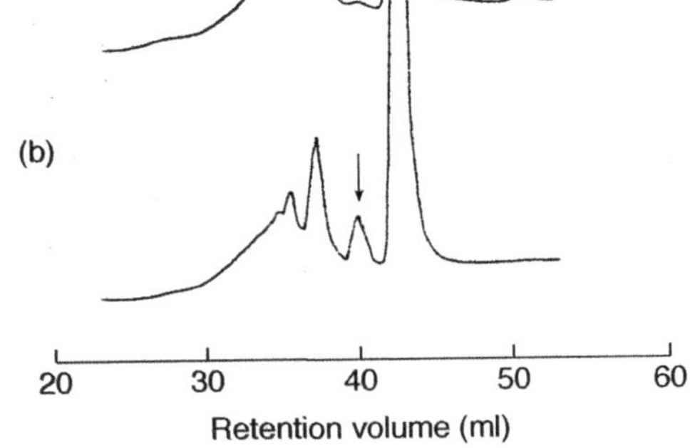

Figure 8.54 Gel permeation chromatography (GPC) comparison of (a) a 'good' batch of isocyanate to (b) one which was too reactive. Reproduced by permission of The Federation of Societies for Coatings Technology, from C. Kuo and T. Provder, *J. Coat. Technol.*, **53**, 69–76 (1961).

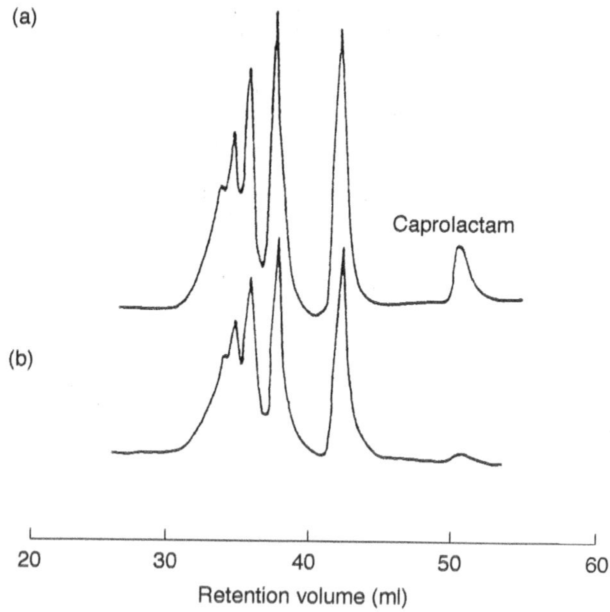

Figure 8.55 Gel permeation chromatography (GPC) comparison of (a) a batch of isocyanate which was too tacky to (b) a 'good' batch. Reproduced by permission of The Federation of Societies for Coatings Technology, from C. Kuo and T. Provder, *J. Coat. Technol.*, **53**, 69–76 (1961).

This component was shown to be caprolactam. In this case, the low-molecular-weight caprolactam acted as a plasticizer and depressed the glass transition temperature of the sample.

Raw materials that go into coating manufacture are sometimes considerably more complicated than even the supplier may realize. Materials that are ostensibly sold as either monomeric or as a 'pure' component dissolved in solvent often have several oligomeric components or by-products. It is possible that one or more of these 'minor' ingredients may play an important role in the chemistry of the coating. Scheuing [14] has demonstrated how GPC could be used to show that a 'pure' sample of an epoxy resin was actually comprised of several components (see Figure 8.56). When using GPC to examine liquid coatings or raw materials, there are a few simple things that must be kept in mind. The necessity of analysing suspect and control samples by the same technique has already been pointed out. It should also be pointed out that relatively dilute solutions prepared in a solvent which is compatible with both the sample and the column's stationary phase are required.

The concentration of a solution injected into a GPC column depends, to some extent, on the molecular weight of the sample. GPC columns have only limited

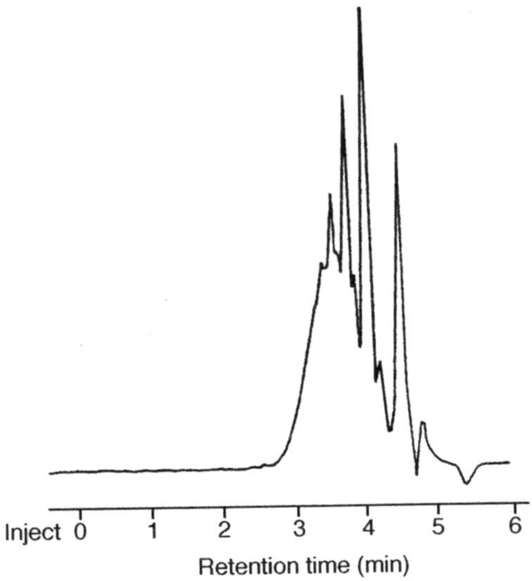

Figure 8.56 Gel permeation chromatogram of a 'good' sample of epoxy resin (Epon 100 1F), showing that it consists of a mixture of several components. Reproduced by permission of The Federation of Societies for Coatings Technology, from D. Scheuing, *J. Coat. Technol.*, **57**, 47–53 (1985).

sample capacity, and artefacts can occur if concentration or viscosities are too high. Concentrations of a few tenths of a percent are typical.

Pigmented samples should never be injected into a GPC system. Pigment can usually be removed either by centrifuging the diluted sample or by filtering it through a membrane filter (0.45-μm filters usually work fine), or both. Centrifuging should be carried out at low speed if one is working with very high-molecular-weight material, in order to avoid physically breaking any long, fragile chains. However, 0.45-μm filters can sometimes clog. In this case, a pre-separation step, using low-speed centrifugation, or a larger filter size, can be performed first.

If the sample in question is water based, one effective technique is to cast it on to a slide or panel and allow the water to evaporate at room temperature (or at a slightly elevated temperature if there is no danger of evaporating a component of interest) overnight. The sample can then be redissolved in a suitable organic solvent (such as THF), filtered and analysed.

In addition to determining the molecular-weight distribution of a suspect sample, it is also possible to analyse samples for low-molecular-weight additives. These could be such things as UV stabilizers, plasticizers, antioxidants and so on. Except for plasticizers, these additives are typically present at levels too low for identification by infrared spectroscopy. A prudent choice of GPC column will usually resolve these components and allow quantitative determinations to be made, at least on a relative level.

8.4.3.2 Investigation of failing samples of cured coatings

Often, the failure investigator does not have the luxury of liquid samples of the batch in question, let alone samples of the individual raw materials that went into the batch. There are many cases where the suspected failure mechanism has nothing to do with a batching (manufacturing) problem, but rather with chemical deterioration as a result of environmental exposure. It is possible that the particular batch of coating was produced within the requirements of the manufacturer's specifications, but failed simply because it could not withstand the environment.

Several years ago, this author worked on a project involving the wholesale delamination of sheets of acrylic coatings from the metal roofs of dozens of buildings. GPC performed on failing and non-failing samples showed a dramatic difference in molecular weight, with the failing samples being much lower than the non-failing ones. Furthermore, the problem was only occurring with a single colour. When new, control samples of two different colours were artificially weathered in a fluorescent UV chamber, there was a marked decline in the molecular weight of the failing colour, compared with the other colour. Apparently, the pigment in the non-failing colour had some UV-blocking capability and was partially protecting the light-sensitive resin from degrading.

As with other analytical methods, GPC should always be performed relative to a control sample. The control could be a batch retain, fresh off the shelf material or a non-failing sample. Extraneous debris, such as dirt, grime and so on, should be removed prior to analysis, if possible. Care must also be taken to only sample the coat of interest. Simply tossing a multi-coat paint chip in a test tube with THF may result in highly irreproducible results. Sometimes, it is not possible to separate a topcoat from a thin primer, but if the primer thickness is consistent between good and bad samples, and if it is thin relative to the coat of interest, the error should be negligible.

GPC is a very useful tool for analysing coating failures. However, it has two major limitations: it cannot be used on insoluble samples and it provides no chemical identification.

The first limitation is the biggest. In order to perform a GPC experiment, the sample must first be dissolved. This is not usually a problem for samples in the liquid state, but it is a serious limitation for cured paint chips. Cross-linked epoxies and urethanes cannot be characterized by GPC, unless one is only interested in additives or unreacted starting materials that can be leached from the cured films. Otherwise, it is restricted to thermoplastic coatings such as acrylics, vinyls, coal tars, certain alkyds and so on.

The second limitation, lack of chemical identification, can usually be overcome by means of a second technique, such as infrared spectroscopy or MS. The peaks are collected as they elute from the detector and analysed by the second technique. It is also possible to interface GPC directly to a mass specific detector, such that GPC–MS can be performed in one step. However, for certain QC functions a known component's retention time may be a reliable enough indication of identity.

8.5 ION CHROMATOGRAPHY

Ion chromatography is an HPLC technique for determining the salt content of aqueous samples. This is important in failure analysis because such salts are sometimes present as contaminants that can shorten the lifetime of a protective coating.

8.5.1 Theory

As with the other forms of chromatography (with the exception of GPC), ion chromatography involves the separation of individual components from a complex mixture based on differential partitioning of the component between a mobile phase and a stationary phase. Like other forms of HPLC, it uses a pump to force the mobile phase and sample through a column, and a detector then identifies the separated species.

The packings used in ion chromatography columns are basically ion-exchange resins and possess charged (ionic) functional groups. The primary retention mechanism is ion exchange between the sample and mobile phase ions and the oppositely charged sites on the packing. The charged sites on columns used to separate anions are positively charged, while the sites on columns used to separate cations are negatively charged.

The sample and mobile phase ions are in competition with one another for the ionic sites on the ion-exchange resin. The situation is usually somewhat more complicated, however, because there is usually more than one ionic species in the sample. For example, if the sample contained chloride, sulfate and nitrates, all of these species are in competition with one another and with the mobile phase anions. Sample ions that interact weakly with the ion-exchange resin, relative to the mobile phase ions, will be weakly retained and have short retention times. Those that interact strongly will have longer retention times.

Ion chromatography as applied to coatings failure analysis is usually used to analyse aqueous samples for inorganic ions such as chloride, sulfate and phosphate, as shown in Figure 8.57.

The technique is not limited to simple inorganic anions, however. As Figures 8.58 and 8.59 show, cations and even charged organic species can be detected by the proper choice of columns and mobile phases.

Commonly, the ion-exchange resins for anion analysis have quarternary amine groups on their backbone, while those used for cation exchange have sulfonate groups. Other stationary phases consist of silica which has ionic functional groups grafted on through short organic connecting groups. Although the primary mech-

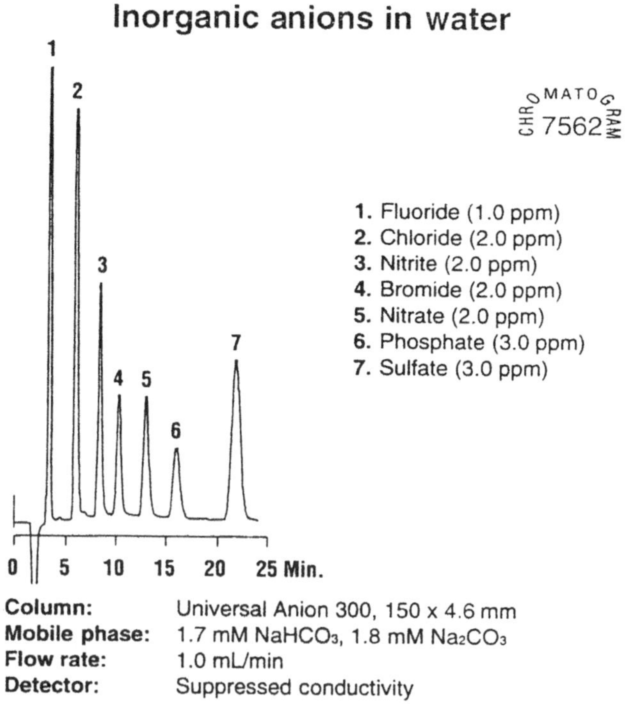

Figure 8.57 Ion chromatogram of some common inorganic anions in water. Reproduced by permission of Alltech Associates, Inc.

anism of separation is competition over the ionic sites on the stationary phase, a secondary mechanism is a classical partitioning between the aqueous mobile phase and the environment within the organic portions of the stationary phase.

For the most part, column packings consist of ion-exchange resins (porous polymeric particles), bonded-phase pellicular particles or bonded-phase porous particles.

The porous polymeric packings consist of porous spheres, typically 5–10 μm in diameter, made from PS–DVB that has been reacted to form pendant ionic functional groups. Sulfonate groups are usually used for cation columns and trialkylammonium groups for anion columns, as shown in Figure 8.60.

The bonded phase particles begin with either silica (porous particles) or glass beads (pellicular), which are coated with a polymeric 'sheath' similar in composition to those shown in Figure 8.60. The primary separation mechanism is therefore essentially the same as that of the conventional ion-exchange resins, but in this case, the slow diffusion in and out of the polymeric phase is replaced by the fast diffusion in and out of this thin organic covering. This achieves a substantial improvement in column efficiency.

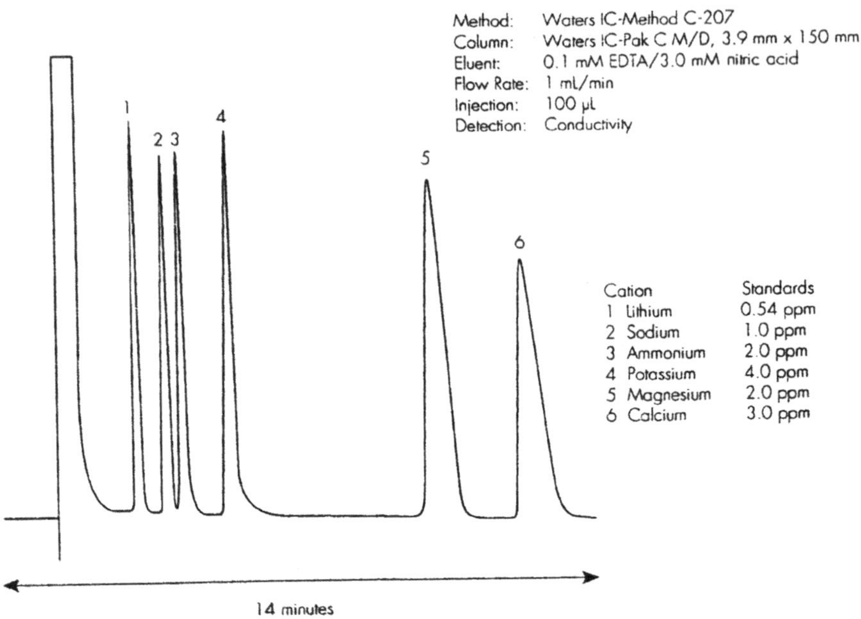

Figure 8.58 Ion chromatogram of some common inorganic cations in water. Reproduced with permission of Waters Corporation.

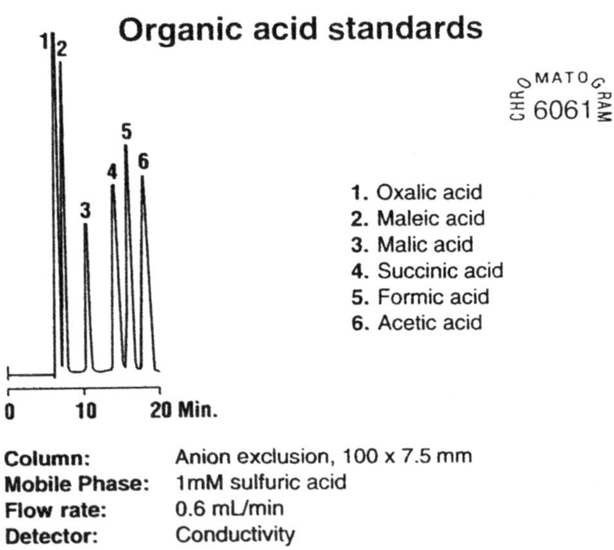

Figure 8.59 Ion chromatogram of some common organic acid standards. Reproduced by permission of Alltech Associates, Inc.

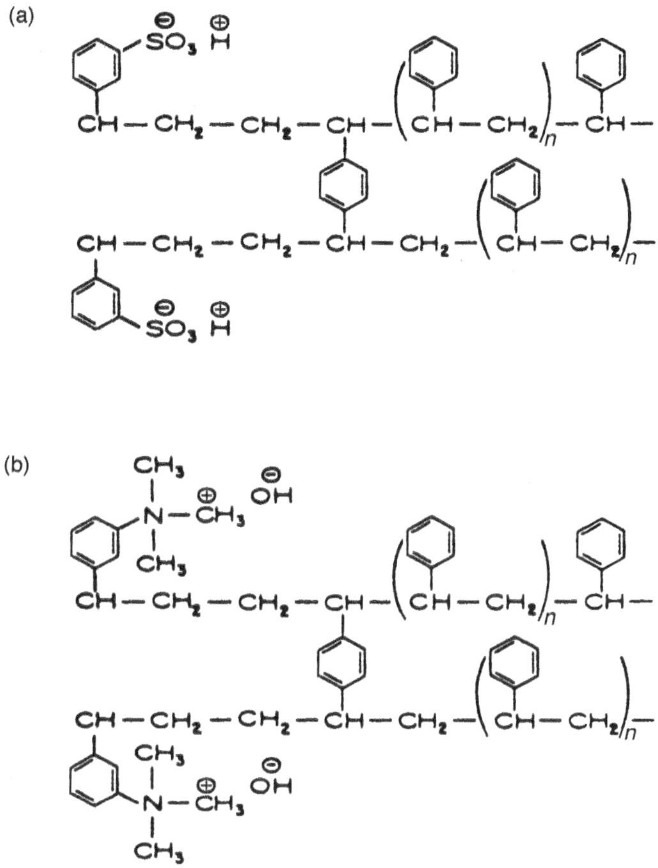

Figure 8.60 Structures of two common stationary phases used in ion chromatography containing (a) sulfonate groups (for cation columns) and (b) trialkylammonium groups (for anion columns). From J. R. Snyder and J. J. Kirkland, *An Introduction to Modern Liquid Chromatography*, 2nd Edn, Copyright © 1979. Reprinted by permission of John Wiley & Sons, Inc.

By far, the most common type of detector used in ion chromatography is a conductivity detector. Such detectors function by sensing the increase in electrical conductivity of the mobile phase when the sample (which is ionic) passes through the detector cell. The greatest sensitivity will be attained when there is a large difference between the conductivities of the pure mobile phase and the ionic species being detected.

Many mobile phases, however, are very conductive, and the difference in conductivity when some of the mobile phase ions are replaced by sample ions is minimal. This results in poor detection limits.

There are two ways around this problem: either use a mobile phase that is not so strongly conducting or alter the nature of the mobile phase after it leaves the column but before it reaches the detector. The first option is the basis for single-column (also called non-suppressed) ion chromatography, while the second is the basis for suppression-based ion chromatography.

Mobile phases for single-column ion chromatography are based on weakly conducting organic acids or bases, such as borate–gluconate, phthalic acid or *p*-hydroxybenzoic acid. Such species can detect anions at the part per million level, while various other mobile phases such as methanesulfonic acid or oxalic acid can be used to detect cations at sub-part per million levels. Figure 8.61 shows an anion analysis performed with a phthalic acid/lithium hydroxide buffer.

In suppression-based ion chromatography, a second column or cartridge is located after the analytical column. This suppression column lowers the conductivity of the mobile phase by substituting weakly conducting ions for the original, strongly conducting ions of the mobile phase. By suppressing the conductivity of the mobile phase prior to the detector, a substantial improvement in detection limits can be achieved. Figures 8.61 and 8.62 are essentially the same analysis, using the same column, but carried out by employing single-column and suppression-based technologies. Note the greatly improved detection limits of the suppression-based system.

Single-column ion chromatography is generally simpler to perform and requires less expensive equipment than suppression-based ion chromatography. The latter

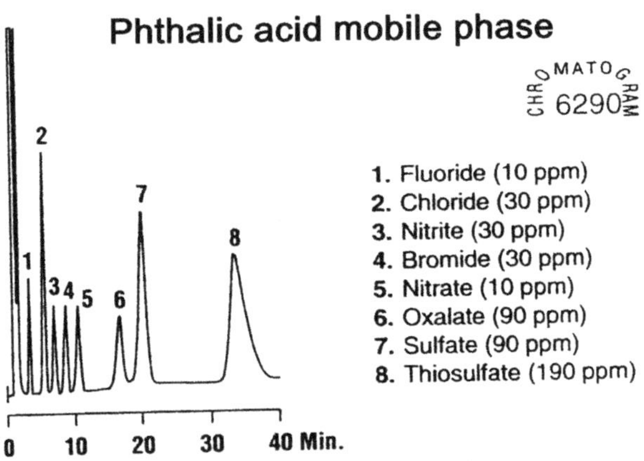

Column:	Alltech universal anion, 150 x 4.6 mm
Mobile phase:	4 mM phthalic acid, pH 4.5 w/LiOH
Flow rate:	1.0 mL/min
Detector:	Conductivity

Figure 8.61 Single-column (non-suppressed) ion chromatogram of some common inorganic anions. Reproduced by permission of Alltech Associates, Inc.

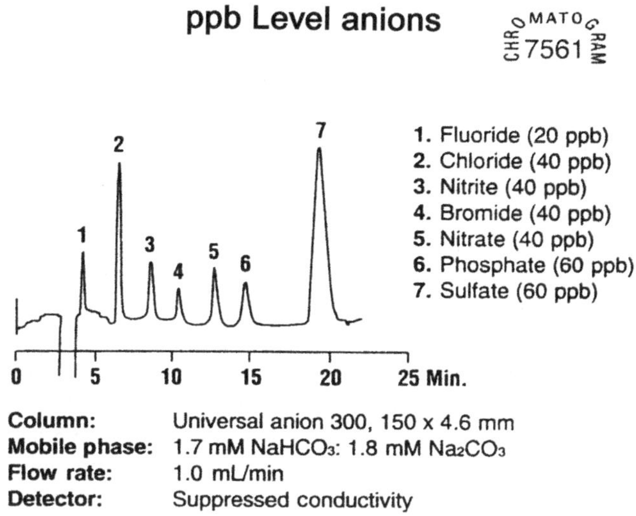

Figure 8.62 Suppression-based ion chromatogram of some common inorganic anions. Reproduced by permission of Alltech Associates, Inc.

technique has much better detection limits, at least for anions. However, the detection limits achievable with single-column ion chromatographs are usually adequate for most failure analysis situations.

8.5.2 Applications

The primary application of ion chromatography in failure analysis is the detection of salts, present either as impurities or contaminants. Salts can result in premature corrosion or blister-type failures. They can be introduced by contaminated pigments or blast-cleaning abrasives or by the environment (road salts, acid rain, salt-laden sea air, etc.).

Salt water is much more corrosive to steel than pure water. One of the requirements that must be met for corrosion to occur is a source of an electrolyte. Without dissolved, charged species, the electrical current required for corrosion cannot exist. Inorganic salts are excellent electrolytes.

In addition to premature corrosion, salts can also cause coatings in immersion service to fail dramatically by aiding in the development of liquid-filled blisters. This process is due to the formation of an osmotic cell, which was previously discussed in Section 2.2.5. When a coating blisters in immersion service, it is extremely important that a sample of the liquid inside the blister be obtained and analysed for soluble species. Water-soluble solvents can be analysed by GC, while ion chromatography can be used to detect salts.

Osmotic blisters can be large or small. If they are small, numerous blisters may have to be 'milked' to even get 0.1 ml of sample. Although this is more than enough for analysis by ion chromatography, it may be desirable to dilute the sample with deionized water to obtain perhaps 0.5 ml. Detection limits will not be a problem with a modest dilution, and sample handling will be simplified.

In investigating a suspected osmotic blister for salt content, the question will inevitably arise as to how much salt is enough to cause blistering. Although a modest amount of research has been published in this area, the question still does not have a simple answer. The detection of any amount of salt in the blister liquid should not be taken as proof that the salt was the cause of the blistering. Even ordinary tap water could have as much as 30 or 40 ppm chloride in it. Levels of 100 ppm or higher should be cause for alarm, but a few parts per million should not be unexpected. It should also be kept in mind that if the blister is relatively large, the measured concentration of salt at the time the analysis was performed is much lower than the concentration of salt when the first tiny amount of water was transported through the coating and dissolved it. Some data do exist on the amount of salt needed to induce osmotic blistering [15, 16], although the levels are generally expressed as microgram per square centimetre, and not as part per million (in the blister liquid). If one knew the area that a particular blister covered, and the volume of the liquid in the blister, one could relate the salt concentration in the blister liquid in part per million to the amount of chloride on the substrate, in microgram per square centimetre. However, neither the area nor the volume can be easily measured. There is much disagreement in the literature on the threshold levels of salt contamination.

Trying to relate salt concentration in the blister liquid to the weight of salt per unit area of substrate also assumes that all of this salt was simply laying on the substrate, waiting to be dissolved by the pure water passing through the coating. While in some cases this may be true, it is also possible that there are soluble salts associated with pigment impurities that are being dissolved by the water as it passes through the coating.

Some workers have cited abrasives contaminated by soluble salts as a potential source of contamination [17, 18], but little experimental data have been produced to verify this. The transfer efficiency for a species associated with the abrasive would be expected to be low, and therefore, abrasives with only very high concentrations of salt would be expected to be a problem. If one has a sample of the suspect abrasive, any soluble salts present can be easily detected by extracting a portion of the abrasive in hot, deionized water and analysing the extract by ion chromatography.

8.6 SCANNING ELECTRON MICROSCOPY

As previously pointed out, a good quality stereo microscope with magnification up to about 50× is indispensable in failure analysis and should almost invariably be the first tool employed in such a study. Sometimes, it is advantageous to examine samples at much greater magnification or to determine the elemental composition of interfaces or inclusions. A scanning electron microscope (SEM) equipped with an X-ray spectrometer can fulfil this need.

A basic understanding of how an electron microscope works is important in terms of appreciating the advantages and limitations of the technique. This section will discuss both imaging theory and elemental analysis by X-ray spectroscopy.

8.6.1 Imaging Theory

The most obvious advantage of an electron microscope compared with a light microscope is its much greater magnification. What makes this magnification possible is the huge improvement in resolution between the two systems. The resolution in a microscope is directly proportional to the wavelength of the light being employed.

In a conventional light microscope, the best resolution is about 0.2 μm [19], which is largely dictated by the wavelength of visible light.

In electron microscopy, electrons are used to form images, not visible light. Since electrons have much smaller wavelengths than visible light, the resolution is usually less than 10 nm [19], easily 20 times better than the best light microscope. Furthermore, electron microscopes have much better depth of field than do light microscopes.

An SEM uses a focused beam of high-energy electrons to scan in a raster-like manner across the surface of the sample. This beam interacts with the sample surface to produce a large number of secondary electrons, which are collected by a detector and eventually converted into a magnified image on a cathode ray tube. The features of the specimen are discernable due to the fact that electrons that were emitted from a surface facing away from the detector are partly absorbed by the specimen itself. This results in shadows, or a gradation in the grey scale, which the human brain processes as 'features'.

The production of the focused beam of electrons is carried out in the electron gun. This basically consists of a cathode constructed of a tungsten filament housed in a specially constructed cylinder, an anode plate and various electromagnetic lenses. The filament is negatively charged by connecting it to the negative pole of a high-voltage power supply, and the anode plate is grounded. This results in a large potential difference between the filament and the anode, called the 'accelerating voltage'. This voltage is usually between 5000 and 30 000 eV. When the filament is heated, a flow of electrons is produced that travel towards the anode plate under the effect of the potential difference.

In a conventional light microscope, glass lenses focus visible light photons. Electrons, however, being charged particles, can be deflected and therefore focused by magnetic fields. The 'lenses' used in electron microscopes are actually electromagnetic lenses. At least two such lenses are employed in the column of the electron gun in order to focus the beam of electrons at a precise spot on the sample. After the last lens, and just before the specimen, will be an aperture to control the spot size, and hence the resolution of the beam. The aperture also helps to reduce spherical aberrations in the lens.

One of the advantages of an SEM when compared with a conventional light microscope is its great depth of field. The depth of field in an SEM may be 100–500 times greater than in a light microscope [19]. This is due in large measure to the extremely small aperture diameters employed in an SEM, which are typically 100–400 μm.

When the focused beam of electrons impinge on the surface of the specimen, numerous possible interactions can take place, thus producing a discrete volume of excitation within the specimen. For imaging, the two types of interaction of most practical significance are the production of either elastically or inelastically scattered electrons. An elastically scattered electron is one that retains all (or nearly all) of its original energy. An inelastically scattered electron is one that has lost much of its original energy due to interactions within the specimen. In scanning electron microscopy, elastically scattered electrons are referred to as 'back-scattered electrons', while inelastically scattered electrons are referred to as 'secondary electrons'. Since back-scattered electrons have more energy than secondary electrons, they have a zone of penetration into the sample that is much deeper than secondary electrons and, hence, are not quite as sensitive to surface features. For both types of electrons, at any given acceleration voltage, the depth of penetration is also affected by the atomic number of the element(s) making up the specimen. Higher-atomic-number elements will provide more 'particles' with which to interact with the electron beams, and the depth of penetration will therefore be reduced.

The nature of the interaction responsible for the production of elastically back-scattered electrons is a simple rebound effect of electrons bouncing off a nucleus or outer-shell electron of one of the sample atoms. An inelastically scattered secondary electron is the result of the primary electron colliding with a specimen electron and transferring some of its energy to that electron, thus resulting in the ejection of the specimen electron and concomittant ionization of the atom. The electrons emitted as a consequence of this ionization procedure have much lower energy than the primary, 'incoming' electrons, and are termed secondary electrons.

Secondary and back-scattered electrons convey different types of information about the sample. It is at least possible to partially discriminate between them in the process of image formation by taking advantage of the differences in energy between the two types of electrons.

In order to actually form an image of the sample, the electrons produced by the interaction of the primary electron beam with the sample must be detected. This is achieved by using a collector that draws the emitted electrons to a scintillator. The latter is coated with a phosphor that generates a short pulse of light when an electron hits it. These photons of light are directed towards a photomultiplier tube, which produces a signal which is then amplified, and which eventually results in an image of the sample on a cathode ray tube.

The collector, which is located above and to one side of the sample, has a positive bias on it in order to attract the electrons being generated at (secondary electrons) or reflected from (back-scattered electrons) the sample surface. Both the secondary and back-scattered electrons are emitted from the sample in all directions, and therefore, some of both will naturally head towards the collector and contribute towards the formation of the image. Since the secondary electrons are of low energy, they will more readily be deflected and collected by the collector than will the high-energy back-scattered electrons. The 'secondary electron image' is formed by a combination of nearly all of the secondary electrons and those few back-scattered electrons that happen to come off at an angle such that they head towards the collector. The secondary electron image is rich in topography and more closely resembles what the human brain has been trained for a lifetime to regard as a 'conventional' image.

If the bias to the collector is switched off, far fewer secondary electrons will be collected, and the back-scattered electrons will be heavily influenced. This 'back-scattered electron image' contains different information than the secondary electron image, because the number of back-scattered electrons produced at the sample surface depends strongly on the atomic number of the elements constituting the sample (this is true of secondary electrons as well, but to a much lesser extent). Higher-atomic-number elements have bigger nuclei and a greater density of electrons around their nuclei. The chances of an elastic collision with a primary electron – or a nucleus – greatly increases. Elements of high atomic number appear white in the back-scattered electron image, while elements of lower atomic number appear grey or even black. Although the image contains less topographical information, it does contain more chemical information.

8.6.2 Elemental Analysis by X-ray Spectroscopy

Being able to examine samples at very high magnification is a great advantage of electron microscopy. However, the power of the technique is greatly enhanced by coupling the visual aspect with elemental analysis, taking advantage of the X-rays produced at the sample surface.

Spectroscopy was previously defined as the study of the interaction of light with matter. Since there are different types of 'light', it follows that there must be different forms of spectroscopy.

Figure 8.13 gave an illustration of the electromagnetic spectrum. From this figure, it can be seen that what we call 'light', or 'visible light', comprises just a small slice of the entire electromagnetic spectrum and is energetically approximately in the middle of the spectral range. X-rays are simply another form of electromagnetic radiation but of much shorter wavelength, and hence much higher energy, than visible light.

In a simplified representation of an atom, there is a positively charged nucleus surrounded by a number of electrons. In a neutral atom, the number of electrons is equal to the number of protons in the nucleus. By convention, the lowest energy shell (the one closest to the nucleus) is called the 'K' shell and can hold up to two electrons. The next shell is called the 'L' shell and can hold up to eight electrons, while the next shell after that is the 'M' shell, which can hold up to 18 electrons. Higher-atomic-number elements (those with atomic numbers 29 or higher) contain even more shells.

X-rays are produced within an electron microscope as a consequence of the collision of the primary electrons produced by the electron gun with the electron cloud surrounding the nuclei of the sample atoms. When a primary electron collides with one of the electrons in an inner-shell orbital of the sample and dislodges it, a vacancy has been formed, thus resulting in an unstable electronic configuration. An electron from one of the outer shells will therefore drop down to fill the vacancy. As a consequence, energy is given off in an amount exactly equal to the difference in energy between these two orbitals. The energy released as a consequence is in the form of X-rays. The energy of the emitted X-ray, then, is characteristic of the element from which it was produced. If these X-rays can be collected and characterized as to either their energy or wavelength, one can identify the chemical elements present in the sample. This is the basis for X-ray spectroscopy.

In this form of spectroscopy, a nomenclature has been developed to describe the source of the X-rays. In this approach, an X-ray is named firstly by the shell where the vacancy occurred and secondly by the number of shells that the 'relaxing' electron fell in order to fill the vacancy. A drop of one shell is referred to as α (alpha), a drop of two shells is referred to as β (beta) and three shells is referred to as γ (gamma). For example, if an electron is ejected from the K shell, and this vacancy is filled by an electron dropping from the M shell, the resulting X-ray would be termed a K β X-ray.

There is a third variable in the production of X-rays, because not all of the electrons in a given shell have exactly the same energy. These energy differences are very small, but as a consequence, there are two alpha energies, four beta energies and several higher-level energies.

It is important to realize that, unless they possess a certain threshold energy, the electrons produced in the SEM cannot remove inner-shell electrons from the sample. This energy is equal to the so-called 'binding energy' holding the particular electron to the element in question. If the energy of the electron beam is lower than the binding energy, no X-rays will be produced except background X-rays, which result from a drop in velocity of the primary electrons as a result of various collisions.

The X-rays produced at the sample surface can be collected and characterized either by their wavelength, using a wavelength-dispersive spectrometer (WDS), or by their energy, using an energy-dispersive spectrometer (EDS). Either method is capable of identifying the elements present in the specimen. The advantages of an EDS include lower cost, rapid simultaneous analysis and low sensitivity to geometric effects. Although a WDS has better resolution and sensitivity and is more quantitative, an EDS is more commonly used for routine applications.

In an EDS, the detection and analysis of X-rays are carried out by using a semiconductor detector. The crystal in a semiconductor converts the electromagnetic energy of the impinging X-rays into purely electronic energy. The magnitude of the current produced by the semiconductor is proportional to the energy of the X-rays striking it. The crystals are usually germanium or silicon, which have been doped with lithium. The crystals and some of the associated electronics must be maintained at liquid nitrogen temperature to be effective.

A critical part of the semiconductor detector is the so-called 'window' separating the crystal from the rest of the SEM. This window is necessary since the semiconductor crystal must be kept in a vacuum, even when the microscope chamber, which is also kept in a vacuum during operation, is vented to allow specimens to be inserted or removed. The vacuum is necessary to prevent contamination, which can seriously degrade the detector.

Usually, the window is a very thin layer of beryllium. A consequence of a beryllium window is that it can absorb some of the lower-energy X-rays produced by low-atomic-number elements. Elements lighter than sodium (atomic number of 11) cannot be detected if a beryllium window is used. Special 'windowless' detectors can be used if certain design criteria are met, which can then lower the detection limit to carbon (atomic number of 6).

In addition to the window problem, low-atomic-number elements are not very efficient producers of X-rays. This is primarily because they contain fewer electrons, and hence, the probability of collision is substantially reduced. Since the energies of X-rays produced by very low-atomic-number elements is low, the emitted X-rays stand a better chance of being absorbed by other atoms in the sample before they get to the detector.

The detection limits for an SEM-EDS depend on a variety of variables, including sample composition, scan time, specimen geometry and accelerating voltage. In practical, day-to-day operation, detection limits of about 0.1% are typical.

An SEM-EDS is a very good tool for carrying out qualitative analysis. However, its ability to do accurate quantitative analysis is limited because there are several factors, not all of which can be controlled, which affect the intensity of a peak. Some of these considerations could be reduced if accurate standards were available in the same matrix as the sample. This is difficult, if not impossible, for the type of unknown samples often encountered in failure analysis. It has been this author's experience that quantitative analytical results on unknowns in an organic matrix are at best accurate to no better than perhaps ±25%. This still means that an SEM-EDS is a valuable semi-quantitative tool, and trends in elemental composition can still be determined.

8.6.3 Sample Preparation

In the area of failure analysis of paints and coatings, sample preparation for an SEM-EDS is usually straightforward. The sample must be small enough to fit into the sample chamber, at least partially conductive to electricity and (except for some of the newer SEMs) extremely dry.

The sample compartments in SEMs vary, but there would usually be no problem in inserting a 2×2 in. sample. However, with a sample this big there may not be enough room to allow the entire surface area to be moved in and out of the focusing area. Some cutting of the sample may be necessary.

If the sample is not partially conductive, bombardment by the electron beam will result in a charge build-up on the sample surface, thus resulting in very poor images. In order to overcome this, non-conductive samples, such as paint chips, must first be coated with a thin layer of some conductive material. The two materials most often used are gold and carbon.

Gold is more conductive than carbon and will typically result in better images. It is more expensive and can degrade the detection limit of the X-ray spectrometer by absorbing some of the lower-energy secondary electrons, especially those produced by the lower-atomic-number elements. Because of these limitations, carbon coating is usually preferred.

The procedure of carbon coating is relatively simple. The specimen is placed in an evacuated chamber, which contains two very pure graphite rods which make point contact with one another. When a large electrical current is passed through the rods, a certain amount of carbon vapour is ablated, so coating the sample.

Traditionally, electron microscopy has been performed under very high vacuum conditions. Gases, particularly water vapour, can interfere with the passage of the electron beams. Air in the chamber would cause the temperature filament in the electron gun to oxidize and burn out prematurely. Although some newer models have options that allow operation at somewhat less stringent vacuum requirements, in most current systems, samples must be dry in order for the operating vacuum to be achieved. This is seldom a problem in failure analysis, although it does prevent studying the drying or coalescing properties of liquid coating samples.

8.6.4 Applications of SEM-EDS

There are numerous instances where the use of an SEM-EDS can provide important information in solving coating failures and numerous instances where it cannot. The ability to examine samples at very high magnification with great depth of field can be a big advantage in problem solving. One may find small voids in a lining where permeation of the tank contents has resulted in crystal growth. One may find that 'staining' or 'discolouration' of a latex coating is due to a thin layer of debris, or that it is due to microscopic differences in the extent of coalescence of the coating. The differences in particle shape and size can cause differences in the way the coating reflects and refracts light, thus causing perceived colour differences. One might detect microscopic cracking in a galvanized substrate, so leading to moisture and salt permeation and hence premature corrosion. Alternatively, the type and uniformity of crystal growth of conversion coatings may be examined.

An SEM-EDS is also very useful for studying aesthetic defects such as cratering. Often, craters are caused by microscopic particles of dramatically different surface tension than the bulk phase of the liquid coating. Sometimes, these particles are contaminants, such as dirt or silicone oils, while at other times, they may be seeds or gel particles in the coating itself. On many occasions, these particles can be seen at magnifications of 500 or 1000× in the SEM and are often characterized by the associated X-ray spectrometer.

An SEM also has the ability to literally 'see samples in a different light' when using the back-scatter mode. Sometimes, no features of interest can be noted when examining a sample with a visible light microscope or even in the secondary electron mode of an SEM. This same sample, however, may take on a totally different appearance when examined in the back-scatter electron mode. In this mode, elements of high atomic number appear white, while those of lower atomic number appear grey or black. In some cases, this allows ready discrimination of phase differences or contaminants that otherwise may not have been visible. It is easy to distinguish rust from iron in the back-scatter mode, since the oxygen content of rust results in an effectively lower 'atomic number' than that of pure iron.

Although an SEM equipped with a wavelength- or energy-dispersive X-ray spectrometer is a powerful tool, it is enhanced by the ability to visualize in both the secondary and back-scatter modes. In the study of the delamination of an epoxy coating from an inorganic zinc-rich primer, it would not be unusual to detect zinc on the back of the epoxy. However, the amount and appearance of this zinc layer is critical in determining the mechanism of failure.

Sometimes, a magnifying glass is all that is required to determine that there is a relatively thick, continuous layer of zinc coating on the back of the epoxy. In this case, one should suspect a curing problem with the zinc primer. However, if little or no zinc is detected at low power, one may still detect zinc by carrying out an X-ray analysis of the back of the failed coating in an SEM equipped with an EDS. If zinc is detected, should one still suspect that there was a curing problem with the zinc primer, or could other mechanisms be at work?

Important information can be gleaned from the appearance and distribution of the zinc. An examination of the back surface of the epoxy in the back-scatter mode may show that the zinc that was detected was present only in widely scattered, discrete islands. Since it would be surprising that absolutely no primer would be on the back of the epoxy, this might indicate that a different mechanism, perhaps a contaminant such as oil or condensation, is responsible for the adhesion failure.

What if the back-scatter examination shows that the zinc is distributed in a pinpoint fashion relatively uniformly over the back of the epoxy? This might indicate that the problem was dry-spray or over-spray of the primer, thus resulting in a very thin, weak interface.

Another possibility is that the back-scatter mode shows an extremely thin, yet nearly continuous, film of zinc on the back of the epoxy. This might be due to a curing problem of the zinc, or it might be due to extreme dry-spray or over-spray. However, there may be a third possibility.

If, under high magnification (1000× or higher), this thin film of zinc has the characteristic appearance of a zinc coating (discrete spheres of zinc pigment embedded in a polymeric matrix), the first two possibilities are likely. However, if the zinc has an amorphous appearance, and if there is other evidence that it is associated with oxygen (different shading in the back-scatter mode, or detection of oxygen via a light element detector in the EDS), it is possible that the failure was due to chemical attack of the zinc by elements in the environment. This would result in a weak, degraded surface layer.

The X-ray spectrometer attached to an SEM greatly extends its ability. There is probably no other analytical tool that in so short a time can provide so much visual and chemical information as an SEM-EDS. If excessive corrosion has formed prematurely behind a coating, such a system may detect chlorine or sulfur, thus suggesting that chloride or sulfate salts have contributed to the corrosion. If an epoxy has disbonded from galvanized steel, the detection of a thin layer of zinc on the back of the epoxy shows that it is not a coating failure, but a problem with the galvanizing.

Sometimes, it is important to know how many coats of paint are present in a sample, but where all of the coats are white, or subtle shades of white. Sometimes, an SEM-EDS is useless for this type of problem, while at other times there is a big enough difference that multiple coats are easily detected in the back-scatter mode, or the elemental analysis capability of the EDS may reveal multiple coats.

An SEM-EDS is very useful in detecting the presence of vinyl butyral wash primers. These primers are very thin (typically 0.1–0.2 mil) and, depending on the nature of the sample, not always easily visualized, even with a good stereo microscope. However, they usually contain a chromate pigment, which can be easily detected by an SEM-EDS, even on the cross section of a sample.

There are numerous advantages to the use of SEM-EDS. There are also several limitations. First of all, the technique is not particularly sensitive, at least when compared with other analytical techniques. Typically, elements should be present at perhaps 0.1% in order to be detected. Contaminants can be present at several hundred parts per million (1000 ppm = 0.1%) and yet still not be detected.

Secondly, unless a light element detector is used, the elements typically associated with organic compounds (C, O, N and H) will not be detected. Even a light element detector will not detect hydrogen, and the detection limit for the other light elements is poor. This, coupled with the fact that an EDS provides no information on how the elements are chemically combined, means that such a technique is incapable of detecting many organic compounds, let alone identifying them. The detection of silicon and oxygen may mean that silicone oil was painted over, or it may simply mean that the paint samples contain talc as an extender pigment.

The above example points out the importance of examining a control sample by the use of an SEM-EDS before jumping to conclusions from data obtained from a failing sample. An SEM-EDS may detect chlorine in a sample, which could be due to salt contamination. Alternatively, it could be due to a poly(vinyl chloride) resin. A variation in the relative amounts of elements associated with various pigments may be observed across the cross section of a particular sample. Perhaps, this means that there is some type of chemical degradation at work on the surface of the sample, or perhaps, this is just the natural layering or segregation of ingredients as a result of different densities. A control sample would answer this question.

Another potential pitfall of the energy-dispersive technique is that sometimes, it allows one to examine too small an area. Often, if one looks hard enough for something, one will find it. However, if the image on the SEM screen represents a fraction of a micron, is it really representative of the sample as a whole? This can usually be answered simply by examining more locations, but often this is not carried out.

Another shortcoming of an SEM-EDS is the fact that the elemental analysis capability is only semi-quantitative. One should be extremely careful about relying too heavily on the quantitative data generated by the instrument's software package. There are too many variables to allow this to be accurate, especially with unknowns, despite the fact that the instrument may report concentrations to four significant figures. Accepting the data as semi-quantitative and looking for trends usually result in more reliable conclusions.

8.7 DIFFERENTIAL SCANNING CALORIMETRY (DSC)

DSC is one of several forms of thermal analysis that involve measuring some property of a material as a function of temperature. DSC, in particular, is useful for analysing paint samples, since it can be employed on solid samples with little or no sample preparation, and only a few milligrams of sample are usually required. It can provide important information about a coating's degree of cure or cross-link density.

8.7.1 Theory

Before discussing the theory of DSC, it is illustrative to discuss the theory of a closely related technique, differential thermal analysis (DTA).

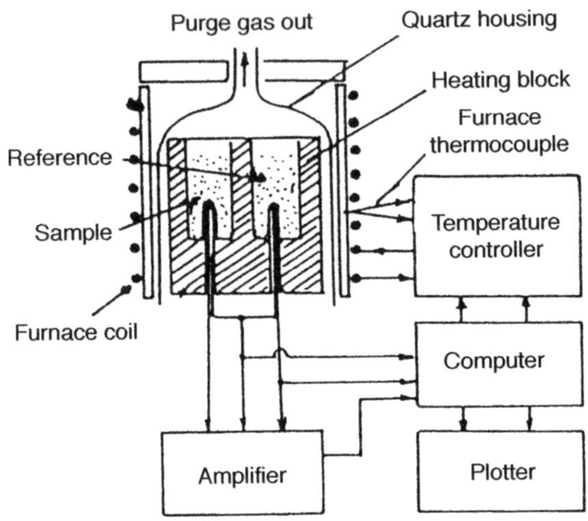

Figure 8.63 Schematic diagram of the layout of a typical differential thermal analysis (DTA) system. From T. Hatakeyama and F. X. Quinn, *Thermal Analysis: Fundamentals and Applications to Polymer Science Strategies*, Copyright © 1994. John Wiley & Sons, Ltd. Reproduced with permission.

DTA basically involves heating a sample, such as a paint chip contained in a small aluminium pan, at a constant, uniform rate, and then comparing the temperature of the sample to that of an inert reference material being heated under the same conditions. The layout of such an instrument is shown schematically in Figure 8.63.

As Figure 8.63 shows, a DTA instrument is essentially a furnace, capable of heating the sample and reference at a controlled rate over a specified temperature range. A variety of reference materials can be used, provided that they do not undergo any phase changes or any chemical reactions over the temperature range of interest. Thermocouples inserted in both the sample and reference holders allow the temperature of both to be monitored as the temperature of the furnace is increased (or, less commonly, decreased).

When no 'events' occur in the sample, the temperatures of both the sample and the reference increase at a uniform rate. However, if an event occurs in the sample that involves either the liberation or absorption of heat, there will be a temperature difference between the sample and the reference. This will be detected and amplified by the instrument, and either an exothermic or endothermic peak will be plotted on the chart paper, as shown in Figure 8.64.

Besides exothermic events, such as chemical reactions or crystallization processes, and endothermic events such as melting, a third possibility exists: there may be an abrupt change in the sample's heat capacity due to reordering on a molecular level. Heat capacity is the amount of heat (energy) required to raise

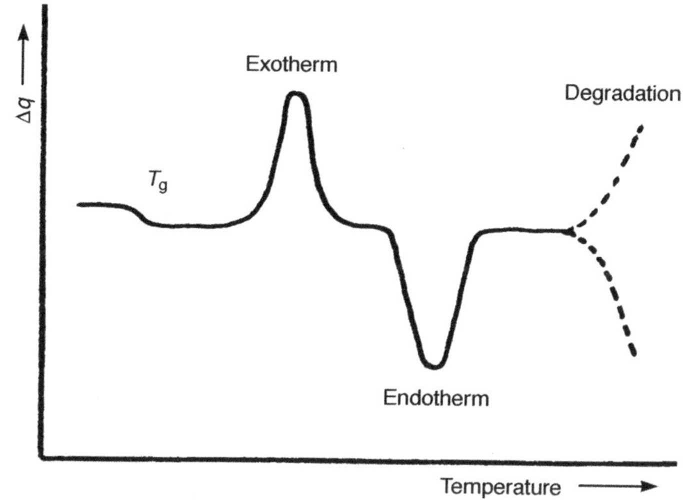

Figure 8.64 Examples of some of the various transitions observable by the use of differential scanning calorimetry. From J. V. Koleske (Ed.), *The New GARDNER–SWARD Handbook (Paint and Coatings Testing Manual)*, 14th Edn, 1995. Copyright ASTM. Reprinted with permission of The American Society for Testing and Materials. T_g = glass transition temperature.

the temperature of 1 g of sample by 1 °C. The change in heat capacity is neither exothermic nor endothermic, but does result in a sudden shift in the slope of the curve, as shown in Figure 8.64, and corresponds to the sample's glass transition temperature.

DSC is really a quantitative form of DTA. The temperature difference between the sample and reference is still detected in much the same way as in DTA, but in this case, the temperature difference is then quantitatively related to the flow of heat from or to the sample. For instance, when a sample melts in DTA, an endothermic peak will result, but the intensity (or area) of this peak can only be measured in some arbitrary unit. In DSC, the area under this same peak can actually be measured in 'real' units such as calorie per gram or joule per gram, thereby allowing one to determine the heat of melting (also known as the heat of fusion) of the sample.

Figure 8.65 shows a schematic diagram of a typical 'heat flux' type of DSC system. In this figure, the thermocouples are attached to the base of the sample and reference holders, with other thermocouples attached to the furnace, and to the conductive, heat-sensitive plate on which the sample and reference pans rest.

During a thermal event, heat flows from the heat-sensitive plate into the sample (an endothermic event) or from the sample into the plate (an exothermic event). In either case, the flow of heat (the 'heat flux') causes a small temperature

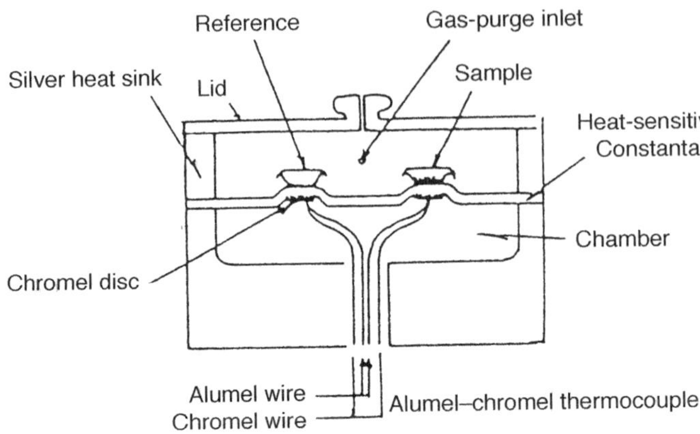

Figure 8.65 Schematic diagram of a typical heat flux differential scanning calorimetry (DSC) system. From T. Hatakeyama and F. X. Quinn, *Thermal Analysis: Fundamentals and Applications to Polymer Science Strategies*, Copyright © 1994. John Wiley & Sons, Ltd. Reproduced with permission.

difference between the plate and the furnace. Since the heat capacity of the plate as a function of temperature is known (having been measured at the factory), the enthalpy, or energy, of the transition responsible for the temperature difference can be calculated.

As was the case with DTA, a sudden change in the sample's heat capacity does not cause a peak, but rather a shift in the baseline. Figure 8.64 shows a few of the possible transitions or events that can occur in a coating sample. At a high enough temperature, the sample will decompose, usually exhibiting a large, uncontrolled exotherm or endotherm.

The type of DSC arrangement shown in Figure 8.65 is known as a 'heat flux' DSC system, because it relies on measuring temperature differences which are proportional to the change in heat 'flux' (energy input per unit time). Another type of DSC system is the power compensation instrument. The set-up in this case involves two furnaces, one for the sample and one for the reference. When a thermal event occurs in the sample, the sample pan will become either hotter or cooler than the reference pan. Power is increased to whichever furnace is associated with the cooler pan in order to bring the temperature difference back to zero. The increase in power can be precisely measured and becomes the y-axis in the plot of power (watts) versus temperature.

8.7.2 Calibration and Sample Preparation

When carrying out quantitative work, it is good practice to calibrate the calorimeter with respect to both temperature and energy. In both cases, this is done by using reference materials.

Table 8.3 Heats of fusion and melting points for some commonly used reference materials.

Reference	Melting point (K)	Heal of fusion ($J\,g^{-1}$)
Indium	429.8	28.5
Tin	505.1	59.7
Zine	692.7	108
Silver	1235.1	107

There are several materials that can be obtained in high purity, the melting points and heats of fusion of which are precisely known. Normal reagent-grade laboratory chemicals are not pure enough for this purpose, and suitable reference materials can usually be purchased from manufacturers of DSC equipment. A few such materials are shown in Table 8.3.

For very precise work, it is desirable to calibrate the temperature scale of the instrument by using two standards whose melting points span the region of interest, although for routine work, a single standard is often sufficient.

Energy (or heat) calibration, on the other hand, is often carried out by using only one reference. When calibrating the energy scale, it is important to have an inert atmosphere, such as nitrogen, in the furnace compartment in order to avoid or at least minimize oxidation of the reference material. This is usually true when analysing samples as well.

There are a variety of sample vessels, or pans, available from the instrument manufacturers. In analysing coatings, aluminium pans work well. These are inexpensive, relatively inert over the temperature range employed, are of low mass and conduct heat reasonably well. The pans are usually on the order of a few millimetres in diameter and can be used open, or with a lid which is crimped shut. The normal crimping procedure does not result in a hermetically sealed container, so if one wishes to trap and hold volatile components as part of the experiment, special vessels will be required.

The size of the sample will vary depending on its nature, but typically runs in the 5- to 10-mg range. It can be in the form of either a powder or slivers. It is important that the sample makes good physical contact with the bottom of the pan, since this will maximize heat flow from the sample to the thermocouple. Thin, flat samples are ideal; thick, chunky samples should be avoided.

One effect of crimping a pan shut with a lid, even though there is no hermetic seal, is to flatten the sample and maximize its contact with the bottom of the pan. It also tends to immobilize the sample, thus preventing artefacts during the scan due to the physical movement of the sample.

Choosing the weight of sample is frequently a compromise. Ideally, a small sample should be used, not only to maximize physical contact with the bottom of the pan, but to minimize the temperature lag between the furnace temperature and the actual temperature of the sample. This results in more accurate measurements. However, especially with pigmented coatings, one must frequently use larger samples because the thermal events are small in magnitude and may be difficult to detect.

It should also be pointed out that thermal lag can be minimized by scanning at a slow rate, such as 1 °C min^{-1}. This will result in very long scan times, and the magnitude of the signal may be smaller. Scan rates of 10–20 °C min^{-1} are common.

Other considerations in sample preparation are the effects caused by either moisture or retained solvents. In failure analysis, one of the main reasons for performing a DSC experiment is to determine the coating's glass transition temperature (T_g). For thermoset coatings, T_g is related to the degree of cure, while for thermoplastic coatings, it is related to monomer composition and, to a lesser extent, molecular weight. In either case, the T_g is an important coating parameter.

The glass transition can be seriously affected by retained moisture or solvents, since these materials can act as plasticizers. A plasticizer depresses a coating's T_g because it provides more mobility on a molecular level. In essence, it results in more free volume. Except for 100% solids coatings, nearly all dried coating samples retain some amount of solvent (or water, in the case of water-borne coatings). If they have been subjected to immersion or other very wet conditions, even 100% solids coatings may have some absorbed water. The amount of solvent or water may be so minimal as not to have any appreciable effect on the glass transition. However, unless one is reasonably certain of this, it is a good idea to take steps to minimize or eliminate this potential problem.

One way to eliminate the effect of solvent or water on the glass transition is to dry the sample in a conventional oven or a vacuum oven. Although effective, there is always the possibility that the heat in the oven may further react the sample, thus resulting in an artificially elevated T_g. Placing the sample under vacuum at room temperature under slightly elevated temperature for several hours, or even overnight, is a safer way to reduce solvent or water content. If water is the only consideration, a desiccator is also a possibility.

8.7.3 Applications of DSC

There are numerous applications of DSC to coatings science. A few recent papers have dealt with the study of intumescent coatings [20], the study of anti-corrosive coatings [21], the measurement of heat capacity [22] and the melting points of polyurethanes [23]. DSC can also be used for kinetic studies of cross-linking reactions and simply to confirm the glass transition temperature of thermoplastic coatings, either as a research tool or for QC purposes. The main use of DSC in failure analysis is to investigate a coating's relative degree of cure.

Previous chapters have discussed the effects of molecular weight and cross-link density on the chemical and physical properties of coatings. In some instances, infrared spectroscopy can provide limited information in this regard, and for thermoplastic coatings, GPC can directly provide molecular-weight information. However, GPC is useless for cross-linked, thermoset coatings that cannot be redissolved. The beauty of DSC is that it can be performed directly on solid samples, with minimal sample preparation.

As the thermosetting reaction proceeds, cross-link density and molecular weight increases, and consequently, the free volume of the coating decreases. This decrease in free volume results in an increase in the glass transition temperature of the coating, which can then be measured by DSC.

If a thermoset coating, such as an epoxy, urethane, vinyl ester or melamine cross-linked acrylic, does not have the physical properties expected of it, or if permeation and blistering is occurring, an obvious area of suspicion is its degree of cure. By applying the coating in the laboratory and curing as recommended by the manufacturer, one can determine if the field sample is properly cured by comparing its T_g to that of the laboratory sample.

If the T_g of the field sample is more than a few °C below that of the control sample, the problem is very likely related to lack of cure. There could be several reasons why the coating is not fully cured. There could be a defective batch of coating. A two-component coating might have been 'mis-mixed'. A heat-cure coating could be under-baked. If it was exposed to particularly harsh service, chemical degradation could have resulted in the lower T_g, and the problem may not be related to degree of cure at all. Thorough background information and additional testing, perhaps involving other analytical methods, would be required to answer such questions.

Powder coatings are ideally suited to characterization by DSC. These coatings contain no solvent, and hence artefacts due to solvent retention are not an issue. Since they are cured at high temperatures, they have a relatively high, distinct T_g (typically 90–120 °C) which is easily detected and measured. The National Association of Corrosion Engineers (NACE) even publishes a standard method that is widely accepted for determining the degree of cure of such coatings by DSC (NACE Standard Recommended Practice RP0394-94, Application, Performance, and Quality Control of Plant-Applied, Fusion-Bonded Epoxy External Pipe Coating).

Figure 8.66 shows a DSC heating curve obtained from a virgin sample of a fusion-bonded epoxy powder coating specified for use on buried pipeline. The sample was scanned from 30 to 280 °C. The data are plotted such that the exothermic direction is 'up'. The first feature in the heating curve is a relatively sharp peak near the beginning of the analysis, at approximately 40 °C. This is purely an instrumental artefact. It is almost always present when doing DSC and should be ignored. The second feature is a 'hump', or shift in baseline, near approximately 60 °C. This is due to the glass transition of the virgin, unreacted powder. The main feature is a very strong exothermic peak centred near 168 °C, which is the result of the reaction exotherm as the virgin powder coating is undergoing cure and releasing heat.

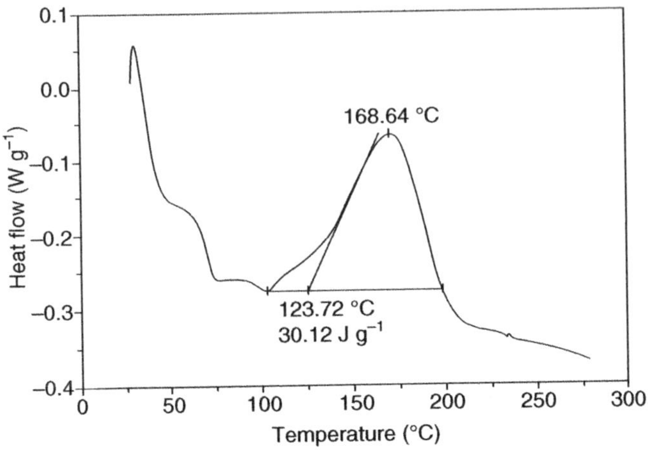

Figure 8.66 Differential scanning calorimetry (DSC) scan of a virgin sample of a fusion-bonded epoxy powder coating.

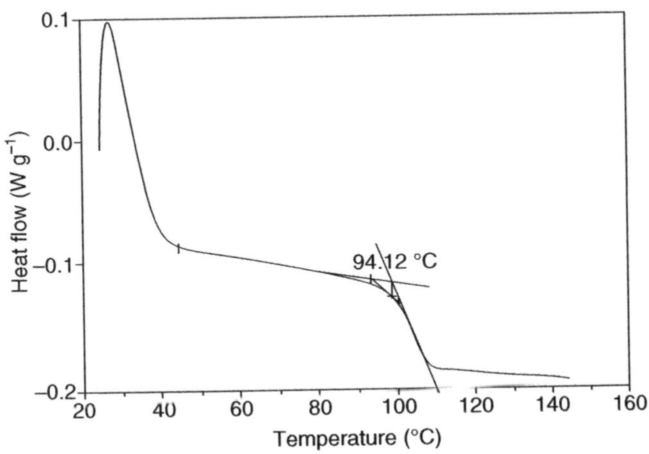

Figure 8.67 Second scan of the same sample of powder coating shown in Figure 8.66. Note the absence of the exotherm (since the sample was cured during the initial run) and the much higher glass transition temperature of the cured coating compared with the virgin powder.

Figure 8.67 is the second scan of the exact same sample which was analysed in Figure 8.66. The second run of this sample, which is now fully cured by virtue of having been 'baked' in the instrument during the initial run, differs greatly in appearance from the first run. There is no exotherm, since the sample was fully cured during the initial run. Furthermore, there is no glass transition near 60 °C, because there is no longer any unreacted, virgin powder. Instead, there is a strong glass transition near 94 °C, which is due to the fully reacted, cured coating. As expected, the glass transition of the cured coating is much higher than the glass transition of the unreacted, virgin powder.

Note that the glass transition in Figure 8.67 was determined by drawing two tangents: one on the stable portion of the curve prior to the transition and one on the steepest portion of the curve after the transition. The point where these two tangents intersect is taken as the glass transition temperature and is usually referred to as the 'glass transition at onset'. This is a very common, although somewhat arbitrary, way of measuring the glass transition.

In addition to measuring the glass transition temperature as a means of determining the degree of cure, some workers also rely on the residual heat of reaction or exotherm. If there are unreacted functional groups in the coating, they can continue to react under the elevated temperature environment of the DSC, resulting in an exothermic peak. By comparing the magnitude of this peak to the magnitude of the exotherm obtained from a virgin, uncured sample, one should be able to determine the suspect sample's degree or percentage of cure. Although this is a valid technique in the laboratory, one should be aware of some potential pitfalls when applying it to 'real-world' field samples.

The most obvious problem with this technique is that, at least for two-component coatings, it assumes that the field sample was properly mixed. Consider, for example, a two-component amine-cured epoxy. In the extreme case, if none of the amine component was added to the epoxy component, no exotherm will be detected, not because the coating is fully cured, but because there is nothing for the epoxy to react with.

Another possible source of error is that one of the co-reactants may have reacted with something that it was not supposed to, thus making it unavailable for continued reaction in the calorimeter. For instance, the isocyanate component of a two-component urethane may have reacted with water from the environment, rather than with the polyol component, and hence, no exotherm will be seen in the heating curve, despite the fact that the coating is under-cured (at least, cured the 'wrong way').

If one can establish, either through good background information or by other techniques, that none of the above concerns are justified, then determining the degree of cure based on the magnitude of the exotherm is viable. It does have a potential advantage over relying on the glass transition, since any retained water or solvent will generally not influence the value of the exotherm.

Another problem, at least for two-component ambient-cured coatings, is allowing laboratory-prepared control samples enough time to reach full cure. Some of these coatings, although dry to touch in hours, require several weeks to attain full cure, and hence a maximum T_g. The common procedure of 'speeding things up' by placing the drawdown in an oven should be practiced with caution, as this may allow reactions to occur that would never occur at normal service temperatures.

Table 8.4 Glass transitions and physical properties of epoxies cured under different conditions.

Coating	Glass transition (°C)	Impact resistance (in.-lb)	% Elongation
Polyamide epoxy, cured at 23 °C	27	40	24
Polyamide epoxy, cured at 48 °C	32	40	7
Cycloaliphatic amine epoxy, cured 23 °C	38	20	<2
Cycloaliphatic amine epoxy, cured 48 °C	35 and 69	20	<2

Measurement of the glass transition temperature is not always done in order to determine a coating's degree of cure. Sometimes, it is done in order to infer things regarding a coating's suitability for a particular environment. Recalling that the definition of T_g is that temperature below which a polymeric material is in a brittle, glassy state, and above which it is in a flexible, rubbery state, knowing the T_g of a coating should provide information on its ability to resist mechanical stress.

Weldon [24] investigated the glass transition temperatures, flexibility (% elongation) and impact resistance of both a polyamide epoxy coating and a cycloaliphatic amine-cured epoxy coating, as a function of curing temperature. Both coatings were ambient cure coatings. In the study, they were cured for either 7 days at room temperature (23 °C), or for 7 days at a somewhat elevated temperature of 48 °C. The results were as follows (Table 8.4):

As the results show, the polyamide epoxy cured at room temperature had a much lower glass transition temperature than the cycloaliphatic amine-cured epoxy (27 ° versus 38 °). As expected, this had very practical significance in that the polyamide epoxy had both much better impact resistance and much better flexibility (% elongation) compared with the higher T_g cycloaliphatic amine-cured epoxy. Interestingly, when panels coated with the cycloaliphatic amine-cured epoxy were heated to a temperature (68 °C) above the coating's T_g and immediately tested for % elongation while still hot, the result was an elongation of 7% – much higher than the <2% value obtained at room temperature. This is a clear example of the practical significance of the glass transition temperature and its relationship to physical properties.

The above data also suggest that the performance of even an ambient-cured coating might vary substantially depending simply on the temperature at which it was cured. The polyamide epoxy was much more flexible when cured at 23 °C than at 48 °C. These temperature ranges could be easily realized in the painting of outdoor structures, since metal surfaces can get quite hot when exposed to the bright summer sun in certain geographic locations. As the above data suggest, DSC could be a useful tool in such circumstances.

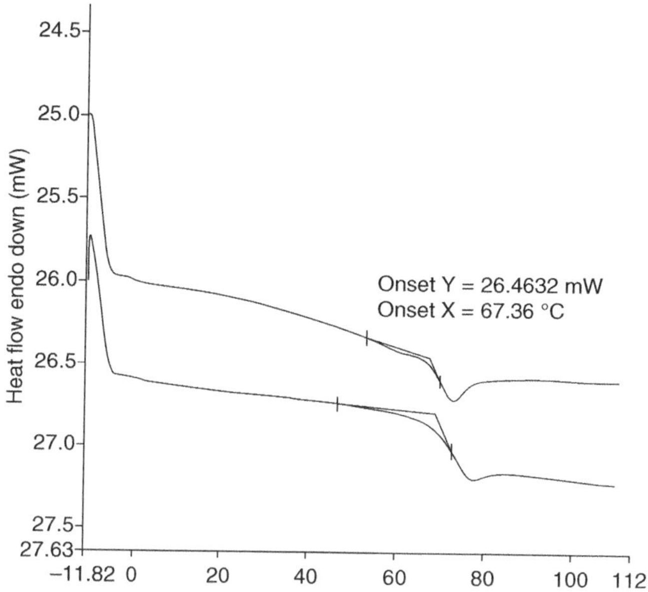

Figure 8.68 Initial differential scanning calorimetry (DSC) scan of a cured epoxy powder coating (upper trace), showing a glass transition near 67 °C and the 'overshoot' (the peak going in the negative direction) shortly after the glass transition, due to physical ageing. The lower trace is the second run of this same sample, obtained after it was allowed to cool down in the instrument. Note that the physical ageing overshoot has been virtually eliminated.

DSC can also be used to investigate possible physical ageing effects in failure analysis. As previously discussed (Section 2.2.2), physical ageing refers to changes in a coating's properties due not to chemical changes, but rather to physical ones, as a result of rapid cooling to temperatures below T_g. It can be detected by DSC, as an endothermic 'overshoot' which shows up in the heating curve at or just after the glass transition (Figure 8.68) and which can sometimes interfere with the accurate measurement of T_g. It can be eliminated by running the identical sample a second time, as the effect of physical ageing can be 'erased' by annealing the sample at a temperature above the glass transition. Figure 8.68 shows two runs of the identical epoxy specimen. The physical ageing overshoot apparent in the initial run (upper curve) is virtually absent in the second run (lower curve), and the glass transition is now easily visible near 67 °C.

A discussion of the detection of and the effects of physical ageing is presented by Skaja, Fernando and Croll [25]. As part of their study, DSC heating curves were obtained from polyester urethane coatings which had aged for various lengths of time, both before and after annealing (Figure 8.69). The data clearly show the typical DSC overshoot due to physical ageing and also demonstrates that the amount of physical ageing increases substantially upon longer storage

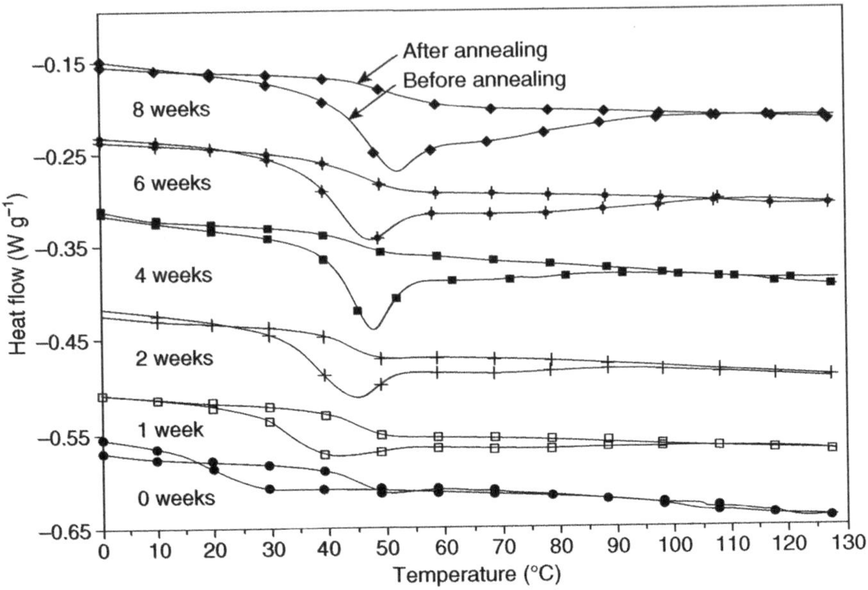

Figure 8.69 Differential scanning calorimetry (DSC) heating curves before and after annealing of polyurethane specimens which aged for up to 8 weeks. From Skaja, Fernando and Croll [25].

times, at least up to a certain point. Again, annealing the specimens in the DSC furnace during the initial run erases the effects of physical ageing.

The extent to which physical ageing contributes to coating failures is unknown but would be expected to be more common with baked coatings than with ambient-cured ones. The author is familiar with at least one case where a powder coating material cracked and failed from a metal substrate subjected to considerable flexing, as a result of densification and embrittlement of the coating due to physical ageing.

8.8 MISCELLANEOUS METHODS OF ANALYSIS

The analytical methods discussed in this chapter do not constitute the limits of analytical chemistry as applied to failure analysis. One of the keys to performing a failure analysis is creativity and flexibility, and if the investigator is aware of another method for accomplishing a specific task, it should be strongly considered. One such 'miscellaneous' method which is very useful in certain cases involving mix ratio problems is nitrogen analysis.

Section 8.2.4.3 discussed the use of infrared spectroscopy as a means of determining whether or not a two-component coating has been applied at the correct mix ratio. While this often works, especially with polyamide epoxies, there are many instances where it will not work, such as amine-cured epoxies, coal tar epoxies and certain urethanes. An excellent alternative for determining mix ratio is nitrogen analysis.

A very old and reliable technique for determining the amount of nitrogen in an organic material is called combustion analysis. Briefly, this involves completely oxidizing the sample using high temperature and an oxidant, sometimes with the aid of a catalyst, and detecting the evolved nitrogen gas. Nowadays, the analysis is usually carried out using automated equipment. For a two-component coating where the only source of nitrogen is the curing agent, the amount of nitrogen in the dried coating is directly related to the mix ratio at which it was applied. Therefore, the technique has much greater applicability than infrared spectroscopy, and, in the author's experience, is more accurate. Occasionally, however, certain pigments which contain nitrogen can interfere with the analysis, if they are in the base component.

There are numerous ASTM test methods for performing chemical analysis of raw materials but very few for the analysis of a formulated paint, and even fewer for the analysis of dried paint chips. In some cases, it may be possible to modify a test method written for a raw material, such that it can be employed on a formulated paint. However, it should never be taken for granted that the 'new' test method will produce accurate results. Most ASTM test methods, whether for physical properties or chemical composition, were developed for a very specific situation or type of sample and may have severe limitations when applied outside this narrow scope. When modifying the procedure, it is always desirable to validate it by analysing a control sample of known composition, or by spiking an unknown sample with the analyte of interest and determining its percentage recovery.

REFERENCES

1. D. R. Brezinski (Ed.), *An Infrared Spectroscopy Atlas for the Coatings Industry*, Federation of Societies for Coatings Technology, Blue Bell, Philadelphia, PA, USA, 1991.
2. J. Haslam, H. A. Willis and D. C. M. Squirrel, *Identification and Analysis of Plastics*, 2nd Edn, Heydon, London, 1980.
3. R. A. Nyquist and R. O. Kagel, *Infrared Spectra of Inorganic Compounds*, Academic Press, San Diego, CA, USA, 1971.
4. J. C. Graham, D. A. Glosky, T. G. Fisher and R. A. Garling, *J. Coat. Technol.*, **60**, 35 (1988).
5. S. G. Croll, *J. Coat. Technol.*, **52**, 65 (1980).
6. D. J. Skrovanek, *J. Coat. Technol.*, **61**, 31 (1989).
7. D. G. Weldon, A. J. Bochan, W. D. Corbett and R. O. Lackey, *J. Prot. Coat. Linings*, **6**, 21 (April 1989).
8. H. A. Dannenberg and W. R. Harp Jr, *Anal. Chem.*, **28**, 86 (1956).
9. M. McMaster and C. McMaster, *GC-MS – A Practical User's Guide*, Wiley-VCH, New York, 1998, p. 5.

10. G. J. Levy and T. P. Wampler, *J. Chem. Educ.*, **63**, A64 (1986).
11. J. M. Washall and T. P. Wampler, *J. Chromatogr. Sci.*, **27**, 144 (1989).
12. L. R. Snyder and J. J. Kirkland, *Introduction to Modern Chromatography*, John Wiley, New York, 1979, pp. 512–514.
13. C. Kuo and T. Provder, *J. Coat. Technol.*, **53**, 69 (1981).
14. D. Scheuing, *J. Coat. Technol.*, **57**, 47 (1985).
15. D. G. Weldon, A. Bochan and M. Schleider, *J. Prot. Coat. Linings*, **4**, 46 (May 1987).
16. B. Appleman, *J. Prot. Coat. Linings*, **4**, 68 (October 1987).
17. R. Garrabant, J. Biesinger, O. Jacinto and M. Melampy, *J. Prot. Coat. Linings*, **16**, 32 (January 1999).
18. W. C. Johnson, *J. Prot. Coat. Linings*, **7**, 54 (September 1990).
19. M. T. Postek, K. S. Howard, A. H. Johnson and K. L. McMichael, *Scanning Electron Microscopy*, 1980, pp. 3, 8 and 20.
20. C. Pagello, F. Raffaghello and D. M. Defaveri, *Polym. Paint Colour J.*, **188**, 16 (1998).
21. E. Kuwano, T. Fujitani and T. Satch, *Proc. Electrochem. Soc.*, **97**, 41 (1998).
22. J. Korus, M. Beiner, K. Busse, S. Kahle, R. Unger and E. Donth, *Thermochim. Acta*, **304/305**, 99 (1997).
23. H. Y. Wuo, H. K. Shim and K. S. Kee, *Macromol. Chem. Phys.*, **199**, 1427 (1998).
24. D. G. Weldon, *Proceedings, PACE 2006, PDCA/SSPC Annual Convention*, Tampa, FL, USA, 2006.
25. A. Skaja, D. Fernando and S. Croll, *J. Coat Technol. Res.*, **3**, 41 (2006).

9

Physical Methods

There are a variety of miscellaneous methods for evaluating the physical properties of coatings, as well as for conducting chemical analysis of coatings and coating materials. The largest single collection of such test methods has been compiled by the American Society for Testing and Materials (ASTM). Some of the ASTM methods that are more directly utilized in failure analysis will be discussed here. In some cases, these test methods can be very useful. In other instances they can produce very misleading information. This is because many of them were developed to fulfil quality control needs for a specific type or types of coatings, under relatively narrowly defined conditions. As such, they are not always appropriate for every type of coating. They may not be able to address all of the variables and environmental factors present in a failure investigation. Therefore, in some cases, it may be desirable to modify certain aspects of such tests to make the results more germane to the problem at hand.

When conducting a failure investigation, it is often desirable to measure those physical properties of a coating that might seem to have a direct bearing on its ability to perform in the environment to which it is subjected. If the coating is applied to a substrate that is expected to move and flex, flexibility would be an important property to determine. The following sections describe some of the more pertinent physical properties of coatings and the standard procedures used to measure them.

9.1 ADHESION

It is easy to argue that adhesion is the single most important property of a coating. Chemical resistance, corrosion resistance, and flexibility will be of little value if the 'protective coating' is lying on the ground rather than being firmly attached to the substrate.

Despite the importance of adhesion, it is this author's opinion that there really are no good, standardized test methods for measuring this property, which work in all instances. Perhaps the most common procedure is ASTM D3359 (Test Method for Measuring Adhesion by Tape Test), often referred to as knife adhesion or tape adhesion. The procedure involves using a cutting tool to scribe a series of cuts into the sample (either an 'x' cut, which is method A, or a cross-hatch pattern, which is method B), followed by the application of a special pressure-sensitive adhesive tape. The tape is then rapidly removed, and the amount of coating that detaches is rated according to either descriptions or illustrations. The best rating is a '5', with the worst being a '0'.

When employed by an experienced investigator, this method can supply useful information. However, it has several shortcomings, and the data are often misused.

One immediate problem with the procedure is that the knife cuts are supposed to go down to the substrate. For thin automotive coatings, this is no problem. For thick, tough, industrial maintenance coatings, 10- or 15-mil thick, this can be a major problem, thus making the method very operator dependent. Another problem is that the tape may adhere much better to one type of coating than to another, meaning that the force transmitted to the coating/substrate interface can vary dramatically. Thus, it is possible to get 'good' results for a coating simply because the tape does not stick to it.

Another problem with the method involves the force of friction between the cut edges of the coating. If the coating is somewhat flexible it can exhibit a 'healing' effect at the cut. Therefore, one is measuring not only the adhesion of the coating to the substrate, but also the force of friction between the 'healed' edges of the cuts. This will be more of a problem with thicker coatings due to the greater surface area contact at the cut edges.

Perhaps the biggest problem with all of the adhesion test methods is that seldom does anybody really know what the results mean. It is tempting to say that a '5' is good and a '0' is bad, but really all that can be said is that a '5' is better than a '0'. The ASTM does not address this in any of their test methods, providing only descriptions of the methods and perhaps a rating scale.

With all its shortcomings, adhesion testing by ASTM D3359 can still provide useful information to an experienced investigator. If a coating has not actually disbonded and failed, it is very difficult to predict the service life based on any adhesion method. However, if delamination occurs, and a survey of the structure shows 'poor' adhesion in some areas and 'good' adhesion in other areas, or if previous experience with this or similar coatings has shown that ratings of '5' or '4' are common versus the actual rating of 'ones' and 'zeros', it is obvious that there is a problem.

This type of testing can also be useful in the laboratory when trying to simulate or reproduce a coating failure. If one has obtained evidence that the coating was 'mis-mixed', it may be desirable to prepare test panels at both the required mix ratio and also at incorrect ratios. If the adhesion properties of the 'mis-mixed' coating are significantly inferior to the properly mixed coating, one has probably solved the problem.

Another common way of measuring coating adhesion is by tensile testing in accordance with ASTM D4541 (Test Method for Pull-Off Strength of Coatings Using Portable Adhesion Testers). In this procedure, a stub or dollie, usually between approximately $\frac{1}{4}$- to $\frac{1}{2}$-in. diameter, is glued to the coated article, and some type of portable device is used to determine the amount of force required to disbond the coating.

At first glance, it would appear as though this method is far superior to tape adhesion. Certainly, it is less affected by operator bias. However, it has its own set of problems.

Like tape, the various adhesives can vary significantly in their ability to bond to certain surface chemistries. In such cases, glue failures occur rather than paint failures. Test surfaces must be flat or nearly flat in order to know the size of the surface area over which the force is being applied. Different types of instruments can result in dramatically different values, and even for a single instrument, readings can vary considerably within a few inches of one another.

It is also possible for results of tensile adhesion testing to vary dramatically from the results of tape adhesion testing. It is not unusual to achieve high tensile adhesion values and yet get low values by tape adhesion.

Another common method of evaluating adhesion is by subjective probing, often with a knife. While not a standardized test method, and hence subject to operator bias, an experienced investigator can learn a considerable amount by such simple probing, particularly if there is a control or non-failing area.

One thing that should be pointed out when subjectively probing or pulling at a coating is the possibility of being misled because of film thickness or film toughness. Consider two coatings that have an identical adhesive strength (if it were possible to accurately measure this property). Coating 'A' is thin and brittle, while coating 'B' is thick and tough. By grabbing a loose or cut edge of coating 'B' and pulling with all one's might, it may be possible to rip large pieces of it from the substrate. One might then say that it has bad adhesion. However, when one attempts to do the same thing with coating 'A', it will break and shatter and only minute fragments can be removed. One might mistakenly conclude that 'A' has better adhesion than 'B'.

9.2 FLEXIBILITY AND IMPACT RESISTANCE

The importance of flexibility has been previously discussed. When a coating is flexed or deformed due to movement in the substrate or because of a forming operation, it must be flexible enough to relieve the stress. If it is too brittle, cracking can occur. If it has poor adhesion, it may disbond. Properties such as percentage elongation, tensile strength and flexural strength, are routinely measured on plastics and other thick, freely supported materials through the use of so-called universal testing machines. Such devices stretch or bend specimens of precise geometry under very controlled, reproducible conditions. Theoretically, there is no reason why these devices cannot be applied to coatings. Indeed, ASTM D2370 (Test Method for Tensile Properties of Organic Coatings) describes one such method. The problem is in sample preparation. It is often extremely difficult to prepare defect-free, unsupported films of coatings at the precise dimensions called for in various test procedures, especially for thin, fragile materials.

As a consequence, the more common way of determining a coating's flexibility, or % elongation, is to bend a thin-gauge coated steel panel around a mandrel, as described by ASTM D522 (Test Method for Mandrel Bond Test of Attached Organic Coatings). The test procedure describes using either a conical mandrel or a series of cylindrical mandrels. When a conical mandrel is used, the test will be more severe at the small diameter of the mandrel and less severe at the large-diameter end. Typically, the coating will crack at the small-diameter end, and by measuring the length of the cracking along the axis of the mandrel, one can then calculate a % elongation value.

The technique is only applicable when one has liquid samples of coating material with which to prepare test panels. If only small- to medium-sized paint chips are available, one will probably have to resort to manually flexing them and then making some sort of subjective conclusion about their brittleness or flexibility.

Coating flexibility can also be examined by using the T-bend test in accordance with ASTM D4145 (Test Method for Coating Flexibility of Pre-Painted Sheet). This is often used to evaluate thin-gauge coil coated stock and involves bending the panel back on itself, coated side out, and examining it for cracking. The most severe test is when the sample is bent back on itself once. This is the tightest degree of curvature that can be obtained and is referred to as a '0T bend'. Making yet another bend results in a somewhat longer curvature and is referred to as a '1T' bend. Such things as temperature and speed of bending can influence the results.

Impact resistance is a measure of a coating's ability to withstand rapid deformation without cracking or disbonding from the substrate. The most common procedure is ASTM D2794 (Test Method for Resistance of Organic Coatings to the Effects of Rapid Deformation (Impact)), where a weight is dropped down a guide tube to impact a hemisperhical tup that rests on the coated panel. There is an opening in the base of the device opposite the tup, such that the thin-gauge panel will deform. Cracking or disbonding in the impacted area is considered a failure, with the results typically reported in inch-pound. If the tup rests on the painted side of the panel, the test is referred to as a direct-impact test. If the backside of the panel is indented, it is called a reverse- or indirect-impact test.

It is important to realize that the substrate plays a major role in the results obtained from impact testing. Thin-gauge panels, which is what the method calls for, will deform more than thicker panels. Therefore, the test will be more severe if carried out on thin panels, as opposed to thick ones.

It must also be realized that the temperature of the test is an important variable. Although normally run at room temperature, if the actual failure at the job site occurred at sub-freezing temperatures, it will be more meaningful to conduct the laboratory test at a similar temperature, where the coating may be significantly more brittle.

9.3 SOLVENT RESISTANCE

For thermoset coatings, the resistance to solvent, usually methyl ethyl ketone (MEK), is often taken as a rough measure of the coating's degree of cure. Two standard methods for performing this evaluation are described in ASTM D4752 (Test Method for Measuring MEK Resistance of Ethyl Silicate (Inorganic) Zinc-Rich Primers by Solvent Rub), which addresses inorganic zinc-rich primers, and ASTM D5402 (Practice for Assessing the Solvent Resistance of Organic Coatings Using Solvent Rubs), which is a more general procedure. Many laboratories have their own in-house procedures for evaluating solvent resistance.

Typically, the methods use a cloth or other suitable material (this author prefers Q-tips) saturated with solvent. The cloth is rubbed back and forth in a defined area under moderate finger pressure, either for a certain number of 'double rubs' (one up and down stroke) or until the substrate is exposed.

Although incapable of detecting minor differences in cure, if carried out by the same operator, the methods are consistent in detecting moderate differences in cure. The major variable is the amount of downward force exerted by the operator. When the data being recorded represent the amount of colour transfer (a subjective enough property in the first place), the operator can be biased by the colour of the paint. Obviously, a small amount of black paint transferred to a white cloth will be much more noticeable than a similar amount of white paint. It is sometimes desirable to use a dark cloth when testing light coatings.

An additional point to consider is that, like all ASTM tests, the solvent rub methods describe the procedure and perhaps provide a rating scale, but do not address the significance of the result. In order to provide useful data in a failure investigation, one must have something to compare the results to. This might be accomplished by a comparison of results from failing and non-failing areas, or to a laboratory-prepared panel of the coating in question or, as a last resort, to the investigator's experience with generically similar coatings.

9.4 WEATHERING RESISTANCE

Many times, a coating is considered to have failed if it significantly changes colour in a shorter than expected time period. Often, this is a consequence of the coating's inability to resist the normal weathering effects of the ultraviolet (UV) component of sunlight, heat and moisture. Sometimes, it is possible to detect such changes analytically, while at other times it is not. In such instances, it may be adequate to simply demonstrate in the laboratory that such changes can occur as a result of exposure to light, heat and moisture.

There are numerous devices on the market that attempt to simulate, and accelerate, the effects of natural weathering. They all employ the same three elements of UV light, heat, and moisture but differ in how they are generated. The most common laboratory devices employ either fluorescent tubes to generate UV light, as described in ASTM G154 (Standard Practice for Operating Fluorescent Light Apparatus for UV Exposure of Non-Metallic Materials), or use a xenon arc, as described in ASTM G155 (Standard Practice for Operating Xenon Arc Light Apparatus for Exposure of Non-Metallic Materials). There are still ASTM methods which describe the use of carbon arc instruments, such as ASTM G152 and G153, but these devices are rarely used anymore. They are awkward and expensive to operate, and do not appear to deliver data which is superior to the xenon arc devices. Carbon arcs are still listed in some old paint specifications, but is becoming common to substitute xenon arcs.

There are several types of bulbs available for the G154 devices, which all vary in spectral output. The various arc sources also vary in spectral output and can be further modified by the use of various filters.

In terms of irradiance (expressed as watts per square metre), sunlight contains only about 6% UV light [1]. The UV spectrum is divided into three regions, namely the UVA region from 400 to 320 nm, the UVB region from 320 to 280 nm and the UVC region below 280 nm. There is no UVC radiation at the earth's surface, since the atmosphere filters it out. There is only a small amount of UVB radiation. Therefore, for an accelerated weathering device to more closely match the effects of natural sunlight, most of its energy should be in the UVA region. Radiation from UVB sources will cause changes in coatings more rapidly than UVA radiation but may give unrealistic results.

It is extremely difficult, if not impossible, to predict a coating's service life from accelerated weathering data. Real-world weathering is so variable that it cannot be reproduced in the laboratory. Nevertheless, if the suspect coating were to discolour or to otherwise deteriorate after a 'short' exposure in an accelerated weathering device in a manner similar to that observed in the field, one has likely shown that the failure is product related rather than application related.

9.5 CHEMICAL RESISTANCE

Coatings are widely used to protect substrates from a variety of chemicals. Sometimes, these chemicals are common and seemingly innocuous, such as water. At other times, they may consist of solvents, caustics or even hot, acidic gases. It is not uncommon for a protective coating to fail because it is being chemically degraded by the very chemicals it is designed to resist.

In the course of conducting a failure analysis, it may be possible to employ certain analytical techniques to detect by-products or other telltale signs of chemical attack. However, in many cases, it may also be desirable to demonstrate a coating's ability, or inability, to resist the chemicals in its environment.

If good background information can be obtained, it is often possible to simulate, or sometimes virtually reproduce, the environment in which the coating failed. If it is known that the particular vessel in question contained 10% sulfuric acid at 50 °C, panels could be coated and cured in the laboratory and subsequently exposed to these conditions for weeks or months. They could then be evaluated for signs of chemical attack, such as blistering, softening, checking, loss of adhesion and so on. Sometimes, the chemical history of a particular vessel is not so well defined. It may have held various materials and been subjected to a variety of temperatures. Although it is likely that the identity of the last 'cargo' is known, the degradation may have resulted from a previous service environment.

In many cases, conducting chemical resistance testing in the laboratory is simply an exercise in ingenuity and common sense. However, there are a few standard test procedures that can also be employed.

The catch-all test for chemical resistance is ASTM D1308, which has the somewhat misleading name of 'Test Method for Effect of Household Chemicals on Clear and Pigmented Organic Finishes'. The name implies that it is used to test the resistance of house paint to detergent, catsup and vinegar, which indeed it is. However, the method is so broadly written that it could be used to test almost any type of coating with almost any chemical, at whatever temperature and time one desires. The method discusses simple spot tests, room temperature immersion and even elevated temperature immersion.

When carrying out immersion testing at elevated temperatures, certain safety precautions must be considered. With chemicals of low volatility, it may be possible to simply immerse specimens in closed jars kept in an oven or a water bath. For higher temperatures or more volatile chemicals, a heated resin kettle fitted with a water-cooled condenser is usually adequate. If caustic or some other chemical aggressive towards glass is being evaluated, a stainless steel container may be more suitable.

One problem with immersion testing is edge effects, especially in prolonged testing. One way to avoid this is to perform what is often referred to as 'one-sided' immersion testing, as described in ASTM C868 (Test Method for Chemical Resistance of Protective Linings) and NACE TM-01-74 (Laboratory Methods for the Evaluation of Protective Coatings and Lining Materials in Immersion Service). In these procedures, a cylinder made of glass (or other suitable material) is employed, and the test panels constitute the ends of the cylinder. The cylinders are typically 4–6 in. in diameter, such that the test panels are usually 6×6 in. or 8×8 in. in size. Holes are drilled in appropriate locations near the corners of the panels, such that they can be firmly attached to the ends of the cells through the use of threaded stock, nuts and bolts. Various gasket materials are used to obtain a leak-proof seal. The cells are then filled halfway to $\frac{2}{3}$ full with chemical, with either heating tape or immersion heaters being used to maintain the temperature. They are also fitted with water-cooled condensers to prevent pressure build-up and minimize evaporation.

These one-sided cells are very versatile and allow testing to be performed simultaneously on both the immersion phase and the vapour phase. They are somewhat awkward, however, in terms of facilitating interim evaluations.

A phenomenon sometimes encountered in failure analysis is the so-called ‘cold-wall effect’. The theory of the cold-wall effect is that chemical permeation and deterioration of coatings will proceed much more readily if the temperature of the tank contents is higher than the temperature outside the tank. This author has seen more than one occasion where this effect has been operative. It is likely due to the result of the chemicals tending to condense on the slightly cooler steel beneath the coating.

It is often stated that one-sided immersion testing reproduces the cold-wall effect. This is not always true, since the temperature differential between relatively thin ($\frac{1}{16}-\frac{1}{8}$ in.) steel panels exposed to a hot chemical on the inside and room temperature laboratory air on the outside may be only 1 °C. Therefore, in order to really reproduce a cold-wall effect, either much thicker steel should be used or the outside surface of the test panel should be jacketed and exposed to cool, circulating water.

When conducting chemical resistance testing, the investigator will often be confronted with the need to ‘speed things up’. This is a legitimate demand, which is often fulfilled by either increasing the temperature or the concentration of the chemicals. However, both of these operations should be carried out with caution.

There is a rule of thumb that states that the rate of a chemical reaction roughly doubles for every 10 °C increase in temperature. If this is the only effect to be considered, a test performed at 120 °C should be four times as aggressive and, therefore, should take only a quarter of the time of that for the same test carried out at 100 °C. This theory ignores two things, activation energy and the glass transition temperature.

Chemical reactions require a certain threshold energy, or activation energy, before they can occur. Once above that energy (or temperature), the 10 °C rule of thumb often applies. In the above example, it is possible that at 100 °C (which might correspond to the actual service temperature of the vessel in question), the activation energy requirement has not been met, while at 120 °C it has. Therefore, chemical reactions can occur at 120 °C that simply would not occur, or would occur very slowly, at the actual service temperature of 100 °C.

The glass transition temperature, or T_g, of a coating has been previously defined as that temperature at which it changes from a glassy material to a rubbery one. It also represents that temperature at which there is a substantial increase in free volume, thus making the coating more permeable. If a coating is below its T_g in the service environment, but above it in the ‘accelerated’ laboratory environment, one runs the risk of failure by artificially increasing the permeability of the coating.

9.6 FREEZE–THAW

Sometimes, coatings fail because they do not expand and contract at the same rate as the substrate. One way to evaluate this possibility is to apply the coating system to the same type of substrate and subject the cured samples to a freeze–thaw regimen, with limits perhaps slightly outside those expected at the job site. It may also be desirable to add a moisture step to simulate the effects of rain, dew, snow and so on. A common cycle which is employed is approximately 7.5 hours at 120 °F (49 °C), 0.5 hour of tap water immersion and 16 hours at 0 °F (−18 °C), for 10 cycles, with adhesion measured both before and after. Occasionally, a coating will crack and disbond when subjected to such cycling, but more often, one will simply observe a decrease in adhesion. The cycle given in ASTM D6944 may be appropriate.

Two problems encountered with freeze–thaw testing include the situations of too 'fresh' a coating system and too small a sample.

Some coatings require considerable time (several weeks) to reach full cure. Others will actually begin to degrade due to the environment, either before or during a period of natural freeze–thaw cycling. In either event, the field-applied coating may be much more brittle than a coating that has only cured for 1 week in the laboratory. Unfortunately, because of the time constraints imposed on most failure analysis work, there is seldom time to age the coatings properly before testing. As already pointed out, simply baking them in an oven can result in misleading results if the coatings subsequently cure via a different pathway than they would typically do at 'normal' temperatures.

For practical reasons, most laboratory work is done with fairly small samples, for example, with dimensions of 3 × 5 in. or 4 × 6 in. Such samples will not expand or contract to the same degree as, say 10 × 20 ft steel plates. As large a sample as can be reasonably accommodated should be used in laboratory testing.

9.7 APPLICATION VARIABLES

It can be safely said that the majority of coating failures are due to poor application. In this sense, the term 'application' is not restricted only to the physical act of mixing and applying the paint, but also to surface preparation. If there is evidence that such an application error has occurred, it may be possible to reproduce, or at least simulate it, in the laboratory. Simply proving that a certain application-related error may cause a failure, with no evidence that such an error actually occurred, is insufficient in assigning the cause of the failure. It can be shown in the laboratory that certain types of coatings will disbond if applied over an oil-contaminated substrate. However, unless oil was detected on field samples, it would not be appropriate to conclude that the failure occurred because of oil contamination.

It is not always necessary to duplicate the failure in the laboratory in order to assign the cause. One may have good analytical evidence that a urethane topcoat peeled from an underlying epoxy because the epoxy was applied at the wrong mix ratio. However, it is possible that laboratory test panels prepared at either the correct or incorrect mix ratio will not exhibit spontaneous disbonding. Should one conclude that 'mis-mixing' was not a factor in the observed disbonding?

It has been this author's personal experience that, more times than not, it is difficult to duplicate a field failure in the laboratory. This is likely due to the fact that it is virtually impossible to reproduce, or even identify, all of the variables involved in the field application, curing, stress, and weathering of a coating system. What can be done, however, is to look for trends in those properties most closely related to the type of failure, as a function of the suspected application variables.

In the above example, it would be highly unlikely that the laboratory-applied urethane topcoat would delaminate from the 'mis-mixed' epoxy while the test panel is simply lying on the laboratory bench. Even poorly adherent coatings will not fall off if there is no stress. It may be, however, that if adhesion testing is performed, one may find that the urethane exhibits much worse adhesion to the 'mis-mixed' epoxy than to the properly mixed epoxy. This would be strong evidence that the failure was due to 'mis-mixing' of the epoxy.

Because of the diversity of coating systems and the variability of natural environments, it is very difficult to simply make a 'punch list' of application variables and how to test for them. Much of this is a combination of common sense and ingenuity. However, some of the more commonly encountered scenarios will be discussed briefly here.

For two-component coatings, the mix ratio is one obvious potential problem. Potential analytical means of identifying a mix ratio problem include infrared spectroscopy, ashing the sample to determine the % pigment (this often works well when one of the components is a 'clear') and nitrogen analysis (if the coating is a urethane, for instance). Other methods may work well for certain additional types of coatings.

If a mix ratio problem is suspected, test panels can often be made at both the correct and incorrect mix ratios, and then evaluated for relevant properties. If the failure is one of delamination, adhesion testing can be performed. If it is one of blistering in water immersion, immersion or humidity testing can be carried out. If it is one of yellowing, accelerated weathering can be employed.

Often, more than one factor is involved in a coating failure. In the above case of a urethane disbonding from a 'mis-mixed' epoxy, the reason for the poor adhesion may be a blush of excess curing agent that exuded to the surface of the 'mis-mixed' epoxy in response to high humidity, or overnight condensation. Although still the result of 'mis-mixing', such panels prepared in a low-humidity laboratory environment may not show any indication of poor adhesion, whereas curing them in a high-humidity environment may duplicate the failure (thus the importance of good background information).

On the subject of epoxies, another potential factor relating to poor inter-coat adhesion is the lack of induction time. Some epoxies, particularly those that are cured with low-molecular-weight amines, require an induction (or 'sweat in') time prior to their application. This gives the curing agent, which may be water sensitive and hence prone to exude or blush under certain conditions, time to partially react with the epoxy resins. This can be simulated in the laboratory simply by painting out panels at various times after mixing the coating.

Somewhat the opposite of induction time is waiting too long to apply the coating. This is a potential problem with any two-component coating and not just epoxies. As the pot life is approached, or exceeded, the coating's viscosity will significantly increase. This high viscosity can make it much more difficult for the coating to flow and intimately wet the substrate, with the possible consequence of a significantly reduced adhesion. Like induction time, this can be evaluated by preparing panels near, or in some cases even after, the expected pot life.

Another way to reduce a coating's ability to flow and wet the substrate is by maintaining too much distance between the spray gun and the substrate. This greater distance allows more solvent to evaporate from the minute droplets of atomized coating, thus resulting in increased viscosity, higher surface tension and greater resistance to flow and wetting. This problem may be exacerbated in warmer weather and for higher solids coatings that can ill afford to lose what little solvent they have.

The obvious way to simulate this in the laboratory is simply to vary the distance of the spray gun from the panel. It is usually more difficult to reproduce high or low temperatures in the laboratory, although the paint itself can certainly be allowed to equilibrate at the desired temperature immediately prior to being applied. Once again, the resulting test panels will need to be stressed appropriately (adhesion testing, flexing, immersion, etc.) in order to observe any performance trends related to the application variables.

Yet another application-related error is working at inappropriate temperatures. There are several ways in which temperature can adversely affect coating performance. One obvious temperature-related problem is condensation on the substrate as a result of the article's surface temperature being below the dew point. This thin layer of moisture can result in peeling of the coatings, especially those which are solvent based. Other temperature-related problems include poor flow and wetting at low temperatures, poor coalescence of latex coatings at low temperatures, and bubbling of thick-film solvent-based coatings at high temperatures. The curing reactions of some two-component coatings can be significantly impeded at low temperatures. Even if the temperature is subsequently raised, much of the solvent will have already evaporated from these coatings, thus resulting in a reduced mobility of the forming polymer chains. This reduced mobility means that the collisions necessary to cause a chemical reaction will be less likely to occur.

It is often difficult to simulate temperature conditions in a laboratory environment. If it is suspected that the problem was due to high temperatures, the paint can be preheated, as can the substrate. Once painted, the panels can immediately be placed in an oven at the appropriate temperature.

If it is suspected that the problem was due to low-temperature application, the paint and substrate can be cooled down prior to application, and the painted panels can then be put in a refrigerated environment. One potential problem is the possibility that the moisture from the laboratory air may condense on panels removed from a refrigerator prior to the paint being applied. If this was a real possibility at the job site, then this is a good thing. If not, the laboratory panels may fail not because of the low temperature, but because the coating was mistakenly applied over moisture. When carrying out such test, the dew point in the laboratory should be monitored. It may be necessary to build a small, temporary, dehumidified environment in the laboratory for this purpose.

Except in the most extreme cases, it is difficult to prove, after the fact, that a coating was over-thinned. However, gas chromatography can readily demonstrate if an unauthorized thinner was used. Such a thinner may have virtually no effect on the coating's performance, or it could prove to have sprious consequences.

Once analytical results have identified an unauthorized thinner, test panels can be prepared with and without this solvent. They will have to be subjected to some logical test regimen to see if the thinner produced the anticipated effect.

Although probably not as common as some people think, failures due to painting over contamination occur on a regular basis. Since identifying the precise contaminant can be difficult, and ascertaining its concentration even more so, it is not surprising that trying to simulate this in the laboratory is problematic. Fortunately, the presence of measurable amounts of a contaminant at the failing interface is often strong enough proof of the failure mechanism so that attempts to reproduce it are often not necessary.

One of the most common problems resulting in paint failures is inadequate cleaning or surface preparation. This can mean failure to blast clean a rusty surface to a white or near-white blast prior to applying a bridge paint, failure to remove dirt and mildew prior to applying house paint or failure to remove mill oils from galvanized strip prior to coating. Sometimes, the evidence of inadequate preparation can be seen with the naked eye, while at other times, it requires the use of sophisticated analytical instrumentation.

In some cases, it may be possible to simulate or produce poorly prepared surfaces similar to what was observed on the sample. Mill scale bearing steel can be easily procured, and rusted, if need be. Although it may be impossible to duplicate the exact surface condition of the job site steel, an approximation can be achieved. If the blast profile was either too shallow, too rounded or too angular, they can likely be approximated in the laboratory by the judicious selection of a blast-cleaning abrasive. However, some types of contamination, such as pollen, tree sap or mildew growth, would be extremely difficult to reproduce in the laboratory.

Another common application-related error that can usually be reproduced in the laboratory is thickness. Sometimes, coatings fail because they are too thin or sometimes because they are too thick. A thorough examination of failing and non-failing areas may reveal such a clear thickness-related trend that additional work is unnecessary. However, varying the coating thickness in the laboratory is one of the easier tasks to undertake.

Another application-related error that can often be simulated in the laboratory is that of over- or under-baking of heat-cured coatings. Under-baking can result in coatings that are poorly cross-linked and soft and lack full chemical resistance properties. Over-baking can result in coatings that are hard to topcoat or that are too brittle.

In many instances, baking effects can be simulated in the laboratory, although perhaps not precisely reproduced. The ovens in some coil coating lines are of extremely high temperature. They have to be, since the moving coil might reside in the oven for less than a minute. Few laboratories have facilities that can reproduce these effects, although useful data can still be obtained with conventional equipment. In many cases, the recommended bake for coil coatings does not specify a time but only a peak metal temperature. In such instances, it is very useful to attach thermocouples to the coated test panels, such that peak metal temperatures can be monitored.

REFERENCES

1. J. V. Koleske (Ed.), *Paint and Coating Testing Manual*, 14th Edn, The American Society for Testing and Materials, Philadelphia, PA, USA, 1995, p. 621.

10

Examples of Coating Failures

The various examples presented in this chapter have been selected not only to encompass a broad spectrum of coating failures, but also to demonstrate the utility of the various analytical techniques which have been described in the earlier chapters.

10.1 URETHANE TOPCOAT DISBONDING FROM EPOXY

The structural steel members for a new bridge were primed, coated with a two-component polyamide epoxy intermediate coat and subsequently top-coated with a two-component aliphatic polyurethane. The specified mix ratio for the white epoxy intermediate coat was 1 : 1, by volume, with a required 20-minute induction time. It was to be applied at a dry-film thickness (DFT) of 4–6 mil. The blue urethane finish coat, which was relatively glossy, had a specified mix ratio of four parts of component 'A' (the pigmented portion which contained the acrylic polyol) to one part of component 'B' (a 'clear' containing the aliphatic isocyanate). It was to be applied at a DFT of 2–4 mil.

Approximately 3 months after the job was completed, it was found that there were areas where the urethane topcoat was peeling from the epoxy intermediate coat. A field investigation showed that while the actual amount of spontaneous peeling amounted to perhaps only 1 or 2% of the painted surface, tape-adhesion testing indicated very poor inter-coat adhesion even in many visually intact areas. Although there were some exceptions, the bulk of the peeling and poor adhesion were on the horizontal and southern-facing exposures, with generally better adhesion in shaded or sheltered areas.

Failure Analysis of Paints and Coatings, Revised Edition Dwight G. Weldon

Several relatively large (up to 1 in.^{2}) chips of delaminating topcoat were obtained. In addition, scrapings of the epoxy coating were obtained after the topcoat was peeled away. Shavings of non-failing, well-adhered topcoat were also obtained, as well as a few relatively thick chips from non-failing areas consisting of the entire three-coat coating system. Although attempts were made at the job site to obtain 'clean' samples of epoxy intermediate coat from non-failing areas, this was very difficult due to the excellent adhesion in these areas.

Since the failure occurred months after the painting was completed, no quantities of liquid paint were left over, and batch retains from the coating supplier were not made available. However, fresh, off-the-shelf samples of coatings were provided.

As previously pointed out, the place to begin a failure investigation in the laboratory is at the microscope. Microscopic examination showed little difference in topcoat thickness between failing and non-failing samples. These ranged from a low of approximately 2 mil to a high of approximately 6 mil. Although the latter was over the recommended maximum of 4 mil, this was not felt to be high enough to be causative. Furthermore, non-failing samples were also found which were 6-mil thick.

At first glance, an examination of the back of the disbonded topcoat failed to reveal more than just a slight amount of dirt or debris, and certainly not enough to account for such poor adhesion. However, when viewed with oblique lighting, there appeared to be a faint, patchy haze on the back of the blue topcoat. Of course, the back of the non-failing topcoat was not available for examination, since it was firmly adhered to the epoxy intermediate coat.

In addition to the microscopic observations, infrared (IR) spectroscopy was also performed. Spectra were obtained of topcoat and epoxy from both good and bad areas, as well as on drawdowns of the fresh batches of coating material. Since, with multiple-component coatings, there is always the possibility of 'mis-mixing', drawdowns were prepared at both the specified mix ratios, and at mix ratios with a 50% excess of each component.

Although the spectrum of an epoxy is usually not greatly influenced by cure times in excess of a day or so, the same is not true of urethanes. Therefore, the urethane drawdowns were cured at a slightly elevated temperature (110 °F (43 °C), a temperature which much of the steel likely reached during the summer afternoons) for 1 week prior to obtaining the spectra.

The results of the IR analysis indicated that 'mis-mixing' of the urethane finish coat was not an issue. The approximate mix ratio of the urethane was determined by using IR spectroscopy because of minor, predictable changes in the relative intensities of some of the silicate-extender pigment bands compared with the ester carbonyl vibration from the polyol component, as well as by variations in the amount of actual carbamate (urethane) formed.

However, the analysis of the epoxy produced mixed results. IR spectroscopy proved to be a viable technique for determining the relative amount of the polyamide component to the epoxy component. This was primarily because of variations in the relative intensity of the amide carbonyl band near 1640 cm^{-1} to that of a band near 1610 cm^{-1}, due to a C–C stretching vibration of the aromatic ring present in the epoxy resin. Some mis-mixed coating was found in failing areas, with no mis-mixed epoxy in the three non-failing areas sampled. However, of the five samples of failing epoxy, three were properly mixed. Therefore, at this point, the results seemed inconclusive.

Although the microscopic observations seemed to indicate that the delamination was a clean one between the urethane topcoat and the epoxy, there was still the issue of what appeared to be a faint haze on the back of the blue topcoat. In order to further investigate this, IR spectra were obtained of both the front and back surfaces of failing paint chips by using attenuated total reflectance (ATR).

The ATR spectra obtained from the fronts of the topcoat samples were typical of urethanes (see Figure 10.1a). Not surprisingly, being reflectance spectra, they differed in several respects from the transmission spectra obtained by the potassium bromide pellet technique. More importantly, however, they differed somewhat from the ATR spectra obtained from the back of the samples. The spectra obtained from the back surfaces (see Figure 10.1b) contained additional bands near 1510, 1180 and 830 cm^{-1}, all of which are typical of epoxies.

Based on the above findings, it certainly appeared that the reason for the intercoat delamination was degradation of the very upper surface of the epoxy intermediate coat prior to topcoating, thus resulting in an extremely thin, loose layer that acted as a 'bond breaker'. This mechanism was consistent with the observation that the bulk of the failure occurred in those areas receiving more sunshine, coupled with the known fact that epoxy coatings chalk upon exposure to sunlight. Two issues, however, seemed unresolved. First of all, inspection records showed that the bulk of the steel was top-coated within 2 weeks of when the epoxy was applied – a relatively short period for surface degradation to occur. Secondly, although not consistent, failing areas were found where the epoxy had been mis-mixed. How much of a factor did this mis-mixing play in the failure? Did it speed up the degradation of the epoxy? Would it, on its own, have accounted for intercoat delamination?

One might be able to predict these answers, given an intimate knowledge of the formulas of the epoxy and urethane coatings, coupled with a detailed understanding of the chemistry of their cross-linking, the chemistry and physics explaining their adhesion to one another and a thorough understanding of the degradation mechanism induced by weathering. However, in addition to the coating manufacturer not wishing to publicize his formulas for the world to see, such depth of knowledge into the behaviour of coatings does not yet exist. Therefore, the best way to answer the questions posed above is by experiment, and not theory.

The epoxy intermediate coat was applied to steel test panels at both the correct mix ratio, and at the off-ratio found on some of the failing field samples (approximately 1.5 : 1 rather than 1 : 1). Three sets of panels were prepared, so that they could be cured in different fashions prior to topcoating. One set was allowed to cure for 2 weeks at laboratory ambient conditions. A second set was cured for 11 days at laboratory ambient temperature and then for an additional 3 days in

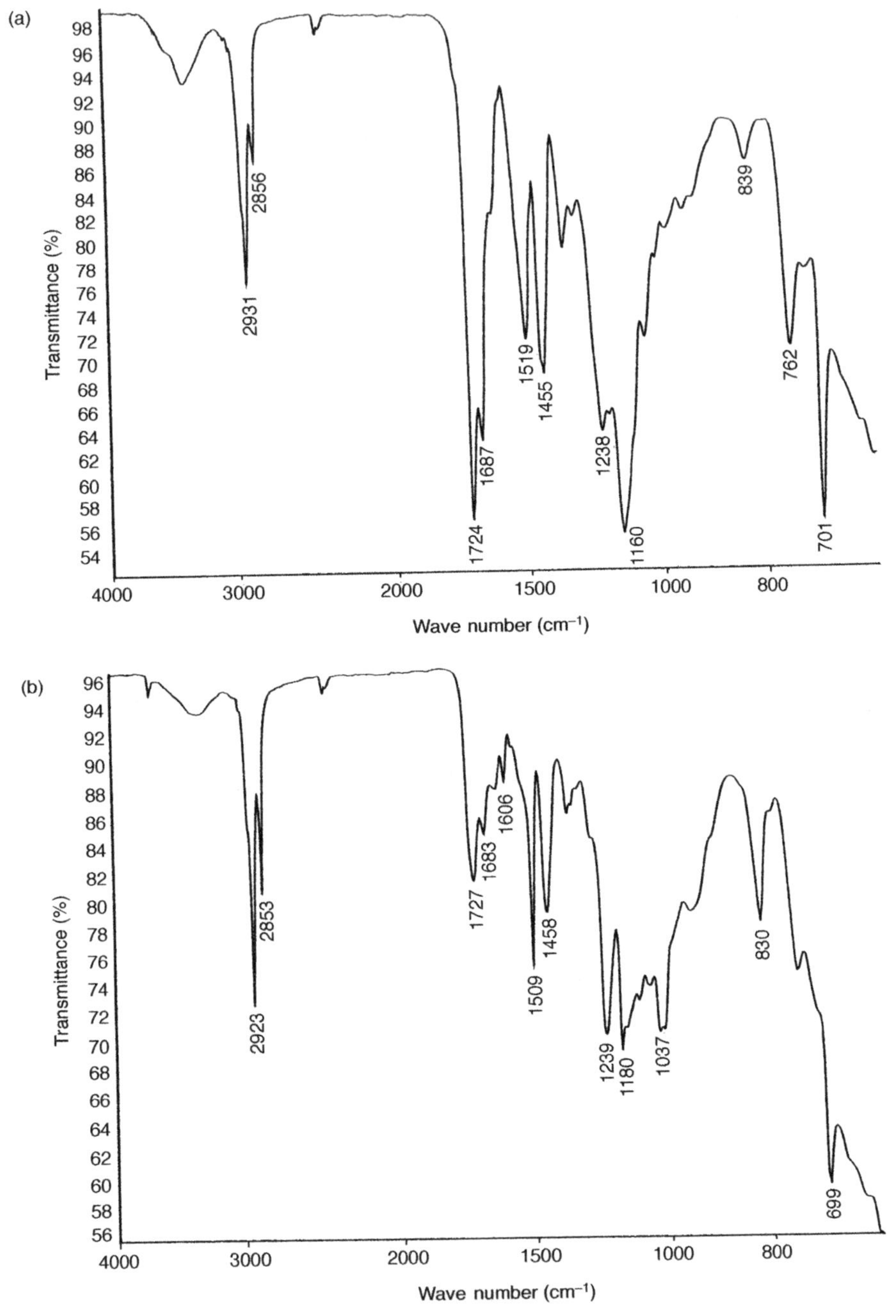

Figure 10.1 Attenuated total reflectance spectra obtained from (a) the front (non-peeling) and (b) the back (peeling) surfaces of a urethane topcoat.

Table 10.1 Adhesion as a function of mix ratio and cure for an epoxy coating.

Epoxy mix ratio	Epoxy cure treatment	Inter-coat adhesion rating[a]
1:1	Ambient	4B
1.5:1	Ambient	4B
1:1	Ambient + 3 days UV	2B
1.5:1	Ambient + 3 days UV	1B
1:1	Ambient + 13 days UV	0B
1.5:1	Ambient + 13 days UV	0B

[a]Measurements made by using ASTM D3359, Method B.
UV = Ultraviolet.

an ASTM G-154 accelerated weathering device equipped with ultraviolet (UV) A bulbs. The third set was cured overnight at laboratory ambient conditions and then cured for an additional 13 days in the accelerated weathering device. Once top-coated with urethane, all of the panels were allowed to cure for an additional week at laboratory ambient conditions prior to being evaluated for adhesion.

Adhesion testing was carried out in accordance with the tape adhesion procedure previously discussed (ASTM D3359). This method involves using a knife or other cutting tool to scribe a cross-hatch pattern into the panel, followed by the application of a special pressure-sensitive adhesive tape. The tape is then rapidly removed, and the adhesion is assessed in accordance with the method's rating system. Ratings can range from a '5B', corresponding to no loss of adhesion, to a '0B', corresponding to 65% or more delamination.

The adhesion testing on the laboratory panels produced some interesting results, as shown in Table 10.1.

The results indicate that mix ratio has no effect on inter-coat adhesion when there is no UV exposure, and only a slight effect when there is UV. However, even the properly mixed epoxy shows a serious drop-off in inter-coat adhesion after only 3 days of exposure to UV light, with more dramatic consequences being observed after 13 days.

Based on the field investigation and subsequent laboratory testing, the coating failure is due to the inability of even the properly mixed epoxy intermediate coat to withstand even a short exposure of heat, light and moisture. This weathering effect caused a visually undetectable deterioration in a very thin surface layer of epoxy. This loose, degraded epoxy resulted in peeling of the subsequently applied urethane topcoat.

10.2 BLISTERING OF AN EPOXY TANK COATING

A shop in the northeast United States fabricated and coated the structural steel for an above-ground potable water tank. The specification called for a near-white metal blast, followed by a two-component polyamide epoxy primer and a two-component polyamide epoxy topcoat. Both coats of epoxy were specified for 3- to 5-mil DFT, for a total coating thickness of 6–10 mil. The manufacturer's product data sheets for the epoxies listed 40 °F (4 °C) as the minimum application temperature. The work was done in the fabrication shop in the month of March, with air temperatures of approximately 50 °F (10 °C). The coated steel was shipped to the job site, where the tank was subsequently erected and put into service. The interior coating blistered within 6 months.

An investigation of the interior of the tank showed extensive blistering over the majority of the coated steel that was below the waterline. Most of the blisters were approximately $\frac{1}{16}$ in. in diameter, although some were as large as $\frac{1}{4}$ in. There was no exposed steel, and there was no visible corrosion. When the blisters were cut open, a clear, odourless liquid was contained inside, and the bare steel was exposed. When examined with a pocket microscope, no rust was present on the steel, and the blast-cleaning profile was evident. Although the adhesion was obviously very poor immediately at a blister, the adhesion between blisters was very good, and the coating was still quite hard.

The coating thickness was surveyed by using a magnetic DFT gauge. Although most of the steel exhibited blistering, there were some isolated areas of several square feet in size that were not blistered. The total coating thickness in the non-blistered areas ranged from 5 to 8 mil, while in the blistered areas, it ranged from 7 to 13 mil. Thus, the non-blistered coating fell within the specified thickness range, while the blistered coating was either within or above the specified thickness.

Samples of blistered, as well as tightly adherent, non-blistered, coatings were obtained. Furthermore, small quantities of the liquid inside the blisters were also obtained by using syringes. Note that when obtaining such samples, it is suggested that the surface of the coating be cleaned prior to puncturing it. Furthermore, some object other than the syringe being used to collect the sample should be used to puncture the blister. This could be a pin or even the needle from another syringe. The reason is to prevent contaminating, clogging or breaking the sampling syringe. Furthermore, in this case, although no paint was left over, fresh samples of primer and topcoat were obtained.

As usual, the investigation began at the microscope. Measurements of coating thickness confirmed the field observation: the blistered coating was either within or above the specified thickness, while the non-blistered coating was always within the specified thickness range. Furthermore, the back of the blister caps had a rough texture, which was apparently the replica of the blast profile. Except for an occasional particle of 'blasting smut' (either minute particles of mill scale dust or broken-down blasting abrasive), the back of the blister caps appeared clean. There was very little cross-sectional porosity in the coating, and there was no evidence of microscopic pinholes at the top of the blisters.

As with most two-component coatings, the possibility of mis-mixing usually needs to be investigated. Circumstantially, mis-mixing seemed an unlikely possibility for two reasons. First of all, even in the space between blisters, the coating was hard, tough and well adherent, with all of the characteristics to be expected of a properly mixed, well-cured epoxy. Secondly, the vast majority of the steel was blistered. While it is certainly possible to mis-mix an occasional kit of coating, it is unlikely that nearly all of the paint needed to coat the entire tank could have been mis-mixed, especially in a shop environment with painters familiar with epoxy coatings.

Despite the circumstantial evidence against mixing errors, this possibility still needed to be investigated. Again, IR spectroscopy proved to be a successful tool in determining the coating's mix ratio. Difficulties were encountered, however, in employing the polyamide region near 1650 cm^{-1}, since traces of moisture in some of the samples contributed to a water band near 1620 cm^{-1}, thus interfering with the analysis. Although thorough oven drying of the samples might have solved this problem, this was unnecessary because another pair of bands in the spectrum varied consistently with the mix ratio. These were the 1245-cm^{-1} band originating from an aromatic carbon–oxygen stretching vibration of the epoxy resin and the 1020-cm^{-1} band coming from the talc extender pigment. As suspected, mis-mixing of the epoxy was not a problem.

The next logical step in the analysis seemed to be a characterization of the liquid contained within the blisters. This was almost certainly water, but additional minor components might explain the blistering mechanism. The liquid was analysed for two components, that is, salts and solvents.

The theory behind osmotic blistering has already been discussed previously in this text. Briefly, such blistering can occur if there is a solution of high concentration behind a coating and a solution of low concentration on the other side of the coating. Since water can eventually permeate any coating, it can dissolve water-soluble contaminants either in the coating or on the substrate. This results in a solution of high concentration, which is separated from the solution of low concentration (the potable water in the tank) by a semi-permeable membrane (the coating). In such situations, the water-soluble material is usually a hydrophillic solvent or a salt.

Salt levels in the blister liquid were evaluated by using ion chromatography, which is very sensitive and which requires very small samples. Only chloride and sulfate were detected, at concentrations of approximately 15 and 5 ppm, respectively. This was somewhat less than the concentrations in the tank water, and it was unlikely that they were responsible for the massive blistering failure.

The presence of solvents in the blister liquid was evaluated by gas chromatography (GC). This showed that, although the blister liquid was almost entirely water, it also contained a small amount (approximately 200 ppm) of a glycol ether solvent. This same solvent was detected in large quantity (several %) in the liquid sample of the epoxy primer and was therefore not an adulterant. Such solvents typically show substantial water solubility and can clearly cause osmotic blistering. Therefore, the cause of the failure was the use of a coating that was not formulated properly for its intended service.

Although a relatively straightforward failure analysis, the above example brings out several interesting factors, one of which is thickness. Although a blistered coating was observed above the maximum recommended thickness, it was also encountered in thinner areas. The non-failing areas were always near the low end of the acceptable range. Therefore, as might be expected, thickness certainly contributed to the coating failure, but it (and therefore the applicator) could not be held as the primary factor in the failure. It simply illustrates that at a low enough thickness, enough of the solvent could escape so as not to cause osmotic blistering.

Secondly, it must be remembered that the steel was painted in the shop and then shipped in early spring to the job site for subsequent erection. Hence, the coated steel had weeks to cure in an outdoor setting with excellent ventilation, and yet enough solvent still remained to cause blistering. This illustrates the fact that, unless solvent escapes during or shortly after application, small quantities can remain trapped in coatings for a very long time (sometimes years). Whether this causes a problem depends largely on the type of solvent and the service environment.

Another interesting aspect of this failure is the fact that the microscopic examination of the blistered coating failed to detect any pinholes in the blister caps. Some people believe that such pinholes play a major role in blister formation since they allow water easy ingress to the coating. However, this is usually not the case.

For a liquid-filled blister to form, there must be a pressure (in this case, osmotic pressure) pushing against the back of the coating and forcing it off of the steel. If water has easy ingress through a pinhole, it also has an easy route back out, and it would be difficult to develop enough pressure to delaminate the coating. This might happen, however, if for some reason the coating had poor initial adhesion or if the ingress of water resulted in the corrosion of the substrate. In the latter case, the formation of the corrosion product might undercut and physically push the coating off of the steel.

10.3 CRACKING AND DELAMINATION OF EPOXY COATING SYSTEMS ON MASONRY WALLS

A new wing housing a testing and quality control laboratory was added to an existing food processing facility. Because of the sterile condition required in such a facility, the masonry block walls had to be effectively sealed with a glossy, hard coating system, resistant to rigorous daily cleaning. The system chosen was a two-component polyamide epoxy block filler and a two-component amine-cured epoxy topcoat. The amine-cured epoxy was selected because it cures to a hard, chemically resistant, glaze-like finish. It also had the advantage of high volume solids (95%), thus emitting minimal odour. The mix ratio of the primer was 1:1 by volume, while that of the topcoat was 4:1 by volume. The primer was to have been applied by airless spray, and then back-rolled to fill in the voids and porosity in the concrete block. It was then top-coated within 2 or 3 days.

Unfortunately, within 6 months, large areas of coating several square inches in size began delaminating from the walls. During the field inspection, it was noticed that, invariably, the delamination initiated at the mortar joints and then proceeded over the face of the block. Delamination occurred on the inside of the exterior walls and on the single interior wall which divided the laboratory area from an adjoining office area.

At the time of the investigation, approximately 5% of the coated surface area had already failed. Subjective probing with a utility knife in visually intact areas showed that poor adhesion was much more widespread. When an 'x' cut was made and the coating was probed at the intersection, it was not unusual to be able to 'pop' off chips as large as one square inch in size. In every case, a continuous coat of block filler was on the back of the disbonded coating. Some block filler also remained on the walls. The amount of remaining block filler varied from a thin residue to what was estimated as a mil or two in thickness.

When the chips were examined in the laboratory, two coats were visible, a highly filled white block filler that ranged from 2 mil up to approximately 12 mil, and a white topcoat at roughly 4–6 mil. The recommended topcoat thickness was 6–8 mil. The block filler did not really have a recommended DFT, but rather a recommended coverage rate. This is because the purpose of the block filler is principally to fill the voids and smooth out the roughness of the masonry block. Its actual thickness over the 'surface' of the block is usually only a few mils, since most of the coating is used in filling the surface roughness. Specifying a dry-film or even a wet-film thickness for masonry block fillers is usually an academic exercise, since there is no practical way for a contractor to actually measure it.

Even the microscopic measurement of masonry block filler thickness in the laboratory can be misleading and sometimes difficult to accurately communicate. With any heavily textured, porous, masonry surface coated with a block filler, huge variations in 'thickness' can occur simply because pits or 'bug holes' in the masonry might be 10- or 20-mil deep. In fact, even the reporting of an 'average' thickness, if it could be readily obtained, can be misleading.

When probing the back of the failing paint chips with a scalpel, it was observed that the block filler was rather soft, especially in the thicker areas. In the thinner areas, the block filler had a very unusual appearance in that it had somewhat of a honeycombed appearance, almost like it had melted and subsequently reformed.

Partly because of these observations, the solvent resistance of the block filler was evaluated. This was carried out in a rather subjective fashion by rubbing with a cotton Q-tip dipped in methyl ethyl ketone (MEK) while observing under the microscope. The block filler, unlike a typical polyamide epoxy, had very poor solvent resistance, with dramatic softening and rapid colour transfer to the Q-tip. The same test performed on the front, or topcoat side, had little effect.

An analysis of the block filler on the back of the failed coating by using IR spectroscopy quickly explained the reason for the poor solvent resistance. Rather than being a polyamide epoxy, the block filler was actually a vinyl acrylic. Such acrylics (sometimes referred to as acrylic-modified vinyls) are thermoplastic water-borne coatings with poor solvent resistance. They are much less expensive, and of lower performance, than the specified polyamide epoxy block filler.

In addition to the poor solvent resistance of vinyl acrylic block fillers, because they are water-borne coatings formulated with considerable amounts of surfactants, they have only a moderate resistance to moisture. When used in a dry, benign environment such as corridors in a school, they are entirely adequate. However, they will tend to soften in moist or humid environments, such as shower areas or areas subjected to repeated daily washings, such as a food laboratory.

In one respect, the failure analysis did not need to go any further. After all, the contractor had substituted a less expensive, inferior product for that which was specified. There was, however, one curious aspect of the microscopic observations that seemed less than satisfactorily explained. This was the honeycombed, melted appearance of the block filler in the thin areas.

Sometimes, when solvent-based thermoset coatings are applied over thermoplastic coatings such as vinyl acrylics, the solvents will partially solubilize, or 'melt', the surface layers of the thermoplastic. When the solvent-borne topcoat subsequently cures, the shrinkage stress can pull the weakened, underlying coating apart, and when the partially redissolved upper layer eventually dries out, the melted, honeycombed pattern is created.

The problem with the above theory in this case is that the amine-cured epoxy topcoat was a 95% solids material. It only contained 5% solvent, to meet both stringent volatile organic content (VOC) requirements and also to produce low odour. It did not seem like such a material could induce the effects in the block filler that were observed microscopically.

One possibility was that, in addition to using a substitute block filler, the contractor had also used a substitute topcoat. However, IR spectroscopic analysis of the failing topcoat and a control sample obtained from the manufacturer quickly ruled out this possibility. 'Mis-mixing' of the topcoat seemed an unlikely explanation, since neither component contained much solvent.

The investigation then shifted to an analysis of the topcoat for residual solvents. This was carried out by using headspace GC.

Headspace GC has been previously discussed (see Section 8.3.4). Basically, this technique involves placing samples in a closed septum vial and then heating the vial. Any solvents that remain in the sample will partially evaporate. A gas-tight sampling syringe is then used to remove some of the air (headspace) from the vial, which is rapidly injected into the chromatograph. The technique is not quantitative but can usually quickly determine if a coating was thinned with an unauthorized solvent prior to application. In this case, not only were material safety data sheets available which listed the solvents in the coating, but liquid control samples were also available for analysis.

The GC analysis showed easily detectable levels of toluene and methyl isobutyl ketone (MIBK) in samples of failed coating from the job site. These solvents were not listed on the material safety data sheets, nor were they detected in the liquid control samples. It is extremely likely that the contractor added some unauthorized thinner to the epoxy topcoat, perhaps as a means of extending its short (20 minutes) pot life. This might also explain why the topcoat was consistently somewhat thinner than its recommended thickness of 6–8 mil. If the contractor applied a thinned coating at the wet-film thickness specified for an unthinned coating, the DFT would be less than expected.

10.4 DISCOLOURATION OF COIL-COATED ALUMINIUM SIDING

Within a period of 9 months, a manufacturer of coil-coated aluminium siding had received numerous complaints about discolouration of the siding. Although records were incomplete, it appeared that the bulk of the siding in question had been coated the previous year, within a period of approximately a month. The complaints came through several contractors who had used the material on houses scattered in several states along the east coast of the United States. Nearly all of the problems were with white, off-white or beige colours.

The process used in the coil coating facility involved a simple pre-cleaning of the aluminium coil, followed by detergent washing and rinsing. A proprietary chromate conversion coating was then applied, rinsed and sealed via a chromic acid rinse. After the coil went through a dry-off oven, the coating was finally applied and baked. The topcoat was a polyester cross-linked with a melamine–formaldehyde resin. The dwell time in the oven was approximately 50 seconds, and the coating supplier had specified a peak metal temperature of 400 °F (204 °C). Although there was a temperature probe located near the centre of the oven, it was not calibrated on a regular basis, nor were there any measurements made of the peak metal temperature.

Samples that were furnished included five pieces of either white or off-white siding taken from different houses in two different states (New Jersey and North Carolina). In one case, the homeowners had actually saved some of the leftover siding, so that a piece of uninstalled, unexposed siding was also furnished. Although it was difficult to know exactly which coil had been used on which house, the manufacturer did provide three small coil-coated retains from runs during the general time period in question, along with a pint of a recent batch of paint.

When the various samples were examined, it was seen that the discolouration was in the form of uneven grey or dark grey splotches, rather than a uniform discolouration. When examined with a stereo zoom microscope, the discolouration appeared to be due to fine particles of embedded dirt or debris. Closer observations with a scanning electron microscope (SEM), equipped with an X-ray spectrometer, confirmed that fine silicon-rich particles were indeed embedded in the surface of the coating. The discolouration appeared to be a simple physical effect and not some type of chemical- or light-induced form of deterioration.

During the course of the observation, attempts were made to remove the discolouration by cleaning. Water, along with a common household cleaner (including bleach), had little effect. A strong, industrial-strength cleaner performed reasonably well. It was also observed that MEK on a Q-tip rapidly removed the discolouration. Closer observations, however, actually showed that the MEK was not so much removing the discolouration as it was the paint. This seemed to be an unusual behaviour for a baked polyester–melamine coating. Similar solvent sensitivity was observed on the unexposed, leftover piece of siding, as well as on the coil retains.

The solvent sensitivity of the various samples suggested two possibilities. Either the coating was not a polyester–melamine or it had been under-baked. A third possibility, that is, that the poor solvent resistance was a consequence of chemical- or sunlight-induced deterioration, was ruled out since unexposed samples were also solvent sensitive.

IR spectroscopy easily confirmed that the coating was a polyester–melamine. In fact, the spectra of the field and retain samples matched the spectrum of the recent batch of coating.

There are certain types of coatings whose degree of cure can be approximated by using IR spectroscopy. However, pigmented polyester–melamine coatings are not one of these. Therefore, differential scanning calorimetry (DSC) was employed in the further analysis.

As previously discussed (see Section 8.7.3), there are two possible approaches to measuring the degree of cure by using DSC, that is, investigating the glass transition temperature (T_g) or the residual curing exotherm. On coatings that are close to being fully cured, the residual exotherm can often be rather small and sometimes difficult to accurately measure. In such cases, the T_g is often a more sensitive and accurate measure of a coating's degree of cure.

One problem with this line of testing was establishing the correct T_g for this particular coating. The only coated samples available were either 'failing' ones or unexposed samples that were suspect by virtue of displaying the same solvent sensitivity as the discoloured ones. The alternative was to produce a control sample from the quart sample of a currently used batch of coating.

The control sample was made by drawing down a film of this coating, at a thickness of approximately 2 mil 'wet', on a 3 × 5 in. thin-gauge aluminium panel. A small thermocouple was attached to the panel, and it was then placed in a laboratory oven at 520 °F (271 °C) until the thermocouple measured a temperature of 400 °F (204 °C) (the recommended peak metal temperature). This does not reproduce the conditions inside a coil coating oven, where air temperatures can be well over 600 °F (316 °C) and dwell times can be very short, but it should provide a reasonably representative control sample.

Table 10.2 Glass transition temperatures (T_gs) of the various coil coating samples.

Sample	T_g^a (°C)
Failing 1	22
Failing 2	24
Failing 3	34
Failing 4	22
Failing 5	25
Homeowner 'retain'	26
Coil retain 1	25
Coil retain 2	27
Coil retain 3	36
Laboratory 'control'	35

[a]Measured at onset.

The DSC measurements of the various samples produced interesting results. Although exotherms were small, broad and difficult to accurately measure, the glass transition temperatures were relatively distinct in all cases. The results obtained are shown in Table 10.2.

The results clearly showed a trend, in that most of the failing samples and associated retains had glass transition temperatures ranging anywhere from 13 to 8 °C below that of the laboratory-prepared control sample. Most of the glass transition temperatures were below 27 °C (81 °F).

The fact that most of the failing samples had significantly lower T_gs than the control sample was a clear indication that they were either under-baked, or perhaps that they simply represented defective batches of coating that would not cure properly even if baked correctly. In either case, not only would the lower cross-link density of these coatings result in inferior chemical and physical properties, but the fact that their T_gs were only in the 22–27 °C range also means that they will soften during hot summer weather. A softer coating is more likely to pick up and retain wind-blown dust and dirt than a harder coating, thus accounting for the splotchy surface discolouration. The examination is also consistent with additional scanning electron microscope–energy-dispersive spectrometer (SEM-EDS) results, which showed fine silicon-rich particles (dirt) embedded in the coating surface.

The above example points out some interesting aspects of practical failure analysis. It would have been desirable to conclusively establish whether the lack of cure was due to defective batches of paint or inadequate baking. Although the lack of reliable temperature monitoring at the coating facility is certainly cause for suspicion, it is not proof of under-baking. If liquid retain samples of the various batches in question were available, either their composition or, more easily, their curing characteristics could have been evaluated against the current batch of coating. Unfortunately, the coil coater did not keep any retains, nor were records kept of the batch numbers of the paint used. An important question could not be answered because of the lack of certain samples. Unfortunately, this is not a rare occurrence.

A second interesting aspect of the above study is that the data are not entirely consistent. One of the discoloured samples had a T_g of 34 °C, essentially equal to the control T_g of 35 °C, and one of the coil retain samples from the time frame in question had a T_g of 36 °C, actually a degree higher than the control.

One possible explanation for this is the experimental error of the analytical technique. If, for example, the T_g results are only accurate to ±2 °C, then it is still possible that the true T_g of the control could be as high as 37 °C, and the true T_gs of the failing sample and the coil retain could be as low as 32 and 34 °C, respectively. In this case, all of the suspect samples have T_gs below that of the control. Although the precision of the analytical method can be established by carrying out a large number of replicate analysis, in many cases, this extra cost burden cannot be justified to the client. In addition, there are times when there is insufficient sample to perform such exhaustive testing.

A second possibility to explain the above data is that the 'control' is not really a true control. The type of oven used in the laboratory was much different to the type used on the coil line. It is possible that the T_g of a true control may have been several degrees higher than the laboratory-prepared version. It is also possible that coil retain 3, with a T_g of 36 °C, did not actually represent a 'failing' batch of coil. Retains were simply selected by the coil coater from runs produced in the general time frame of the reported failure. This particular retain may actually have represented a run of coil that was satisfactory.

Another thing to consider is that coating failures are really statistical in nature. Things have a certain probability of failing, and this probability can be increased or decreased due to certain variables (paint thickness, bake, % of catalyst, etc.), some of which may be totally unknown (a brief wind storm, for instance, or some chemical excursion). To use an analogy, while it is well known that cigarette smoking greatly increases the risk of lung cancer, not everyone who smokes develops cancer. Therefore, in failure analysis, as well as in medical studies, there will always be exceptions to the general rule.

10.5 DISCOLOURATION AND DELAMINATION OF PLASTISOL

A new, large warehouse facility in the south-west United States was constructed by using roofing panels consisting of coated galvanized steel. The panels, which were light blue in colour, were installed in the fall. After approximately 1 year, the light blue topcoat was starting to take on a greenish hue, and by the following summer, a little over a year and a half after installation, some sporadic flaking and disbonding were observed.

Since the panels were purchased prefabricated, there was very little information available concerning surface preparation or even the type of coatings. It was discovered, however, that the light blue topcoat was a plastisol. Plastisols are basically highly plasticized poly(vinyl chloride) (PVC) material (see Section 5.10.2).

A visit to the warehouse showed that the extent of spontaneous delamination was probably at most only 1% of the total roof area. However, when tape-adhesion testing was performed, pieces of coating as large as the tape itself could be easily disbonded. In every case, it appeared as though the failure was between an off-white primer and the galvanized steel.

The warehouse roof had only a slight slope to it, and there did not appear to be any pattern to either the poor adhesion or the discolouration that, although somewhat splotchy, was evident to varying degrees on most of the roof. There was, however, a smaller office building on the north side of the warehouse. This building was physically attached to the warehouse, but was only a single-storey construction, compared with the two stories of the warehouse. As a consequence, the roof of this structure was generally shaded. Not only were the roof panels over the office annex only slightly discoloured, but the coating adhesion was also dramatically better. In addition to flaking paint chips, sections of two panels, one from the warehouse and one from the office annex, were provided. A piece from a leftover, uninstalled panel that had been stored in the warehouse was also provided.

When examined in the laboratory, the three pieces of siding were dramatically different in appearance. The uninstalled panel was a semi-glossy light blue, with excellent adhesion. The panel from the warehouse roof was badly discoloured, with one large (approximately 2×3 in.) area where the coating had already peeled, and with poor adhesion in the remaining apparently intact areas. The panel from the office annex was intermediate in both discolouration and adhesion.

Relatively large paint chips could be easily disbonded from the warehouse panel, thus revealing a clean, shiny galvanized substrate. When examined microscopically, the back of the off-white primer appeared clean, and there was no evidence of whitish zinc corrosion product. The primer was approximately 0.5-mil thick, while the discoloured topcoat was approximately 2-mil thick. There were also numerous small microscopic cracks in the topcoat.

Only very small chips could be disbonded from the office panel. The thickness of the primer and topcoat were similar to those of the warehouse panels. There was less cracking of the topcoat.

Although shavings or scrapings could be removed from the uninstalled panel, it was extremely difficult to remove any intact paint chips. No microscopic cracking was observed in the topcoat, and the total coating thickness was similar to those measured on the other samples.

Based on the visual and microscopic observations, it appeared very unlikely that the cause of the failure was any type of contamination on the galvanized steel prior to painting. There was no visual or microscopic evidence of contamination (although many types of contaminants are not visually detectable). Furthermore, had contamination been present, it seems almost impossible that it would have been restricted only to those panels that happened to have been installed on the warehouse roof. At least some of the office annex panels would also likely have been contaminated and would have shown similar poor adhesion. It would also be difficult to explain how a surface contaminant could have caused cracking and discolouration of the topcoat.

One possible explanation for the poor adhesion that could not be entirely ruled out at this point was that of either a failure within the galvanizing or within a thin layer of oxidized galvanizing. If the galvanizing was somehow defective, it could possibly be splitting within a thin upper layer, in which case the delamination was really not a paint failure, but rather a galvanizing failure. Again, however, defective galvanizing would hardly have been confined only to those panels that were installed over the warehouse, with no defective panels having been installed over the office annex.

If for some reason the galvanizing was oxidizing prematurely (perhaps as a consequence of the micro-cracking causing the coating to be more porous to water and oxygen), a thin layer of zinc corrosion product might be responsible for the poor adhesion.

Neither possibility seemed very likely, primarily because the surface of the freshly exposed galvanizing was clean and shiny and apparently not deteriorated. However, the possibility of such a failure mechanism was explored by utilizing an SEM-EDS system.

The procedure involved examining both the back of a disbonded paint chip from the warehouse panel, as well as the galvanized surface of a small specimen cut from this panel. The result was not unexpected. The only elements detected on the surface of the metal panel were zinc and trace amounts of oxygen, titanium, aluminium and silicon, due to residues or islands of primer. The elements associated with the primer were detected in abundance on the back of the delaminated paint chip. Except for a few small isolated specks, no zinc was detected on the back of the failing paint chip. Therefore, the failure was not related to a galvanizing problem or to premature corrosion of the galvanizing.

Based on all of the observations and analytical results obtained thus far, it clearly appeared as though the discolouration, cracking and disbonding were due to an environmentally induced degradation of the coating system. IR spectroscopy was employed to investigate the composition of the failing and non-failing samples.

Since the primer was only 0.5-mil thick and very well bonded to the topcoat, its spectrum was obtained by using ATR spectroscopy. Although large chips were readily obtainable from the failing warehouse sample, only chips no larger than perhaps $\frac{1}{16}-\frac{1}{8}$ in. could be obtained from the office sample, and almost no usable, intact chips could be obtained from the uninstalled sample. A semi-micro ATR accessory was used to obtain spectra of primer from failing warehouse chips and 'non-failing' office annex chips. The IR analysis showed that the two primers were extremely similar and consisted of acrylic containing a silicate-extender pigment and likely some titanium dioxide. This was consistent with the SEM-EDS results.

It was possible, working at a stereo zoom microscope, to obtain sufficient scrapings of topcoat from all three samples such that transmission spectra could be obtained via the potassium bromide pellet technique. As expected, the main features of the topcoat were consistent with PVC. There was also a strong ester carbonyl band, near 1730 cm^{-1}, that is not a feature of PVC and which was probably due to a plasticizer.

The intensity of the ester carbonyl band near 1730 cm^{-1} varied somewhat with respect to some of the characteristic PVC bands, although the differences were not entirely consistent among replicate spectra obtained from the various samples. Although these variations suggested that chemical changes were going on in the coating, they were not conclusive. Gel permeation chromatography (GPC) was therefore performed.

As previously discussed (see Section 8.4), GPC is a form of high-performance liquid chromatography (HPLC) that separates molecules based on their hydrodynamic volume or effective size in solution. This technique is extremely useful for characterizing the molecular weight distribution of polymers and for separating low-molecular-weight additives, such as plasticizers, from the high-molecular-weight polymeric binder. It is an ideal choice for characterizing thermoplastic coatings such as the plastisol used on the roof panels.

A very common solvent that is compatible with most GPC packing materials is tetrahydrofuran (THF). The latter is also a very good solvent for most thermoplastic coatings, especially considering that in order to obtain good chromatographic results, the sample concentration should only be approximately 1.0% or less. It was easy to disbond sufficient paint chips from the failing warehouse panel to dissolve in THF, although the coatings were dissolved directly off of the office panel and the uninstalled panel, due to their better adhesion.

One source of error in the above procedure was the inability to dissolve the topcoat in THF without also dissolving the primer. However, since the primer was relatively thin compared with the topcoat, it was felt that it would have a minor influence on the results. Furthermore, the error imparted by the primer would be relatively constant for all three samples.

The GPC results turned out to be remarkably simple to interpret (which is not always the case). The three samples produced chromatograms that displayed a large, broad peak with a single maximum, along with a relatively sharp peak that eluted several minutes after the broad first peak. These fractions were collected, the THF was evaporated and IR spectra were obtained from the residues. The spectra showed that the broad peak was a PVC resin and that the lower-molecular-weight component was a phthalate plasticizer.

Although the three chromatograms had the same general appearance, they were not identical. They differed in both the shape and retention time of the PVC peak and in the relative intensity of the plasticizer peak. It was important to determine both the molecular weight of the PVC component and the relative amount of the plasticizer component.

Prior to injecting the samples, the chromatograph had been calibrated for molecular weight by injecting polystyrene standards of known molecular weight and relying on the instrument's software to construct a linear plot of the logarithm of the molecular weight as a function of the retention time. The software was then able to calculate the average molecular weight of the PVC sample. As pointed out in the above discussion of GPC, this results in only an approximate molecular weight, since the standards employed were not PVC standards, but rather polystyrene standards. However, the importance of this investigation is not so much the sample's absolute molecular weights, but rather their relative molecular weights.

Likewise, the actual concentration of the plasticizer was not determined, but rather the relative amount of the plasticizer. This was done simply by measuring the area of the plasticizer peak as a percentage of the total area of both the PVC peak and the plasticizer peak. The results of the analysis are shown in Table 10.3.

As the results show, two things were happening to the plastisol topcoat. The PVC binder was 'losing molecular weight' and the coating was losing plasticizer. Furthermore, based on the location of the samples, the deterioration was either sunlight induced, heat induced or a combination of sunlight and heat. This conclusion is based on the fact that the warehouse roof received the full, unmitigated effects of sunlight and heat, while the roof on the office annex was shaded for much of the day. The uninstalled panel, of course, was kept indoors.

PVC resins are inherently brittle and are therefore seldom if ever used for coatings without some kind of modification. In this case, the modification was the large amount of plasticizer, which accounts for the generic name of 'plastisol'. However, the phthalate plasticizer used in the coating on the warehouse panels apparently had sufficient volatility to partially evaporate from the coating as it baked in the hot summer sun. This would cause the topcoat to embrittle with age and explains the microscopic cracking seen in the samples.

The second problem, that is, the loss of molecular weight of the PVC binder, is also heat and light related. It is well known that, unless properly stabilized,

Table 10.3 Molecular weight and plasticizer content data obtained from gel permeation chromatography analysis of plastisol topcoat samples.

Sample	Molecular weight[a]	Plasticizer content[b] (%)
Uninstalled	207 000	31.2
Office annex	189 000	23.3
Warehouse	151 000	18.7

[a] Weight-average molecular weight (M_w).
[b] Relative values.

PVC can deteriorate by an 'unzipping' mechanism, resulting in the generation of hydrochloric acid (HCl).

Depending on conditions, this thermal breakdown can begin at temperatures as low as 230 °F (110 °C) [1], or even lower with the combined effect of UV light. Both heat and UV light are in plentiful supply on an unshaded metal roof in the south-west United States.

The unzipping reaction results in the formation of an extended structure of conjugated double bonds. Not only does this structure allow the molecule to absorb light in the visible region of the electromagnetic spectrum, thus imparting colour to the resin (and probably accounting for the discolouration of the paint), but it also makes it much more susceptible to oxidation. The latter can result in additional cross-linking and embrittlement, or even lead to bond breaking, thus accounting for the observed decrease in molecular weight.

The combined result of a deteriorated, embrittled coating is its inability to flex with movement in the thin-gauge galvanized steel substrate. This apparently resulted in both cracking and loss of adhesion, since the acrylic primer had better adhesion to the plastisol topcoat than it did to the galvanized substrate. Another possibility is that the micro-cracking as well as the general deterioration of the coating, resulted in increased permeability, thus allowing water to more easily penetrate, undercut and affect the primer–substrate bond. This seems an unlikely mechanism in light of the very dry environment in the south-west United States.

10.6 DELAMINATION OF FLOOR COATING

The concrete floor of a new corporate aircraft hanger was coated with a medium gloss, two-component, water-borne epoxy. The specification called for acid etching of the floor, followed by two coats of the water-borne epoxy, at a thickness of 2–4 mil per coat. The water-borne epoxy was recommended for use on floors and was advertised as having good impact, abrasion and stain resistance. Its mix ratio was 1 : 1 by volume, with no induction time required. The data sheet further stated that it would reach full cure in 10–14 days, depending on the temperature and humidity.

The facility was not put into service until nearly 3 weeks after the painting was completed. Unfortunately, within 2 months, sporadic instances of disbonding began occurring. This was first noticed where a forklift had gouged the coating, but other areas that were not subjected to such obvious physical abuse also began to peel. Once peeling had started, maintenance employees at the hanger were able to easily pry off pieces of coating several square inches in size by using only dull putty knives. Samples of disbonded coating were submitted, along with two small core samples with coating still attached, taken within inches of where the coating had been peeled away.

Although no professional paint inspector had been on the job site, the general contractor had personally witnessed some of the surface preparation and painting activities of the painting subcontractor. He was able to personally vouch for the fact that the subcontractor had acid etched and rinsed the floor, and that it had been allowed to dry for at least 2 days prior to paint application. Although he could not guarantee that all of the paint had been mixed in the proper proportion, in at least one instance, he clearly remembered seeing a painter adding the two components together and blending them with an air-driven power mixer.

When examined microscopically, the paint chips varied in thickness from approximately 5–10 mil, generally within or slightly above the recommended range. Because the application consisted of two coats of the exact same material, it was difficult to determine if this was one thick coat or two thinner coats. There were a few voids in the cross section of the chips, but nothing out of the ordinary. The coating was tough and slightly flexible, with physical characteristics similar to those of a drawdown of the coating prepared in the laboratory.

The most interesting finding was that the back of the disbonded chips was smooth and almost glossy. When chips were removed from the concrete core samples, the surface of the underlying concrete was very smooth, with no evidence of etching.

When coatings are applied to concrete floors, the concrete must be clean, dry and abraded, in order to obtain maximum adhesion. The two most common ways to roughen up a concrete floor are blast cleaning and acid etching. The latter technique, which was employed on this job, involves applying a solution of muriatic (hydrochloric) acid to the concrete. When the acid reacts with the concrete surface, vigorous bubbling occurs. When carried out properly, this results in a surface with a texture similar to medium sandpaper (Figure 10.2). The residues are then thoroughly rinsed off with copious amounts of fresh water. In some cases, a vacuuming step is also performed in order to remove any residual dust-like debris not removed by rinsing.

As stated above, the general contractor personally saw the painting subcontractor applying the acid and vaguely remembered noticing at least some amount of bubbling. Therefore, why was the concrete surface so smooth? There are a number of possibilities for explaining this.

It is not uncommon for substances known as membrane-curing compounds to be used on concrete floors. Concrete requires moisture to cure, and sometimes these compounds, which are often simple organic oils or waxes, or acrylic polymers, are applied to the concrete surface to prevent the surface from drying out. If present, they can often interfere with the etching action of muriatic acid and need to be physically removed prior to acid etching.

The contractor was not aware of any membrane compounds having been used at the job site. Nevertheless, a light surface scraping of one of the concrete cores was analysed by IR spectroscopy. Only inorganic materials were observed, along with some other bands that could be accounted for by residues of the water-borne epoxy. Although it was unlikely that the coating, being water-borne, would have had enough solvency to lift any membrane-curing compounds from the floor, the back of it was also examined by IR spectroscopy, using ATR.

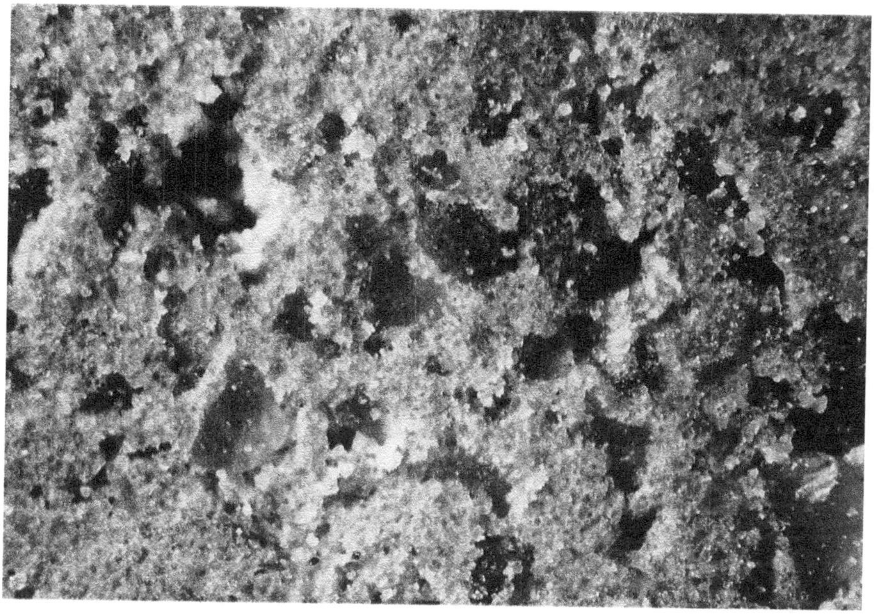

Figure 10.2 Photomicrograph showing the typical appearance of acid-etched concrete when examined at approximately 15× magnification.

Detecting an organic contaminant on the back of an organic coating is not always an easy thing to do, especially if the contaminant is a simple alphatic hydrocarbon, the characteristic bands of which often overlap or even coincide with those coming from the coating material itself. An important first step, therefore, is to obtain a reliable control spectrum of the coating.

A drawdown had been prepared in the laboratory of a properly mixed sample of the water-borne epoxy. Initially, the ATR spectrum of this coating, although quite similar, did not match the spectrum of the back of a disbonded chip. Although there were not any new or missing bands, some of them were present at different relative intensities. Did this mean that there was an organic contaminant on the back of the coating, contributing to the greater intensity of some of the organic bands relative to the pigment bands?

As it turned out, the spectrum of the drawdown was obtained by slicing through the coating to remove a small chip, the front of which had been analysed. Occasionally, the spectrum of the front of a chip will differ from that of the back of the chip, due to a natural tendency for some of the ingredients to stratify, perhaps because of density or solubility differences. When a second spectrum was obtained of the back of the coating removed from the drawdown, it was virtually identical to that obtained from the field sample. There was no reason to believe that a membrane-curing compound had impeded the acid-etching process.

Another piece of evidence was obtained that also tended to prove that no membrane-curing compound had been used. A section of coating was peeled away from one of the core samples to reveal the smooth underlying concrete. While observing under the microscope, a medicine dropper was used to apply two or three drops of 10% muriatic acid to the smooth concrete surface. Vigorous bubbling began to occur immediately and continued for about 2 minutes, after which time, it subsided noticeably. A squirt bottle was used to rinse the area, which was then patted dry with a paper towel. The concrete surface to which the acid had been applied was dramatically rougher than the neighbouring area. This was a clear indication that there was no curing compound present on the concrete, or at least not to such an extent that it materially hindered the acid-etching action.

Therefore, why was the concrete surface so smooth? Further discussion with the contractor revealed that the floor had been finished by steel towelling. Steel towelling can result in a very smooth, dense, hard concrete surface. The effect is very aesthetically pleasing, but it does make etching somewhat more difficult and certainly more critical. It is also possible that the crews doing the etching diluted the muriatic acid too much, did not apply enough of it or did not leave it on the floor long enough. Any or all of these circumstances, coupled with the glossy steel trowel finish, could have combined for a poor etching job. It must be remembered that the goal of acid etching is not simply to pour acid on a concrete floor. The goal is to create a rough, sandpaper-like finish to promote paint adhesion. In this case, the floor should have been inspected after the acid was applied and rinsed, to confirm that it had done its job. If not, a second application of acid should have been performed.

10.7 DELAMINATION OF DRY-FALL ALKYD

The owner of a tool and die shop wished to have the structural steel columns, which were old and dingy grey in colour, painted white. He also decided to put a coat of white paint on the old, bare, galvanized ceiling decking. Not only would this give his shop a more pleasant appearance, but it would also improve the lighting for the workers. After talking with various painters, the owner decided to use a dry-fall alkyd coating for this work.

Dry-fall alkyds are usually highly pigmented and have a relatively fast blend of solvents. As a consequence, any over-spray dries very quickly and, after falling a few feet, will end up as a loose dust which can be easily vacuumed or otherwise removed from floors and machinery.

An inspection of the underside of the galvanized metal ceiling panels showed that they were in good condition, and the old, existing paint on the columns, thought to be an oil-based coating, appeared to have good adhesion to the underlying steel. The only surface preparation employed was steam cleaning with a detergent to remove any soil or oils, followed by a clear water rinse. The whole cleaning and painting operation was conducted over a brief 3-day shutdown of the shop. Approximately 1 year after the job was completed, problems began occurring with the dry-fall alkyd disbonding from the ceiling. Sometimes, chips 1 or 2 $in.^2$ in size would fall to the floor or land on equipment. There was also an area on one of the structural steel beams where some of the paint was lifting from the original grey coating. However, this seemed to be an isolated failure, as the owner reported that the paint on all of the other columns appeared well adherent, even when pried at with a knife.

Various samples were submitted to the laboratory for analysis. These included coatings that could be peeled from one area on the steel column, plus some small shavings from another steel column that had good adhesion. Some chips that had fallen to the ground from the ceiling panels were also submitted along with other large chips that the owner had freshly peeled from a ceiling panel. When the painter had finished the job, he had about a half gallon of paint left over, which he left with the shop's maintenance man in case any minor touch-up work needed to be done. The laboratory was provided with a quart of this paint.

As usual, the investigation began at the microscope. The chips that had fallen or been peeled from the ceiling panels appeared clean and smooth on the back, and ranged from 3 to 6 mil in thickness (the recommended thickness was 2–3 mil per coat). The front of these chips had a rough, pebbly texture, with numerous small globules of white paint (Figure 10.3). This appearance is typical of dry-spray. The latter, as the name implies, is the consequence of the atomized paint drying out to such a degree that by the time it hits the substrate, it has insufficient solvent to allow it to flow and level properly. This can be caused by several things, such as excessively high ambient temperatures, too great a distance between the spray gun and the substrate, a poor solvent blend or improper atomization. The latter is not usually a problem with airless application, which was used at this job. As might be expected, dry-spray is more commonly encountered with dry-fall coatings than with other types.

The cross section of the chip did not show any appreciable porosity, with only an occasional void being present. The chips were, however, quite brittle. The samples from the steel column varied little in appearance compared with the samples from the ceiling, except that the back was not quite as smooth, and the thickness was only 2–4 mil. Whenever a coating fails from galvanizing, it makes sense to look for zinc on the back of the coating. The presence of a thin layer of zinc could mean that the failure actually occurred in the galvanizing itself, or it could indicate the presence of zinc corrosion products that are responsible for the poor adhesion. One way of conducting such an analysis is via the use of an SEM-EDS.

When the sample from the ceiling was examined in this way, substantial amounts of carbon, calcium, titanium, aluminium and silicon were found, along with a lesser amount of oxygen. This indicated the presence of calcium carbonate,

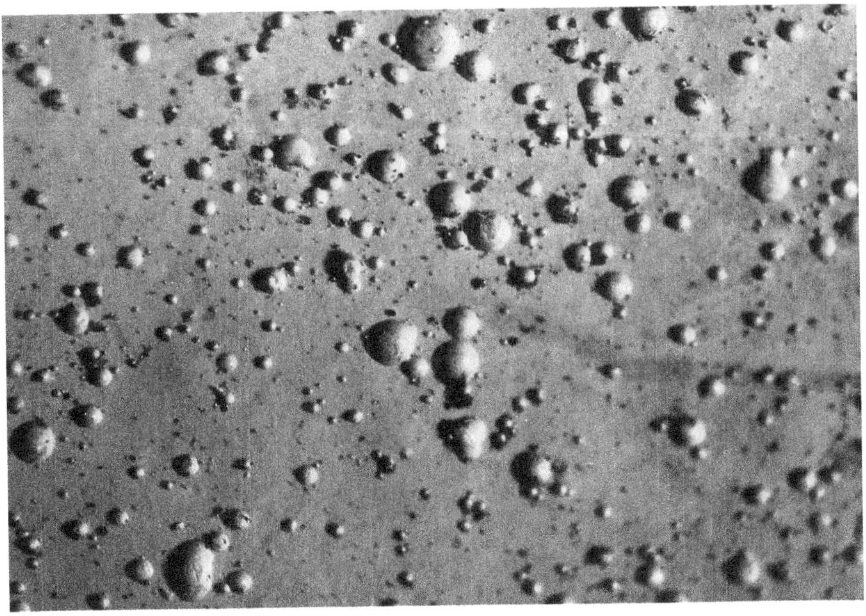

Figure 10.3 Photomicrograph of globules of dry-spray on the surface of a dry-fall alkyd coating.

titanium dioxide and aluminium silicate pigments. Only a very slight trace of zinc was found. Except for the latter, the composition was extremely similar to both the front of the chips and to a drawdown of the leftover paint. Analysis of the back of one of the chips that had disbonded from the previously painted structural steel column showed only those elements associated with the dry-fall alkyd.

In addition to the SEM-EDS investigation, the samples were also analysed by using IR spectroscopy. Initially, this was carried out by combining sample scrapings with potassium bromide and forming them into pellets. It was quickly found that this yielded very limited information. Because of the relatively small amount of resin and large amount of pigment, all this testing did was to confirm what the SEM-EDS analysis had already indicated, that is, that the paint contained large amounts of calcium carbonate, silicates and titanium dioxide. Calcium carbonate, when present in large amounts, is particularly troublesome to a spectroscopist, as its large, broad band near 1430 cm^{-1} can obscure many other important, weaker bands coming from the resin component.

Based on the results of the potassium bromide pellet spectra, there seemed little chance of obtaining useful data by running ATR spectra. However, the possibility of organic contamination on the back of the chips had not been ruled out.

At this point, it seemed fruitful to consider what type of contamination and/or failure mechanism might be at work, and then try to devise experiments that would prove or disprove these theories. Since 99% of the problem was with the ceiling, this became the first consideration.

In contrast to the steel column, which may have been contaminated with an isolated splash of machine oil or other fluid, it seemed rather unlikely that such contamination could have occurred on a wholesale basis on the ceiling. In addition, visual and microscopic observations, along with SEM-EDS investigations, ruled out common contaminants such as dirt or zinc corrosion products. Two likely possibilities remained, that is, oils from the original galvanizing process or saponification of the alkyd coating. In both instances, because of spectral interference from the large amounts of pigment, solvent extractions seemed the most likely avenue of success.

In order to check for oil contamination, a solvent extraction was performed using hexane. The latter is a very mild solvent that can dissolve grease and common oils, but which would dissolve very little of the alkyd resin from the topcoat. Since only the back of the chips was important, a rather large, irregular chip was angled over a small beaker. A disposable pipette was then used to apply approximately 2 ml of hexane to the back of the chip, such that it ran down the chip and collected in the beaker. This same quantity of hexane was used to wash the back of the chip several times. Eventually, the hexane was carefully evaporated onto a potassium bromide crystal. No residue was visually apparent, and the resultant spectrum showed only the weakest of bands, which were very similar to what was obtained via a similar procedure carried out on a laboratory-prepared drawdown. Apparently, there was little or no grease or oil contamination on the back of the chip.

Hexane is a non-polar solvent and is a good choice when looking for oil or grease. However, what if the alkyd resin had been saponified? Would hexane have revealed the presence of a saponified alkyd?

Since alkyd resins are basically oil-modified polyesters, they are unstable when exposed to moisture and alkalinity. Under such conditions, the ester linkage can hydrolyse (saponify), which results in the formation of both alcohols and carboxylic acids. Since this reaction occurs under alkaline conditions, the salt of the acid is formed, rather than the free acid. Since these acid salts are polar, they will not dissolve in weak hydrocarbon solvents such as hexane, but require stronger polar solvents. Therefore, MEK was chosen.

When the extraction procedure was repeated using MEK, the solvent clearly removed material from the back of the chips. It also removed material from the lab drawdown. However, the two spectra differed dramatically. Although both had features characteristic of alkyd resin, such as bands near 2900, 1730, 1280, 1120 and 1170 cm^{-1}, the extract from the disbonded chips had additional weak bands near 1580 and 1400 cm^{-1}, which are characteristic of carboxylic acid salts. It appeared as though at least a small amount of saponification had taken place. However, this was detected in only two of the three chips examined.

Since the facility was not air conditioned, humid summer days could account for the moisture needed for saponification. However, what was the source of the alkalinity? The answer is the galvanizing.

Virgin galvanizing, or zinc, is not an alkaline surface. However, when it begins to oxidize, it forms alkaline species such as zinc hydroxide and zinc carbonate. These species have sufficient alkalinity, even when present in low amounts, to saponify sensitive binders such as alkyds. The result is a weak interfacial layer that dramatically compromises adhesion.

Saponification was not observed on one of the delaminating samples. Does that mean that saponification was not the failure mechanism, and if not, what caused the coating to delaminate?

Totally apart from saponification, using a dry-fall alkyd to coat a smooth, thin-gauge substrate is not a good idea. The coatings have so little resin and dry so quickly that they have little ability to flow and wet the substrate. Thus, initial adhesion is often only fair. Furthermore, these coatings eventually reach a very brittle stage and cannot flex without either cracking (if their adhesive strength is stronger than their cohesive strength) or delaminating (if their adhesive strength is weaker than their cohesive strength). This is not a problem when applied to heavy elements such as structural steel columns, but can be a major problem when applied to thin-gauge steel that can move in strong winds, expand and contract due to temperature changes or flex when walked upon.

Did the coating fail because it could not wet well and became brittle, or did it fail because it saponified? The answer is both. It is not uncommon for coatings to fail by more than one mechanism. Indeed, it is sometimes necessary for several forces to act in concert to cause a coating failure.

An interesting question is whether the above failure was due to defective coating or defective application. Although it is possible that the lack of initial wetting could have been due to poor application technique, not only would this be difficult to prove, but it is also unlikely that this would have occurred over the entire ceiling. Was the paint defective? There was nothing to suggest that this was the case.

The root cause for the failure was neither defective paint nor defective application. It was simply a case of using the wrong type of coating for the job at hand. The alkyd saponified when exposed to the humid, alkaline conditions, as would any alkyd. It did not wet well and embrittled with time, as any dry-fall alkyd would. It was simply the wrong coating for the job, a failure category that is probably larger than commonly realized.

10.8 COAL TAR EPOXY FAILURE

Due to increased demand, a large municipal waste-water treatment plant was enlarged by the addition of a new clarifier. Both the concrete clarifier and the steel rake arm were to be coated with an amine-cured coal tar epoxy coating. The specification called for the concrete to be sweep-blasted and for any 'bug holes' over 0.25 in. in diameter to be opened, such that the coal tar epoxy could flow into these defects, rather than simply bridging them. The steel of the rake arm assembly was to be abrasive blast cleaned to achieve an SSPC-SP-5 'white metal blast', with an anchor profile of 3–4 mil.

After surface preparation, both the concrete and the steel were to be coated with two coats of two-component, amine-cured coal tar epoxy. The coating had a volume solids content of 65%, a mix ratio of 18 parts of component 'A' to one part of component 'B' (by volume), an induction time of 30 minutes and a recommended DFT of 6–8 mil per coat, for a total system DFT of 12–16 mil. The coating was applied by airless spray and could be thinned up to 5% with the manufacturer's recommended thinner. The mixed material had a pot life of 4.5 hours at 72 °F (22 °C). The recommended re-coating interval was 24–48 hours, with 24 hours recommended in very hot weather. If the re-coat interval was exceeded, the product literature called for sweep blasting of the first coat to impart surface roughness.

The coating work was carried out in the summer, with typical daytime temperatures in the 80–90 °F (27–32 °C) range. In at least one area, it was known that the 48-hour re-coat interval had been exceeded, and that the contractor had reportedly sweep blasted that area prior to applying the second coat.

In the spring of the following year, after approximately 9 months of service, problems began showing up. Corrosion of the rake arm was coupled with spotty paint delamination, as well as peeling of the coating from some areas of the concrete. No such deterioration was being experienced on the other clarifiers, which were exposed to essentially the same waste streams and which had been coated with the same coal tar epoxy several years ago. An inspection of the clarifier revealed that corrosion had broken through the coal tar epoxy in several areas of the rake arms, primarily on the top and on one side. In one particularly bad area, the coating was missing from several square inches of steel, and corrosion was quite heavy.

A closer examination showed that there appeared to be extensive inter-coat delamination associated with the top and side of the rake arm that exhibited spot corrosion. Measurements with a magnetic DFT gauge showed coating thicknesses of 4–6 mil immediately adjacent to the rusted spots. Where it appeared obvious that two coats were present, the total coating thickness was 10–13 mil. Probing with a knife in these areas revealed that the topcoat of coal tar epoxy could be easily pried from the first coat. On the side of the rake arm that showed little corrosion, both coats were largely intact, inter-coat adhesion was better (although small chips of topcoat could be disbonded with some difficulty), and the total thickness was approximately 10–14 mil.

In general, the concrete areas were in better condition than the rake arms. However, some inter-coat adhesion failure had occurred there as well. Samples that were brought back to the laboratory included delaminated paint chips as well as scrapings of the underlying coat of coal tar epoxy. Small chips and scrapings of well-adhered, non-failing coating were also obtained, as was a 5-gallon kit of fresh coal tar epoxy.

The investigation began at the microscope. In addition to dirt and debris, the front of the original black coal tar epoxy had a brownish or bronzed appearance. This is typical of such coatings exposed to heat and sunlight. The back of the disbonded topcoat was very smooth and clean, as was the front surface of the first coat from where the topcoat had disbonded. The smooth nature of both sides of this interface made it quite clear that, if they corresponded to areas where the 48-hour re-coat window had been exceeded, no sweep blasting had been performed to roughen the first coat.

An examination of the cross section of several chips confirmed the thickness readings obtained at the job site. In general, the coatings had been applied at, or even slightly below, the recommended minimum thickness. The examination of the cross section also revealed another finding, namely porosity. Although not so prevalent in the topcoat, chips of the first coat of coal tar epoxy were extremely porous, with large numbers of small to large voids spanning the cross section. This extensive porosity dramatically reduced the effective thickness of the first coat, which would make it much more susceptible to permeation from the waste water. The porosity was more prevalent in chips removed from the steel rake arm than from the concrete.

At this point in the investigation, it was apparent that one of the major factors in the corrosion of the rake arm was the poor inter-coat adhesion of the topcoat of coal tar epoxy to the first coat of coal tar epoxy. When the topcoat delaminated, this effectively reduced the coating thickness in half. Under ordinary circumstances, even a single coat of coal tar epoxy should have been able to last considerably longer than 9 months, but to make matters worse, this first coat was extremely porous and full of voids. Thus, instead of a 6- to 8-mil thick first coat, as specified, only 4–6 mil was present, with so many voids that the effective thickness was probably only 2–3 mil.

As previously stated, it was common knowledge that the 48-hour re-coat window had been exceeded in some areas. In the hot summer temperatures, steel surfaces can reach well over 120 °F (49 °C). The coating product data sheet cautioned that in very hot weather, the re-coat interval should be only 24 hours. Therefore, it was possible that a much larger percentage of the coated surfaces exceeded the re-coat window than initially thought. This may not have been a problem had the first coat been mechanically roughened, but the smoothness of the interface showed that this step had been omitted.

It was interesting that the bulk of the inter-coat delamination occurred on only certain surfaces of the rake arm and hardly at all on the concrete. Although there was no way of proving it, the likely explanation was that the concrete, being a much bigger heat sink, never got as hot as the steel rake arm, thus extending the re-coat window. The fact that one face of the rake arm had better inter-coat adhesion than the top and opposite side could have been due to its solar orientation at the time of painting, or perhaps simply because its re-coat window was not exceeded.

Another important factor in the failure was the porosity of the coal tar epoxy, which allowed water and other chemicals easier access to the underlying steel. Voids within a coating are caused by gases trying to rapidly escape from a coating that in turn is rapidly drying. The vapour pressure of the gas is pushing against the thick, viscous coating, thus resulting in trapped bubbles as the coating's viscosity becomes too great too soon. The gases could be entrapped air, moisture vapour or low-boiling solvent vapours.

Since the coating was applied by airless spray, and since it would be difficult to over-agitate and whip air into a viscous coal tar, the chances of the voids being caused by entrapped air seemed remote. The dry, hot temperatures during the job tended to rule out surface condensation and hence water vapour. The likely cause of the porosity was therefore entrapped solvent.

The coating manufacturer's product data sheet made it clear that the product could be applied under the environmental conditions at the job site. Therefore, either the porosity was due to a poorly formulated product that was prone to void formation at the temperatures encountered, or it was either over-thinned and/or thinned with the wrong thinner.

In order to determine if the formulation was prone to porosity, test panels were prepared in the laboratory. The steel panels and the paint material itself were both pre-heated to temperatures of 80, 100 and 120 °F (27, 38 and 49 °C, respectively), and the coating was applied by airless spray, both unthinned and thinned by the maximum amount (5%) with the recommended thinner. The painted panels were immediately placed in ovens maintained at the appropriate temperatures. Panels were prepared to achieve DFTs of 4–6 and 8–10 mil.

When the cross sections of the test panels were examined microscopically several days later, only the most minimal of porosity was observed, with one exception. This was the thinned panel applied at 120 °F (49 °C) and 8–10 mil. Even here, however, the extent of void formation was considerably less than what was observed in the field samples. It seemed apparent that, if properly thinned and applied, the coating was not prone to excessive void formation. This left the prospect of over-thinning or the use of the wrong thinner.

Over-thinning would be a very difficult thing to prove with only dried paint chips for evidence. However, if the wrong thinner had been used, there was a good chance that residues could still be detected by GC. This possibility was therefore investigated by using headspace GC.

This chromatographic technique has been described previously. Basically, it involves placing paint chips in small septum vials and heating the vials for a few minutes at elevated temperatures. Residues of solvent will partially evaporate from the chips into the air, or headspace, within the vial. A special gas-tight sampling syringe is then used to remove a quantity of this headspace, which is immediately injected into a GC unit. The technique is not quantitative but is a simple and reliable qualitative method for many common paint solvents.

Based on the material safety data sheets for the paint and thinner, what should have been detected was primarily xylene, with smaller amounts of ethyl benzene and butyl alcohol, perhaps with some traces of various hydrocarbons related to the coal tar pitch. An analysis of shavings removed from laboratory test panels was consistent with this prediction.

An analysis of failed paint chips from the rake arm showed two strong peaks in addition to those corresponding to the expected solvents. A little additional detective work in the laboratory showed that these peaks corresponded to MEK and toluene. It was obvious that the contractor had thinned the coating with thinner that was not recommended, perhaps in an attempt to extend the coating's relatively short 4-hour pot life (which would have been even shorter at the hot summer temperatures encountered at the job site). MEK, in particular, is a fast-evaporating solvent that could have developed considerable vapour pressure and contributed to excessive void formation. The lack of such porosity on the concrete surfaces could have been due to lower surface temperatures there as opposed to the steel rake arm, or maybe simply because the coating applied to the concrete had not been thinned as much.

10.9 SPLITTING OF INORGANIC ZINC-RICH PRIMER

A large quantity of new steel intended for a bridge in midwestern United States was shop coated with a three-coat system consisting of an ethyl silicate zinc-rich primer, a polyamide–epoxy intermediate coat and a urethane topcoat. The primer had a specified DFT of 2–4 mil and, according to the product data sheet, could be top-coated after 24 hours at 72 °F (22 °C). The epoxy, which had a mix ratio of 1 : 1 by volume, had a specified DFT of 3–5 mil and could be top-coated after 16 hours. The urethane finish coat had a 4 : 1 mix ratio and a recommended DFT of 2–4 mil.

The blast-cleaning and painting operations were carried out in the early summer, and the steel was subsequently shipped to the job site. When it was time to touch-up and erect the steel, the job site painting contractor pointed out that there was much more handling damage than usually expected and insisted on being paid extra for the extensive amount of touch-up work required. When the field crew began the touch-up work on the first steel member, they found that when they attempted to feather back an area of damaged coating, they were unable to feather it back to sound coating. They were able to easily scrape the coating off of the entire steel beam.

A site inspection confirmed that the coating system had extremely poor adhesion. Although in most cases, scraping with a knife resulted in chips ranging up to approximately 0.5 × 1 in. in size, in some spots chips, several square inches in size could be readily peeled off. In every case, the coating system was splitting within the zinc-rich primer, with zinc on both the back of the chip and on the underlying steel. Some members were found where the coating system was well adherent, and only small fragments of coating could be disbonded after considerable effort.

Thickness measurements taken with a magnetic gauge showed that the total three-coat system varied dramatically in thickness.

If all three coatings had been applied as specified, the total DFT should have ranged from 7 to 13 mil. In actuality, it ranged from 6 to 19 mil. There was definitely more spontaneous peeling in thicker areas (approximately 12 mil and above) and most of the areas that were approximately 8 mil or less had relatively good adhesion. However, there were exceptions to this trend. It was possible, in fact, to find thin areas with poor adhesion and thick areas with good adhesion.

It was not possible to bring a steel beam into the laboratory. However, very good samples of peeling coatings were obtained, as well as scrapings of the underlying primer that was still adhered to the steel. Smaller fragments of non-failing coating were also obtained, as were fresh samples of the liquid coatings from the manufacturer.

When samples were examined microscopically, three coats were visible: the zinc-rich primer, the epoxy intermediate coat and the urethane topcoat. As expected from the field measurements, the thickness of these coats varied. In general, failing samples tended to be thicker than non-failing ones. Most of the variation in thickness was due to the epoxy, which ranged from a low of 2 mil to a high of 9 mil.

Failing samples were as large as approximately 2 × 3 in. At 20× magnification, the surface of the zinc primer attached to the back of the epoxy was very rough and uneven, consistent with a cohesive 'splitting' mechanism. When lightly scraped with a knife, the primer 'powdered' up very easily – a sign of a poorly cured coating of low cohesive strength.

In addition to the powdery nature of the primer, its relative degree of cure was determined by its resistance to MEK. The resistance of inorganic zinc-rich primers has been shown to correlate with their degree of cure and can be evaluated according to ASTM D4752 [2]. The ASTM procedure involves rubbing a sample with cheesecloth saturated with MEK until the substrate is exposed or until 50 double rubs (one complete forward and back motion) have been reached. This author has found that using cotton Q-tips, rather than cheesecloth, gives more consistent results and is much easier to carry out on small samples.

When the various samples were tested, it was found that the zinc primer on the back of failing samples had extremely poor solvent resistance. In every case, there was very heavy colour transfer to the Q-tips within 50 double rubs, and in some cases, the primer was completely removed to reveal the back of the epoxy intermediate coat. Because only small fragments of non-failing coating were available, it was not possible to make a direct comparison of solvent resistance between failing and non-failing samples. However, it appeared as though the non-failing samples were more resistant to MEK than the failing ones, and they did not crumble or powder as easily when probed with the tip of a knife.

At this point, it seemed evident that the root cause of the coating failure was insufficient cure of the primer prior to the application of the epoxy and urethane coatings. Inorganic zinc-rich primers, because of their high loading of pigment (the coatings are actually formulated above their critical pigment volume concentration), do not have very good cohesive strength to begin with. If not adequately cured, this property is compromised even more. It has been shown that when such under-cured primers are top-coated with solvent-based, thermoset coatings such as epoxies, cohesive 'splitting failures' can result [2]. It is thought that the failure is due to the solvents in the epoxy penetrating the porous, under-cured primer and softening it, followed by the stress imparted when the epoxy begins to cure and shrink. Such failures are aggravated by thicker coatings, as observed on this project. There are two simple reasons for this, that is, thicker epoxy means more solvent and thicker coatings impart more shrinkage stress than thinner ones.

Although the root cause of the failure appeared obvious, the question remained as to whether this was a product problem or an application problem. Due to a lack of documentation, there were conflicting reports of how long the zinc-rich primer had been allowed to cure prior to the application of the epoxy intermediate coat. Although there was little doubt that in some cases more than 1 day had elapsed, there was the possibility that some members may have been coated sooner than the specified 24-hour cure time. Furthermore, since a large amount of steel was involved, the entire operation spanned more than 2 weeks, with considerable variations in temperature and humidity. An examination of weather reports showed that daytime temperatures varied from 60 to 80 °F (16–27 °C), with relative humidity (RH) values from 40 to 60%.

A characteristic of inorganic zinc-rich primers is that their rate of cure can be greatly influenced by the amount of moisture in the air. Some manufacturers acknowledge this on their product data sheets and will list a 'cure to topcoat' at specific combinations of temperature and humidity. Unfortunately, the product data sheet for this particular primer was vague and simply listed a re-coat time of 24 hours at 72 °F (22 °C), with no admonition concerning humidity. Based on the scant documentation, there was practically no way to prove how much curing time the zinc primer on any particular beam had received prior to application of the epoxy intermediate coat, nor the temperature and humidity to which it was exposed. However, it was possible to prepare a series of test panels at different conditions, assess their degree of cure, topcoat them and then carry out adhesion testing.

In order to keep the number of panels manageable, it was decided to carry out an initial testing at only two conditions of temperature and humidity, that is, 72 °F (22 °C) and 40% RH, and 72 °F (22 °C) and 60% RH. The temperature of 72 °F (22 °C) was selected because this was the temperature listed on the product data sheet. The RH was selected to span the minimum and maximum conditions at the job site. Enough panels were primed such that they could be coated with the epoxy intermediate coat after 8, 24, 72 and 168 hours. The epoxy was applied at the specified DFT of 3–5 mil, and also at 8–9 mil, which correlated with the highest readings seen on field samples.

In addition to the physical testing described above, additional primed panels were tested for degree of cure by both the solvent rub test and by IR spectroscopy.

The solvent rub test has already been described. Another way to follow the cure of such primers is by IR spectroscopy [2]. In order to appreciate how this works, it is necessary to understand the curing mechanism of inorganic zinc-rich primers. This mechanism, as well as the IR spectroscopic technique, have previously been described in Section 8.2.4.5. Briefly, the loss of ethyl groups from the silicate pre-polymer as it reacts with moisture is detected and expressed as a 'cure ratio'. The larger the value of this cure ratio, then the more fully cured is the sample.

The test panel project yielded very interesting results, which are summarized in Tables 10.4 and 10.5. Although the data are not perfectly consistent (a fairly common occurrence in the field of real-world failure analysis), some very clear trends are evident. At 72 °F (22 °C) and 40% RH, the zinc-rich primer cures very slowly, as seen by both its poor solvent resistance and sluggish increase in IR cure ratios. Although the manufacturer stated that it could be re-coated at this temperature in 24 hours (no humidity requirement was listed), at this point, it still had very poor solvent resistance, and its IR cure ratio had barely changed from what it was after only 8 hours of cure. At this humidity, there was really no substantial improvement in the coating's degree of cure until after 168 hours (a full week). The situation was considerably better at 60% RH, but even here, the degree of cure was marginal after the recommended 24 hours.

Not surprisingly, the adhesion results mirrored the cure results. Panels coated with epoxy intermediate coat after the recommended 24 hours had very poor adhesion (ratings of '0B' or '1B'), with all failures occurring cohesively within the zinc-rich primer. A substantial improvement in adhesion was not observed until after 168 hours of cure at 40% RH or 72 hours at 60% RH. There was a slight trend towards poorer adhesion at higher epoxy thickness.

Table 10.4 Methyl ethyl ketone (MEK) resistance and cure ratios of zinc primer as a function of the cure conditions.

Conditions[a]			
Time (h)	**RH (%)**	**MEK resistance**	**Care ratio**[b]
8	40	Very poor	23.2
24	40	Very poor	26.8
72	40	Poor	35.3
168	40	Fair	47.9
8	60	Very poor	22.9
24	60	Poor	31.1
72	60	Fair	45.3
168	60	Good	61.8

[a] Curing carried out at 72 °F (22 °C) for all samples.
[b] Determined by infrared spectroscopy.
RH = relative humidity.

Table 10.5 Adhesion of epoxy–zinc sample as a function of the cure conditions and thickness.

Conditions[a]			
Time (h)	**RH (%)**	**DFT**	**Adhesion[b]**
8	40	Thin	0B
		Thick	0B
24	40	Thin	0B
		Thick	0B
72	40	Thin	1B
		Thick	0B
168	40	Thin	3B
		Thick	2B
8	60	Thin	0B
		Thick	0B
24	60	Thin	0B
		Thick	1B
72	60	Thin	3B
		Thick	2B
168	60	Thin	3B
		Thick	3B

[a] Curing carried out at 72 °F (22 °C) for all samples.
[b] Measurements made by using ASTM D3359, Method B; all failures cohesive within the zinc.
DFT = dry-film thickness.

To return to the original question: was the primer under-cured because the contractor re-coated it too soon, or because it was a defective product? There is still no way of knowing for sure if the contractor always waited the required 24 hours at 72 °F (22 °C) prior to applying the epoxy. However, would it have mattered? Based on the laboratory test panels, apparently not. According to these results, to obtain adequate cure (and hence adequate adhesion), the contractor would have to wait not 24 hours, but between 3 and 7 days, depending on the humidity.

Is the product then defective? In a sense, yes, although this is somewhat an argument over semantics. The problem is not that the product cannot work. Good results were seen not only in the laboratory, but at the job site as well. The problem is that the product does not perform 'as advertised'. The claim that it can be re-coated in 24 hours at 72 °F (22 °C) is not entirely accurate, since its cure time is very humidity dependent. Perhaps it could be successfully top-coated in 24 hours at 72 °F (22 °C) if the humidities were high enough, such as perhaps

80%. However, in most cases, such high humidities will not be encountered, and much longer curing times will be required.

10.10 DEFECTS ON ELECTRO-COATED PANELS

Veneri and Kramer [3] were able to identify the cause of defects occurring on electro-coated galvanized steel panels for an automobile parts manufacturer. The sheet steel was cleaned and then pre-treated with a conversion coating of zinc phosphate, prior to the application of an electro-coat (E-coat) primer.

Zinc phosphate conversion coatings for use on steel have been discussed previously (see Section 2.1.3.1). They are typically formed by exposing the steel to a solution of primary zinc phosphate ($Zn(H_2PO_4)$) and excess phosphoric acid. When carried out correctly under carefully controlled conditions, this produces a very dense, very thin film of either $Zn_3(PO_4)_2 \cdot 4H_2O$ (hopeite) or $FeZn_2(PO_4)_2 \cdot 4H_2O$ (phosphophyllite) on the metal surface which results in improved paint adhesion and greater corrosion resistance.

The E-coat primer used was based on an epoxy resin. The parts were dipped in the bath of coating, and an appropriate electrical charge was applied which caused the oppositely charged coating to deposit on the part. The parts were then removed from the bath and were rinsed and baked. When examined, however, a series of small, raised defects were often encountered.

Initially, a microscopic examination of the defects indicated that there were small particles apparently embedded in the paint. Although at low magnification, the particles had the appearance of dirt, this seemed an unlikely failure mechanism since the cleaning, pre-treating and baking operations were all carried out under stringent conditions.

A very careful cross-sectioning of defect areas, followed by an examination via the use of an SEM-EDS, clearly showed that the defects were due to large accumulations of zinc phosphate crystals, sometimes referred to as 'nubs'. These nubs begin at the substrate and extend through the entire thickness of the primer, thus resulting in a 'volcano-like' defect. The process actually resulted in the formation of a small pit in the galvanized substrate. Slight changes in this phosphating bath chemistry were able to eliminate this problem.

10.11 FAILURE OF COATED LIGHT POLES

A local municipality had contracted for a number of light poles to be installed along one of the main thoroughfares leading through town. The poles, which were over 30 ft in height, were constructed of steel which had been blast cleaned and subsequently coated with a powder coating. The powder coating was of the polyester variety, which was to be baked to a peak metal temperature of 204 °C.

The poles were installed in the spring and summer. Within 2 years, problems had developed with rusting, blistering and peeling of the paint.

The investigator travelled to the municipality for a first-hand look at the problem and to take samples for subsequent analysis. In addition to observations and sample taking, coating thickness measurements and adhesion tests were also performed. Thickness measurements were made using a non-destructive magnetic thickness gauge. Adhesion was performed both subjectively by cutting and prying with a utility knife and in accordance with ASTM D3359, Method B ("Standard Test Method for Measuring Adhesion by Tape Test"). In this procedure, a knife is used to cut a cross-hatch pattern into the sample, followed by the application of a pressure-sensitive tape. The tape is then rapidly removed, and the amount of coating detachment is assessed based on the method's visual rating scale. This goes from a 5B, corresponding to no loss of adhesion, to a 0B, corresponding to 65% or greater loss of adhesion.

The first pole examined showed considerable failure. Not only was there extensive rusting and lifting of the coating from the base plate at ground level, but patches of coating were also missing from the pole itself, often with associated rusting of the exposed steel (Figure 10.4). There were also dozens of spots were the coating was lifted or blistered but was still loosely adhering to the pole. The areas of blistered or lifted coating were typically 1–2 in. in diameter. Most of the failure started at ground level and extended about 7–8 ft up the pole and was concentrated on that side of the pole facing the street. In fact, the failure was principally on that quadrant of the pole which was facing oncoming traffic. There was no failure on the opposite quadrant, facing away from the street, and only very few isolated failures higher up on the pole. Beyond about 15 ft, there were no failures.

Figure 10.4 Coating failure on a light pole.

When these blistered or lifted areas were cut into with a knife, the coating fell off of the steel, which had a slight amount of rust staining on it directly beneath the loose coating. When examined with a magnifier, it was evident that the steel had been blast cleaned. The coating thickness within the general area of failure was 7–9 mil. About 180 degrees opposite the failing area, where the coating was in very good condition, the thickness was 5–6 mil. The coating adhesion near the blistered coating (but not immediately on a blister) was poor, with a rating of 0B. The steel beneath where the adhesion test was performed was found to be clean and shiny and had a distinct blast profile. Rubbing at the nearby coating with a cotton swab saturated with MEK showed that the solvent had very little effect on the coating. Indeed, 100 double rubs (one double rub corresponds to one back and forth motion, using moderate finger pressure) showed only slight to moderate colour transfer and no apparent softening of the coating.

Adhesion testing was conducted on this pole at non-failing locations. While one rating of 1B was found, which is usually considered poor, the other tests showed good adhesion, with ratings of 4B or 5B.

Several other poles were examined, with similar results. The same distinct pattern of failure was found, with the great majority of it occurring on that portion of the pole which faced oncoming traffic and which extended from the base of the pole to roughly 7–8 ft in height. This was true regardless of which side of the street the poles were on, so the pattern was not related to sunlight exposure or weather. Coating adhesion in non-failing spots was generally good, and there was no relationship between coating thickness and failure. Typical coating thickness was 5–8 mil. Sometimes, examination with a 15× pocket magnifier showed fine, hairline cracking near the centres of the blisters.

The laboratory investigation of samples from the poles consisted of visual and microscopic observations, IR spectroscopy and SEM-EDS.

The visual and microscopic observations showed that the chips from the various poles consisted of what appeared to be a single black coat, with thickness usually in the 5- to 8-mil range. Some chips had areas of rust or rust staining on the back and also replicated a blast profile, confirming that the steel had indeed been blast cleaned (Figure 10.5). Some of the chips had several small, hairline cracks on the front side. When one such chip was further examined by carefully slicing off layers of coating with a sharp artist's knife, it was found that the crack was very deep and appeared to go virtually all the way through the coating. Indeed, other chips were found where the cracking was visible through the entire cross section.

The cracks in the chips would seem to suggest that the coating was brittle. However, this was not the case, as failing chips could be bent and flexed considerably without breaking and without the development of visible cracks. Indeed, one of the larger chips (some were as large as 1 × 2 in.) was bent about a $\frac{1}{4}$-in. diameter steel mandrel without cracking. However, when a similar chip was placed for a few hours in a freezer at roughly 0 °C, along with the mandrel and

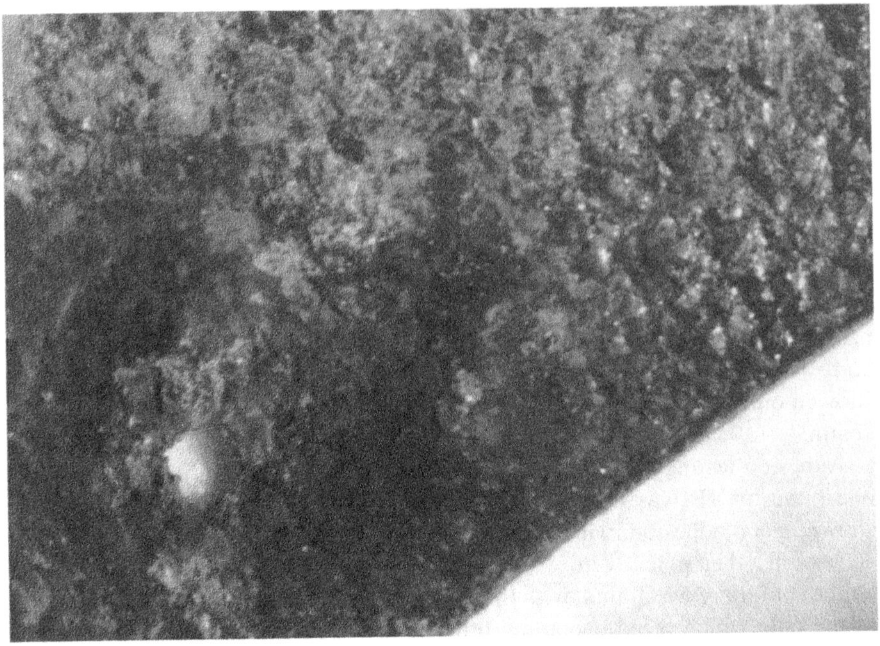

Figure 10.5 Back side of a chip which was removed from an area of failure on a light pole. Note the rust and also the replication of a blast-cleaning profile.

then immediately tested while still cold, it was very brittle and snapped in half almost immediately upon beginning the bend test. Therefore, although the coating was flexible at room temperature, it was brittle at near-freezing temperatures.

IR spectra were obtained of coating samples from both failing and non-failing areas. The spectra showed that the samples were virtually identical to one another and were characteristic of an isophthalate polyester (Figure 10.6), as seen by characteristic bands near 2900, 1725, 1470, 1375, 1305, 1240 and 725 cm^{-1}. Thus, the coating was consistent with the type said to have been applied. Furthermore, no obvious chemical degradation was indicated, since spectra of failing and non-failing samples were virtually identical.

SEM-EDS was performed on both the front and back side of coating samples, including those with rusted areas. The analysis revealed elements typically associated with coatings, such as carbon, oxygen, aluminium, silicon, barium and sulfur. The aluminium and silicon indicate an aluminium silicate pigment, while barium and sulfur are likely due to barium sulfate pigment (barytes). No chloride was detected on two of three rusted areas on the back sides of the chips, and only a trace of chloride was detected on the third. Thus, salt contamination appeared minimal.

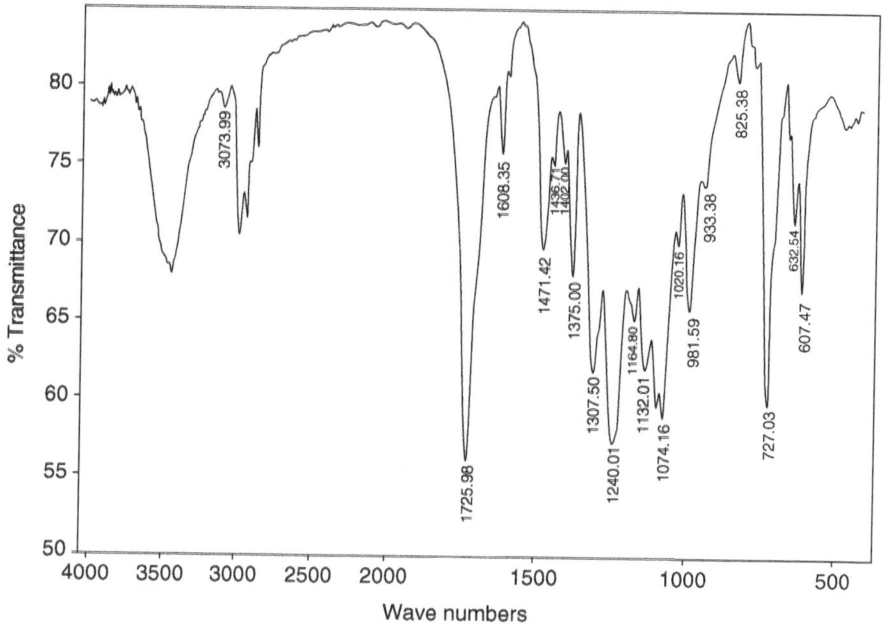

Figure 10.6 Infrared spectrum of coating removed from a light pole. The spectrum is typical of an isophthalate polyester.

Based on the above, it appeared as though the proper coating had been applied (IR spectroscopy confirmed a polyester) and adequately cured (resistance to solvent was good). Furthermore, it had been applied to a clean, blast-cleaned substrate at a relatively uniform thickness and apparently had good initial adhesion. However, several observations supplied clues as to the cause of the failure. The first clue was the pattern of failure, which almost always occurred on that side of the poles facing oncoming traffic, and within a few feet of ground level. The second was the observation of cracks in the failing coating, which would suggest brittleness, even though the chips could be bent and flexed in the lab at room temperature without cracking. The third clue was the finding that the coating became very brittle when cooled to roughly 0 °C.

The above findings show that the coating failed because it became brittle at low temperatures. This brittleness, and hence lack of impact resistance, resulted in cracking of the coating, likely when it was impacted by debris thrown up by snow plows in the cold, snowy winter months. This would account for the orientation of the failure, which was always on that portion of the poles facing oncoming traffic (including snow plows) and usually only a few feet off the ground. The cracking allowed water, and perhaps even a trace of road salt, to breach the coating and initiate rusting of the underlying steel. The undercutting reaction, over time, resulted in the lifting or blistering of the coating.

10.12 BLISTERING OF COATING ON CONCRETE FLOOR, NUMBER 1

The existing concrete floor in an older facility was coated with an epoxy primer and an epoxy topcoat. The facility had changed hands several times over the years, but most recently had been used for the manufacture of perfume. The concrete floor had never been previously coated. It was prepared by abrasive blast cleaning and vacuuming, prior to the application of the two-coat epoxy system. Both the epoxy primer and the epoxy topcoat were 100% solids materials. The primer was specified to be applied at 5–9 mil and the topcoat at 7–10 mil. The coating manufacturer's representative made several visits to the jobsite during the preparation and painting of the floor, and indicated that everything appeared to have been done satisfactorily, and that no thinners were used.

However, about 1 year later, problems developed with blistering of the coating system from the concrete. These blisters were as large as 1 in. in diameter, and, where they had cracked or were broken open, a watery dark liquid was often observed.

Samples of disbonded floor coating, blister liquid and even a concrete core sample with blistered coating were submitted to the laboratory. The investigation consisted of visual and microscopic observations, IR spectroscopy and gas chromatography–mass spectroscopy (GC-MS).

When the sample of blistered, disbonded coating was examined, it was found that the light grey primer was 8- to 9-mil thick, as specified, but that the dark grey topcoat was 40- to 50-mil thick, which was much heavier than the specified 7–10 mil. The front side was smooth and slightly glossy, while the back side appeared to be clean when examined at 30× magnification, except for an occasional speck of concrete. The microscopic examination also showed that the back side of the primer was rough, likely due to the replication of a blast profile in the concrete.

The coating on the 2-in. diameter concrete core sample, said to have been removed from a blistered area, had very poor adhesion. Once the edge of it was lifted using a knife, the entire disc of coating could be removed from the concrete using one's fingers. The back side of the coating was similar in appearance to the previous sample, and it appeared to have been applied to a roughened surface. Indeed, the concrete from which it had been removed was rough, with exposed aggregate, indicating that it had been abrasive blasted. The surface of the concrete was slightly brown in colour. A microscopic examination of the cross section of the coating removed from the core sample showed two coats: a light grey primer 10- to 12-mil thick and a dark grey topcoat 9- to 11-mil thick. Therefore, in contrast to the first sample, both the primer and topcoat on this sample had been applied very close to or at the specified thicknesses.

The sample of blister liquid was dark brown, and had some fine, powdery material in it. The pH was 12. When two or three drops of it were evaporated, an amber waxy material was obtained.

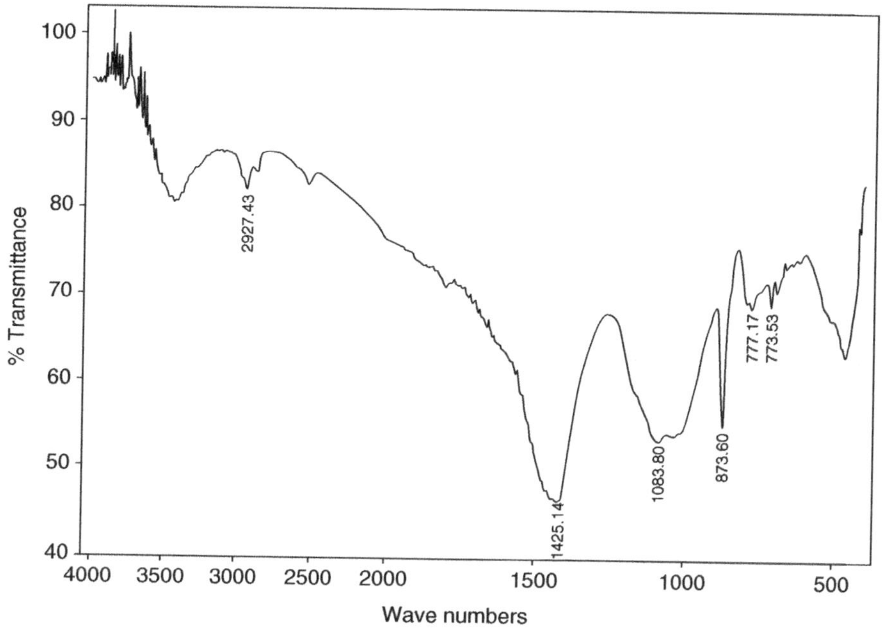

Figure 10.7 Infrared spectrum of scrapings taken from the surface of a concrete core sample immediately beneath the failing coating.

An IR spectrum was obtained of the surface of the concrete core sample by very lightly scraping a portion of it and combining the scrapings with potassium bromide (KBr) in order to make a pellet. The spectrum (Figure 10.7) showed only calcium carbonate (bands near 1425 and 875 cm^{-1}) and silica and/or silicates (broad region from 1100 to 1000 cm^{-1}), materials which would be expected from concrete. However, an additional spectrum obtained by extracting a portion of these scrapings in MEK and carefully evaporating away the solvent was much more revealing (Figure 10.8). This spectrum showed the presence of a long-chain aliphatic hydrocarbon (bands near 2900, 1455 and 1375 cm^{-1}) and weak epoxy bands (1510, 1245, 1180 and 830 cm^{-1}). The aliphatic hydrocarbon bands are likely due to an oil or wax-like contaminant, while the epoxy bands are likely due to isolated residues of coating material still present on the surface of the concrete.

Even more interesting was the spectrum of the waxy amber material obtained from evaporation of the blister liquid (Figure 10.9). While some calcium carbonate was present, likely due to traces of fine particulate debris which had not been completely removed prior to evaporation, the two strongest bands were those near 2900 and 1565 cm^{-1}. Weaker bands were also present near 1230, 1195 and 1130 cm^{-1}. The very strong bands near 2900 cm^{-1} suggest the presence of a long-chain aliphatic material, such as oil or wax. However, this did not seem likely,

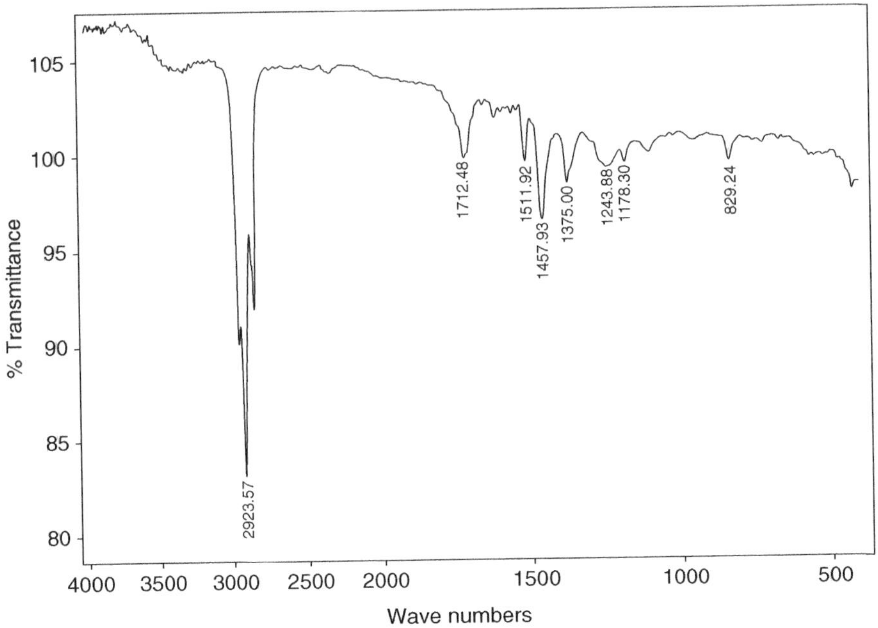

Figure 10.8 Infrared spectrum obtained from the solvent extraction of scrapings taken from the surface of a concrete core sample immediately beneath the failing coating.

since such materials would not be soluble in the aqueous blister liquid. However, the strong, broad band near 1565 cm^{-1} is typical of salts of carboxylic acids. The strong 1565- and 2900-cm^{-1} bands are consistent with the presence of a saponified fatty acid. Such acids, which are organic carboxylic acids, are unstable under the moist, alkaline conditions of a concrete floor (remember, the pH of the blister liquid was 12) and react or degrade by a process known as saponification (alkaline hydrolysis) to produce carboxylic acid salts. These salts are water soluble and explain the presence of this material in the aqueous blister liquid.

In addition to IR spectroscopy, the blister liquid was also analysed using GC-MS. The resulting chromatogram (Figure 10.10) had over 30 peaks corresponding to various volatile or semi-volatile organic compounds. Many of these peaks were identified using a computerized library search routine based on their mass spectra. The major peaks corresponded to 1-butanol, 1-methoxy-2-propanol, 2-butoxyethanol, benzyl alcohol and either tetradecanol or tetradecene. Two smaller peaks were identified as fluorinated compounds.

The spectroscopic and chromatographic analysis described above detected several materials which likely were contaminants. GC-MS in particular detected several somewhat unusual compounds which may have permeated into the previously bare concrete floor during its use as a perfume manufacturing facility, or possibly even from prior use. Additionally, a large amount of benzyl alcohol was

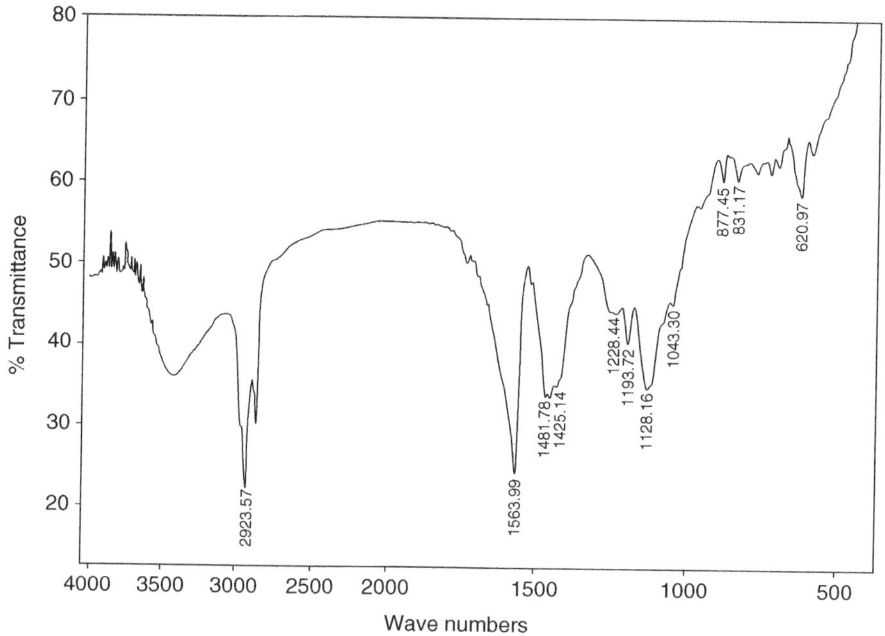

Figure 10.9 Infrared spectrum of waxy amber material obtained from the evaporation of blister liquid.

detected, which is a common solvent for epoxy coatings. However, both of the epoxies used on the project were of the 100% solids variety and should have contained no solvent, and reportedly no thinners were used. Thus, the source of the benzyl alcohol was something of a mystery, although perhaps it too was a contaminant.

In addition to contaminants found by GC-MS, IR spectroscopy found organic material in both the upper surface of the concrete core sample and in the blister liquid. Indeed, a visual examination of the evaporated blister liquid showed a waxy, amber material, which was shown to be a saponified fatty acid. Such fatty acids are not normally used in epoxy coatings, suggesting that a fatty acid got on or in the concrete floor at a much earlier date and saponified over time to form the salt of the acid. A spectrum of the solvent extraction of the concrete surface also showed the presence of a simple long-chain aliphatic hydrocarbon, such as grease or oil.

The presence of all of these extraneous materials, which apparently got on or in the floor during the many years in which it was uncoated, has resulted in the present coating failure. Hydrocarbon materials such as oils and fatty acids can certainly interfere with the bond of the epoxy to the concrete, despite the fact that the concrete had been roughened by abrasive blasting. While the blast cleaning would have removed surface contamination, it would not have removed materials

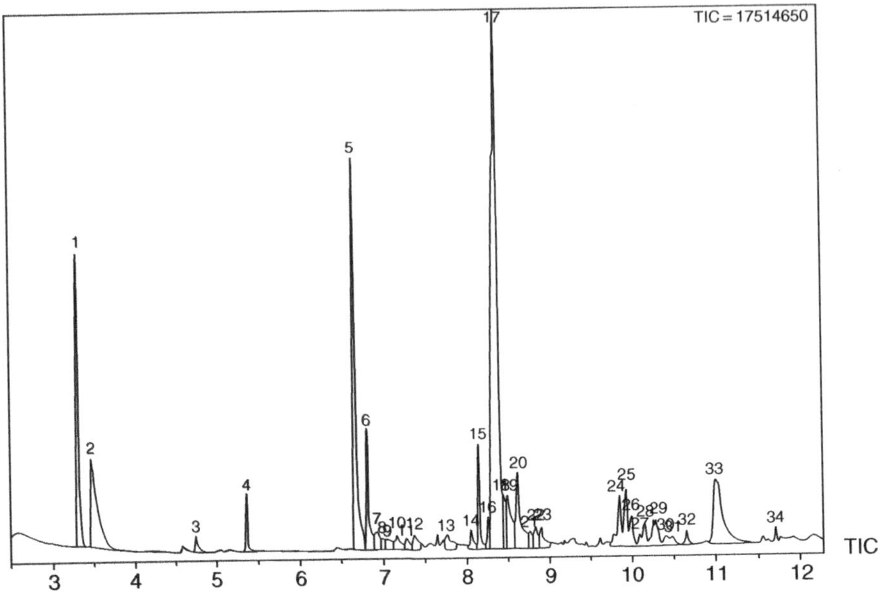

Figure 10.10 Gas chromatogram of blister liquid.

which had permeated farther into the concrete over a period of many years. Furthermore, some of the contamination consisted of water-soluble materials. This includes the carboxylic acid salts, which probably originally entered the floor as fatty acids or even fats, and which formed over time via saponification. As previously explained (Section 2.2.5), water-soluble materials can lead to osmotic blistering. In this case, the source of the water is likely ground water beneath the concrete slab, and it is the concrete itself which is acting as the semi-permeable membrane.

10.13 BLISTERING OF COATING ON CONCRETE FLOOR, NUMBER 2

A new manufacturing facility was constructed in south-west United States. The main area of the facility contained a roughly 35 000 ft^2 concrete floor. Even though this area of the country is warm and dry, a vapour barrier had been installed beneath the floor. The concrete floor was poured without the use of any membrane-curing agents or dust suppressants. It was allowed to cure for approximately 2 months, after which time it was prepared by water blasting with sand injection. Several days later, the floor was primed with a 100% solids two-component epoxy primer with a recommended thickness of 4–6 mil, followed by

Figure 10.11 Clear, viscous gel on concrete surface underneath the blistered coating.

a 100% solids epoxy topcoat with a recommended thickness of 10–15 mil. About a year after the project was completed, blisters began to develop in the floor coating. Reportedly, there was no pattern to the blistering.

Two roughly 3-in. diameter concrete core samples were received from the facility. Both had blisters which ranged in size up to about $\frac{1}{2}$ in. in diameter. When blisters were cut into, it was found that the entire two-coat coating system had blistered from the concrete. Furthermore, there was a substantial quantity of a clear, viscous gel underneath the blistered coating (Figure 10.11), which was found to be highly alkaline. When one such broken blister was examined a few days later, the viscous gel had dried to a crusty, clear, crystalline material (Figure 10.12). The back side of the blistered coating also had some of this material on it but was otherwise clean and rough, with only some small fragments of concrete on it. The cross section showed two coats, each close to the recommended thickness.

An IR spectrum was obtained from the viscous gel after it had dried and hardened (Figure 10.13). The spectrum was typical of a silicate, as seen by the characteristic strong, broad band near 1020 cm^{-1}, along with the 460-cm^{-1} band. Moisture was also evident by the water bands near 3400 and 1650 cm^{-1}. SEM-EDS analysis of this dried gel (Figure 10.14) showed that it was composed mainly of silicon, sodium, potassium and oxygen.

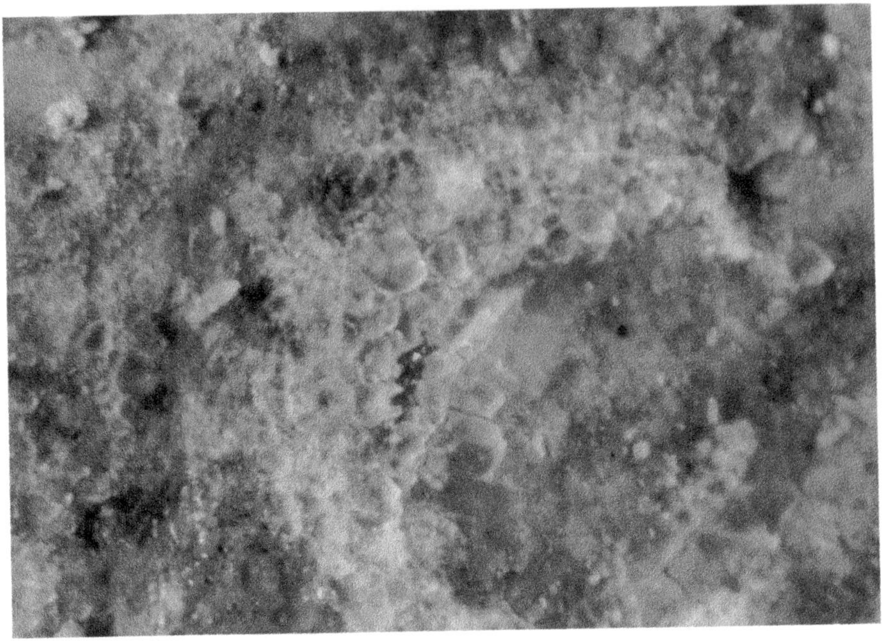

Figure 10.12 Clear, viscous gel examined a few days after the blistered coating was removed. Note that the gel has now crystallized.

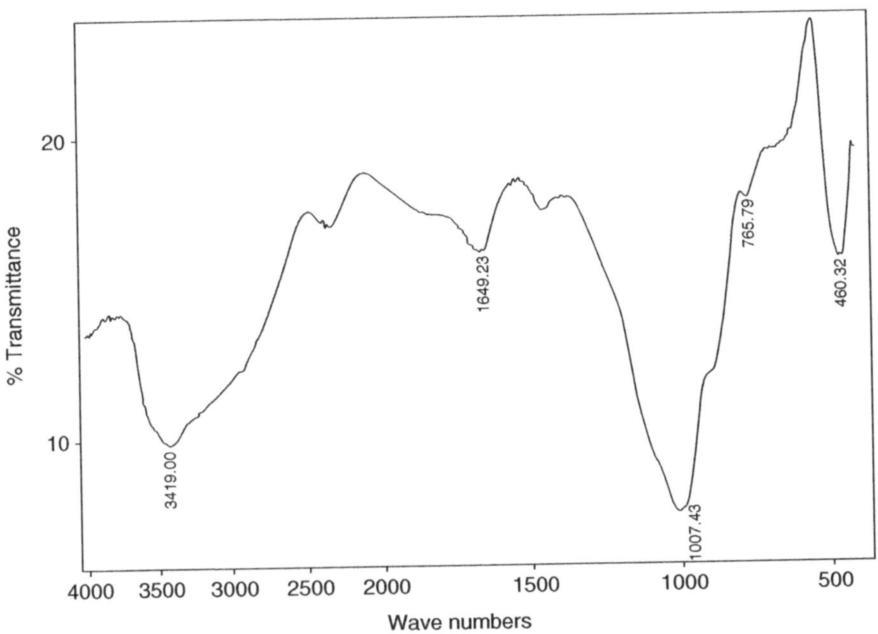

Figure 10.13 Infrared spectrum of crystallized gel shown in Figure 10.12.

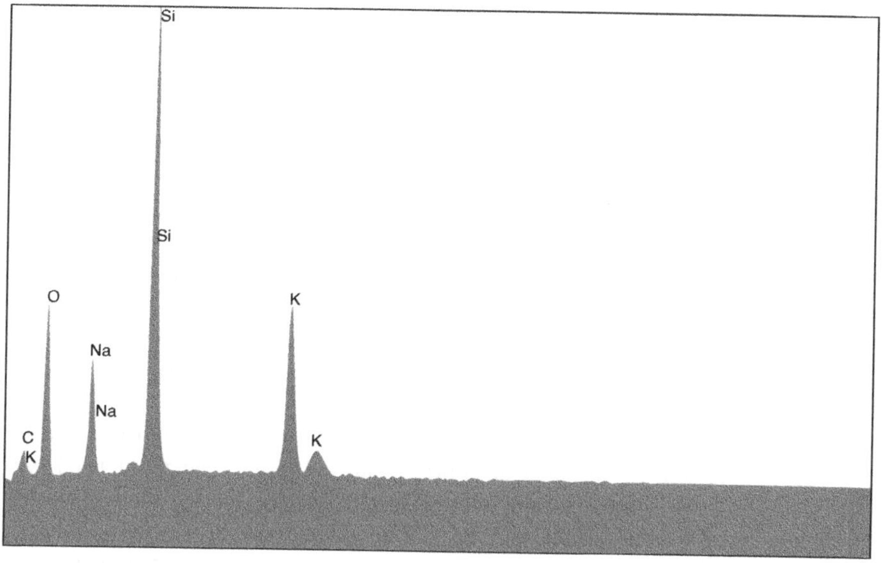

Figure 10.14 Scanning electron microscope–energy-dispersive X-ray spectrum of dried gel found on concrete beneath the blistered coating.

Sodium and potassium silicates are water-soluble materials which are sometimes used as hardeners and dust suppressants for concrete floors. Because they have a strong affinity for water when top-coated with thick, relatively impermeable coatings, they can draw up excess, unreacted water from the concrete mix (or ground water, in some cases) and essentially develop an osmotic cell, as discussed in the previous example (Section 10.12). However, in this project, all the involved parties were certain that no such materials were used. Therefore, there must be a different explanation for the presence of this gel.

Certain types of aggregate sometimes used in concrete are known as alkali-silica reactive aggregate. This type of aggregate can react with moisture, sodium and potassium in concrete to form a gel-like material which has a very strong tendency to expand [5–7]. The reaction itself is referred to as alkali-aggregate reactivity (AAR) or alkali-silica reactivity(ASR), and it is ASR which is responsible for the blistering of the floor coating.

ASR is well known in the concrete industry, where it is of concern due to its ability to cause cracking of concrete. It was first recognized in California during the 1940s, where work showed that expansion and cracking of concrete could result when specimens containing certain combinations of high-alkali cement and aggregate were stored in high-humidity conditions. The distribution of ASR in a concrete slab can be highly variable and requires water, reactive silica and a high concentration of hydroxyl ions. In a dry, interior environment, even if reactive silica and a high pH are present, ASR may not occur. However, applying a relatively impermeable coating can trap in moisture, with the potential of accelerating the reaction. Below about 80% internal RH, the reaction will cease [6]. Initially, it was thought that there were a limited number of aggregate constituents which could cause ASR, such as opal, chalcedony and some volcanic rocks. It is now thought that it can occur with a wider range of aggregates [6].

10.14 PEELING OF PAINT FROM METAL CHAIRS

A manufacturer of folding metal chairs received a complaint from one of their biggest customers: the paint on about 200 of the nearly 1000 chairs recently purchased was flaking and peeling. The bare metal chairs are run through a three-stage washer, oven dried and then cooled to room temperature prior to painting. The painting is done in a spray booth using a series of rotary atomizers. The paint is given a 10-minute 'flash off' time to allow the bulk of the solvent to evaporate, and then baked for 12 minutes at approximately 165 °C. The paint used was described by the coating manufacturer as a water-borne baking enamel with a recommended DFT of 1.0 mil and a recommended bake of 12–14 minutes at 165 °C. It was also described as having very good hardness and impact resistance, and listed solvent resistance as passing '50 double rubs with MEK'. After receiving the complaint, the chair manufacturer replaced the 200 failing chairs but used a different paint which was advertised as being glossier and harder than the original paint. No complaints were received with the replacement chairs.

The laboratory was supplied with two failing chairs, along with one of the replacement chairs with the new paint. The investigation consisted of visual and microscopic observations, IR spectroscopy and DSC.

When the first of the failing chairs was examined, it was obvious why the customer had complained. There were two relatively large patches of missing paint, each about 1 × 2 in., on the seat of the chair, along with several smaller spots of missing paint. All of the failing areas were located in the centre portion of the seat area, corresponding to where there would have been physical contact with a person. There was no visible failure on the legs, cross braces, back rest, underside of the seat or even on the edges or perimeter of the upper side of the seat.

Adhesion testing was performed at several locations on the first of the failing chairs. Adhesion was found to be very poor near the centre of the seat area and on the back rest (even though there was no missing paint on the back rest), but fair to good elsewhere on the chair.

The back sides of chips removed from the seat area of the failing chair were examined with a stereo microscope at up to 60× magnification. The chips were found to be very clean and replicated faint striations indicative of the surface roughness of the underlying metal. The metal surface itself was also examined microscopically and found to be very clean. The cross section of the failing chips showed a single coat of paint approximately 0.6-mil thick. The solvent resistance of the paint on the seat of the failing chair was also evaluated, using cheesecloth saturated with MEK. After 50 double rubs (the requirement in the coating manufacturer's technical literature), there was moderate colour transfer, but the coating had not been removed down to bare metal. This was considered to be a 'pass'.

When the paint on the failing chair was examined microscopically, it was found to be rough and badly 'orange peeled' in the failing area of the seat, as well as having numerous small pinholes. Most of the pinholes were relatively deep, and some went all the way down to bare steel. This texture could be found on other areas of the chair, but some spots were found where the paint was smooth and not pinholed.

The second of the failing chairs was very similar in appearance to the first, including the pattern of failure. Adhesion testing produced similar results, and solvent resistance was perhaps slightly better than the first chair. The thickness of the failing paint was approximately 0.8 mil.

The third chair examined was the non-failing chair which had been coated with the new paint. This paint was considerably smoother and more glossy than the paint on the two failing chairs. Microscopic observations showed it to be free of both orange peel and pinholes, and the paint had a more 'slippery' feel compared with the other chairs. The coating passed the 50 double rub MEK test with no colour transfer, and adhesion was found to be good. Gloss reading was taken from this chair, as well as from the two failing chairs. The paint on the failing chairs had 60° gloss of 20–30% versus 60–65% on the new chair. Gloss readings were taken on areas which would not have been subjected to abrasion.

IR spectra were obtained from the paint on both failing and non-failing areas of the two failing chairs, by the potassium bromide pellet technique. The spectra (Figures 10.15 and 10.16) were virtually identical and showed that the original coating was an isophthalate polyester cross-linked with a melamine resin and pigmented with china clay and titanium dioxide. The isophthalate polyester was evident by the characteristic bands near 2900, 1725, 1230 and 730 cm^{-1}, while the band near 1550 cm^{-1} was due to a melamine or possibly some other type of nitrogenous cross-linking agent. China clay could be seen by bands near 3690, 3615 and 1000–1030 cm^{-1}, while the broad area from 700 to 500 cm^{-1} was due to titanium dioxide. The composition was typical of what one might expect for such a coating.

DSC was performed in a single step, under a nitrogen purge, typically from about −5 to 110 °C at 10° per minute. Samples were analysed from failing and non-failing areas and also from the non-failing chair with the new paint on it. The results were as follows (Table 10.6):

For a thermoset polyester–melamine coating, the glass transition temperature is related to the coating's degree of cure, or cross-link density. The cross-link density, in turn, is related both to the chemical composition of the coating and its baking temperature. Since the glass transition temperature was found to be the same (35–37 °C) at several locations on the failing chairs, the problem was not one of uneven baking in the oven.

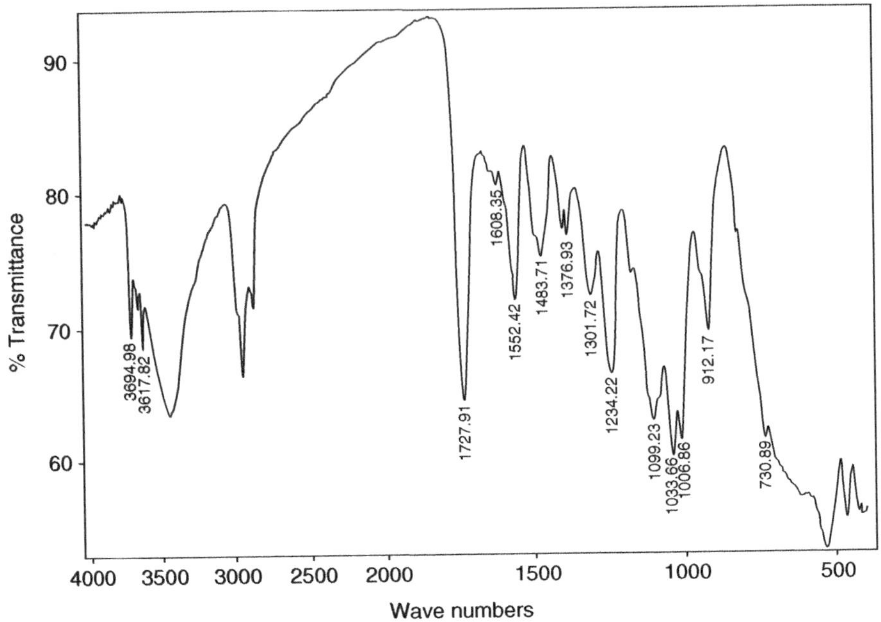

Figure 10.15 Infrared spectrum of coating from a non-failing area of a chair.

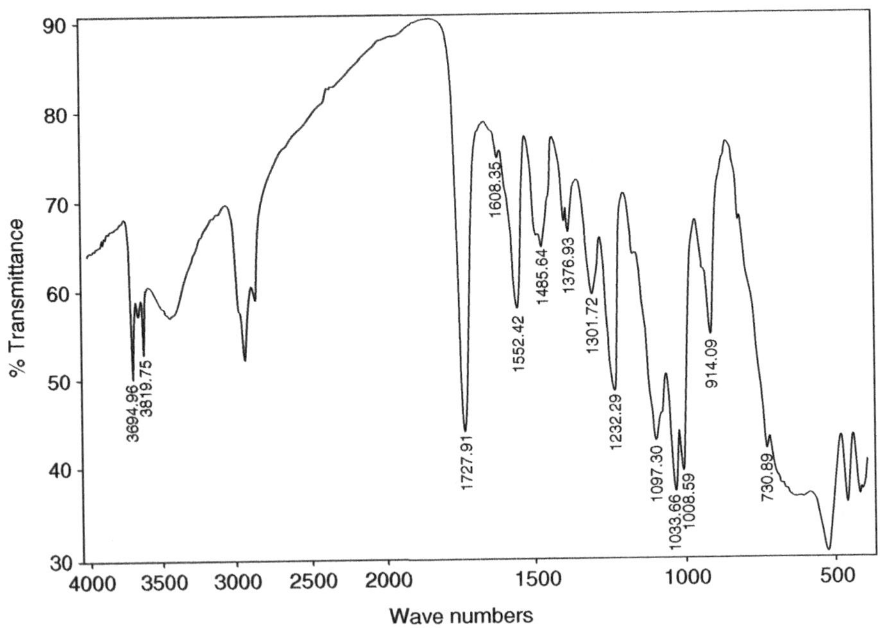

Figure 10.16 Infrared spectrum of coating from a failing area of a chair.

Table 10.6 Glass transition temperatures of failing and non-failing areas on chairs.

Sample	Location	Glass transition (°C)
First failing chair	Failing seat area	37
First failing chair	Non-failing seat area	35
First failing chair	Non-failing brace	36
Second failing chair	Failing seat area	36
Second failing chair	Non-failing brace	35
New chair/new paint	Seat area	Two transitions, at 31 and 79 °

The most obvious finding when the chairs were examined was the fact that nearly all of the failure was located in that area of the seat corresponding to physical contact with a human being. Indeed, not only was there little or no failure on other parts of the chairs, but there was also no failure on the back 3 or 4 in. of the seat, where there would be no physical contact. Furthermore, even on a failing chair, adhesion testing near the back of a seat showed good adhesion despite the fact that a few inches away, there were large patches of missing paint, and chips could be easily flaked off of the metal. It was also noted that the adhesion on the back, or underside, of the seat was also very good, even in the centre of the seat (which was bad on the opposite, or front, side).

This pattern of failure was not due to the large part of the chair (the seat) being a heat sink, thus not reaching the same temperature as the smaller parts. If this were true, one would expect poor adhesion everywhere on the seat, not just in the area exposed to physical contact. Furthermore, DSC testing indicated that there was no difference in the degree of cure, or cross-link density, between failing and non-failing areas on the same chair, since all the areas had similar glass transition temperatures.

The pattern of failure also suggested the possibility that the coating had been softened or otherwise attacked by body oils, sunblock or other personal lotions. However, not only would one expect to see a lower glass transition temperature due to the absorption and plasticizing effects of such agents, which was not the case, but IR spectroscopy showed no difference between failing and non-failing areas on the same chair. IR spectroscopy is very good at detecting the presence of organic chemicals such as oils and lotions.

Coatings can also fail if they are applied over contaminated surfaces, but this was not likely here. First of all, microscopic observations showed that both the back side of the failing paint and the surface of the underlying steel were free at least of any visual contamination. Secondly, it is difficult to see how contamination of the underlying steel would always have been confined to the same areas of the chair, and that these same areas would always correspond to where there would later be human contact.

Since the coating failure did not appear to be due to painting over contamination or to body oils or lotions permeating the paint, and since presumably the initial coating adhesion in the failing areas of the seat was just as good as in the non-failing areas of the seat, it was apparent that the coated chairs could not, over time, withstand the stress due to a human being sitting on them. This stress can come in several forms, including body heat, moisture from perspiration and physical stress from the torque applied to the coating as a person shifts and twists while seated.

Visual observations, along with thickness measurements, showed that not only did failure occur in those areas subjected to human contact, but also that those areas had a rough surface, commonly referred to as orange peel. Orange peel is a consequence of the coating not flowing out properly. In addition to poor flow, the coating contained numerous pinholes, some of which went completely through the coating down to the bare steel. This porosity would allow moisture from perspiration to permeate the coating and come in contact with the steel/coating interface, thus weakening adhesion over time. This would be exacerbated by the fact that the coating was only about 0.7-mil thick in the failing areas, rather than the 1.0 mil listed on the coating's data sheet. Although orange-peeled and pin-holed coating could be found in non-failing areas such as the cross braces, these areas would not have been subjected to body contact. Interestingly, the new paint applied to the more recent chairs was much smoother and had little or no orange peel or pinholes compared with the older, failing chairs.

In addition to the failing coating being somewhat thin and porous, its rough, orange-peeled texture also likely played a role in the failure. This surface roughness means that there would be a greater force of friction between the coating and a human being. Therefore, torque forces caused by people sliding or twisting around on the chair will be more efficiently transferred or coupled to the coating and hence to the steel/coating interface, resulting, over time, in an adhesion failure. Interestingly, the new chair with the new paint was not only free of pinholes, but also had a much smoother surface. It was higher in gloss and smoother and more 'slippery' to the touch. This slipperiness would make it much more difficult for any torque or force generated by human movement to be transmitted through the coating to the coating/steel interface, thus resulting in better adhesion and longevity.

Yet another factor related to the adhesion failure was the glass transition temperature of the coating. Although DSC testing did not indicate that the failing areas were any less cured than non-failing areas, it did show that the coating on failing chairs had a glass transition temperature in the range of 35–37 °C (about 97 °F). Since this is very near the normal human body temperature, it is possible that at times, the coating was being subjected to moisture and torque forces when it was close to its glass transition temperature. As a coating approaches its glass transition temperature, it will become softer, and the force of friction between it and a person will increase. Therefore, the torque forces exerted by a person will be even more effectively transmitted through the coating, putting additional stress on the bond between it and the steel. Furthermore, not only does a coating become softer near its glass transition, but it also becomes more permeable. Thus, not only are the forces intensified, but the amount of moisture permeation will also be increased, resulting in reduced adhesion at the very time that the stress due to torque is being magnified.

The DSC analysis of the new paint on the new chair showed that it had a glass transition similar to, or even slightly below, the 35–37 °C of the failing chairs. However, the analysis also showed that it had a weaker, second glass transition near 79 °C. The reason for the second glass transition was not clear, but it could have been due to an additive or a second resin which was simply harder than the original resin. At any rate, it would likely tend to maintain a higher degree of hardness at higher temperatures than the old coating.

In summary, the paint failure on the chairs was not due to contamination on the steel or uneven baking in the oven, or even to poor initial adhesion of the paint to the steel. Rather, it occurred over time due to the combined effects of somewhat thin, relatively porous paint, a rough surface which increased the amount of stress being transferred to the coating/steel interface and a glass transition temperature which was possibly close to the service temperature of the chairs.

10.15 FAILURE OF RAILROAD CAR LINER

A railroad tank car had been blast cleaned to an SSPC SP10 near-white degree of cleanliness and coated with one coat of a red epoxy primer, one coat of a beige modified epoxy intermediate coat and one coat of a blue modified epoxy topcoat. The intermediate and topcoat were the same coating type but pigmented differently. Prior to the car having been put in service, isolated areas were noted where the beige intermediate coat had delaminated from the red primer. This prompted the owner to perform adhesion testing, which showed that the intercoat adhesion of the intermediate coat to the primer was suspect throughout the car. Delaminated paint chips, shavings of the underlying primer and kits of the primer and intermediate coat were supplied to the laboratory. The laboratory testing consisted of visual and microscopic observations, IR spectroscopy and physical testing.

The delaminated paint chips were as large as 2×5 in. and were blue on the front and beige on the back. When examined with the unaided eye, there was a slight reddish haze on the back side of the beige intermediate coat. When examined microscopically, this haze was found to be due to small, red, thread-like segments on the back surface of the intermediate coat. In some cases, the fragments seemed to be of random orientation, while in other areas, they formed geometric patterns which ranged from ring-like to almost hexagonal (Figure 10.17). In some instances, they were mostly red in colour, while in other instances, they consisted of clear material with reddish primer or pigment particles poorly dispersed in them. There were also many small craters or pinholes at the back of the beige coating. A microscopic examination of the chips of red primer removed

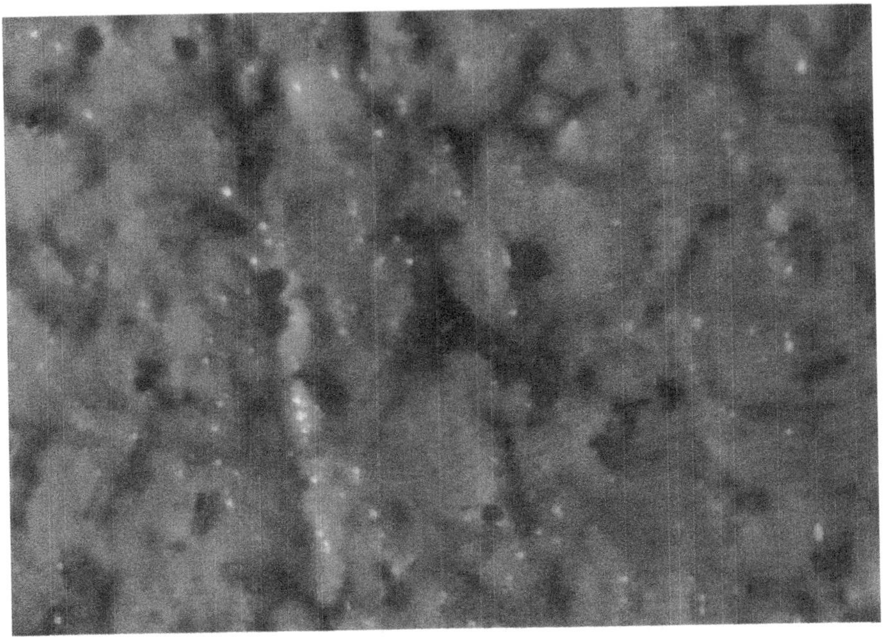

Figure 10.17 Reddish tendrils on the back side of the modified epoxy intermediate coat which delaminated from the red epoxy primer.

from the steel beneath the delaminated intermediate coat showed spots of a clear or milky material on the coating surface.

IR spectroscopic analysis showed that the control sample of red primer, applied and cured in the laboratory, was an epoxy cured with a polyamide curing agent and pigmented with calcium carbonate, iron oxide and a silicate. The beige intermediate coat consisted of an epoxy containing barytes (barium sulfate) extender pigment (bands near 1180, 1115, 1075, 635 and 610 cm^{-1}) and apparently modified with a plasticizer, which resulted in a strong ester carbonyl band near 1730 cm^{-1}.

A spectrum was obtained of the beige intermediate coat from a failing paint chip, by using a razor blade to scrape away a mil or two from the back of the chip (which had the reddish tendrils) and sampling from the interior of the coating. This spectrum was very similar to the spectrum of the control sample of beige intermediate coat prepared in the laboratory, except for some slight differences in resin to pigment ratio.

The spectrum of the clear/reddish tendrils on the back side of the failing beige intermediate coat had several features similar to the spectrum of the red primer but had notable differences as well. The major differences included a strong carbonyl band near 1725 cm^{-1}, a strong, broad band near 1605 cm^{-1} and bands near

1375, 1110, 1070 and 1040 cm^{-1}. It was thought that perhaps these differences were due to the applicator having applied the red primer at the wrong mix ratio, resulting in an under-cured surface of different composition which was partially lifted by the intermediate coat. Therefore, additional spectra were obtained of the red primer applied in the laboratory at different mix ratios and even of the individual components of the red primer. In no case was a spectrum obtained which matched that of the tendrils on the back of the beige intermediate coat. Therefore, the red tendrils were not due to mis-mixed red primer.

Another possible explanation for the spectral features of the reddish/clear tendrils is that they were caused by residues of the beige intermediate coat which were inadvertently sampled along with the tendrils. However, with the exception of the carbonyl band, the unusual spectral features of the tendrils were not present in the spectrum of the intermediate coat, or even in the spectra of the two individual components of the intermediate coat.

In another attempt to account for the composition of the tendrils, a spectrum was obtained by mixing the pigmented part A of the red primer with the clear part B of the intermediate coat, thinking that perhaps the painter had mistakenly mixed these materials at the jobsite. However, this was quickly found to have been impossible, as the two components were incompatible with one another and gelled within a few seconds.

At this point in the investigation, it was decided to prepare a series of test panels in the laboratory to investigate the inter-coat adhesion of the intermediate coat to the primer. Variables included thickness and re-coat time. The test panels consisted of $4 \times 6 \times \frac{1}{8}$ in. commercial-grade hot-rolled carbon steel panels which had been blast cleaned. The coatings were applied by conventional (pressure pot) spray technique. The primer thickness varied from 1–2, 2–3 and 3–5 mil. It was allowed to cure for 0.5, 1, 3 or 24 hours prior to being coated with the beige intermediate coat. The intermediate coat was applied in two coats, each at about 4–5 mil.

The panels were allowed to cure for approximately 1 week, at which time they were tested for tensile adhesion (ASTM D4541, Standard Test Method for Pull-Off Strength of Coatings Using Portable Adhesion Testers). This procedure uses a high-strength adhesive to attach a fixture or 'dollie' to the test panel, and the force required to disbond the dollie/coating is measured.

The results of the adhesion testing were as follows (Table 10.7):

In every case, the location of failure was between the intermediate coat and the primer. Although giving the primer 24 hours to cure resulted in marginally better inter-coat adhesion than shorter re-coat times, the above results showed that in all cases, the intermediate coat had poor adhesion to the primer. Although there is no universal agreement on such things, with the type of instrument used in this testing, values of at least 500 psi could certainly have been expected. Furthermore, cutting into some of the test panels with a knife also showed poor adhesion of the intermediate coat to the primer.

Table 10.7 Adhesion of beige epoxy to red epoxy primer as a function of primer thickness, and recoat time.

Primer thickness (mil)	Re-coat time (h)	Inter-coat adhesion (psi)
1–2	0.5	240
1–2	1	230
1–2	3	200
1–2	24	270
2–3	0.5	200
2–3	1	200
2–3	3	210
2–3	24	240
3–5	0.5	260
3–5	1	210
3–5	3	200
3–5	24	300

Many of the adhesion dollies from the above testing showed a reddish haze on the back side of the beige intermediate coat. When examined microscopically, this haze was found to be due to short reddish/clear tendrils, virtually identical to those observed on the back side of the paint chips which failed from the rail car. Indeed, in some areas, even the geometric pattern of the tendrils was reproduced. Furthermore, the surface of the red primer from where the adhesion dollies were pulled had varying amounts of the same clear, milky haze which were present on chips removed from the rail car. In short, testing in the laboratory reproduced the failure which occurred in the rail car.

The above testing, particularly the physical testing which reproduced the failure, clearly demonstrated that the inter-coat adhesion failure was not due to mis-mixing of the coatings, re-coating too soon, external contamination or any other application-related factors. What it did show was that the failure on the rail car was due to inherently poor adhesion between the red epoxy primer and the beige modified epoxy intermediate coat. An interaction apparently occurs between these two coatings which results in an altered surface chemistry and an associated weak interfacial region. The exact chemical mechanism responsible for this interaction was not elucidated, but for the purposes of the client, it was sufficient simply to discover that the failure was due to an inherent incompatibility between the primer and the intermediate coat and not the result of misapplication.

This particular failure analysis demonstrates the benefits which can sometimes be obtained by preparing test panels. Although the observations from the test panels did not offer a precise explanation as to the mechanism of the formation of the reddish/clear tendrils, it was able to reproduce it, thus demonstrating that this was a ‘normal’, albeit undesired, event. Extensive analysis of these tendrils by IR spectroscopy offered no such information.

10.16 FAILURE OF CALCIUM SULFONATE MODIFIED ALKYD TOPCOAT FROM A BRIDGE

An older bridge which had originally been painted many years ago with an alkyd system was repainted by cleaning and spot priming of any rusted areas. Once the spot priming was completed, all of the structural steel was given a full prime coat and a full topcoat. The primer was a grey calcium sulfonate modified alkyd with a recommended thickness of 2–3 mil, and the topcoat was a metallic green calcium sulfonate modified alkyd, with a recommended thickness of 2–4 mil. The re-coat interval between priming and topcoating varied considerably, from approximately 1 week to as long as 10 weeks.

The following year, numerous areas were noted where the topcoat was peeling from the primer. Most of these areas were on the south side of the bridge. Although the specification called for full priming of the bridge, there were some areas where this had not been done and where the new topcoat had been applied directly to the old alkyd topcoat. There were no adhesion problems in these areas.

The laboratory was supplied with several paint chips from the bridge, as well as gallon cans of the primer and topcoat. The laboratory testing involved visual and microscopic observations, IR spectroscopy and physical testing.

The first chip examined was roughly $\frac{1}{2} \times 3$ in. and consisted of a relatively flexible metallic green coating. The back side of this chip was extremely smooth and somewhat shiny. Very small specks were noted on the back side. When examined microscopically, these specks were due to particles of both white and orange over-spray. Apparently, other painting operations were going on besides the priming and topcoating of the structural steel, and over-spray from this activity had settled on the new primer. The over-spray particles subsequently sheared off when the green topcoat disbonded from the grey primer. The cross section showed a single green coat, 6- to 8-mil thick.

A second sample consisted of chips of metallic green topcoat attached to adhesive tape, which had been used to disbond the topcoat from the primer. These chips were up to about 1×2 in. in size. Again, the back side was smooth and shiny, but there were no over-spray particles embedded in it. The green topcoat was 2- to 3-mil thick. There were also smaller chips in this sample container which consisted of the metallic green topcoat, grey primer and some older coats of paint. Probing at these chips with the tip of a knife caused the topcoat to flake off of the grey primer. The surface of the primer was clean and smooth.

IR spectra were obtained of both the primer and the topcoat samples from the bridge and compared with the spectra of the liquid coatings applied in the laboratory. In both cases, the spectra of the paint chip samples matched those of the liquid samples, showing that there was no substitution or adulteration of the coatings.

Because large samples of failing topcoat were examined which were very clean microscopically on the back side, because the failure was reportedly much worse on the south side of the bridge, and because no failure was reported where the topcoat had been applied to the old, existing paint and not the new primer, it seemed very unlikely that the failure was due to painting over contamination. Why would contamination only be present on the south side of the bridge? Why would it only be present on the newly applied primer and not on the old, existing paint? The observations and findings suggested that perhaps re-coat interval and/or sunlight exposure may have been responsible for the poor adhesion.

In order to investigate this possibility, test panels were prepared in the laboratory. These primed panels were allowed to cure for different lengths of time and under different conditions, prior to topcoating them. In some cases, the primer was allowed to cure at laboratory ambient conditions for varying times, while in other cases, they were exposed for a short length of time in an accelerated weathering device. These devices use UV light, heat and moisture to accelerate the effects of natural ageing.

Once the panels were top-coated, they were allowed to cure at room temperature for a day, followed by 1 week at 130 °F, prior to evaluating them for inter-coat adhesion. The adhesion testing was done via the X-cut method of ASTM D3359, previously described. The best adhesion rating using this method is a 5A, and the worst is a 0A. The results were as follows (Table 10.8):

The inter-coat adhesion after 1 week of accelerated weathering was even worse than the 0A rating implies, in that the topcoat could actually be peeled from the primer using one's fingers.

Based on the above testing, the failure of the calcium sulfonate alkyd topcoat to the calcium sulfonate alkyd primer was not due to any type of contamination. Although in some cases, the topcoat had been applied over over-spray debris from other painting operations going on at the bridge, which might have interfered with adhesion, in other cases, the back of the failing topcoat was smooth and clean. Furthermore, it stuck when applied to the old, existing paint, which if anything would be expected to have more environmental-related contamination than the newly applied primer.

Rather than a contamination issue, the observations and testing clearly demonstrated that the poor inter-coat adhesion was related to the ageing of the primer prior to topcoating. Not only was this theory consistent with the observations that most of the failure was on the south side of the bridge, where the primer would be exposed to higher temperatures and more sunlight, but the failure was repro-

Table 10.8 Intercoat adhesion of calcium sulfonate alkyds as a function of primer cure.

Primer cure	Adhesion, topcoat to primer
7 days, lab ambient	Good (5A)
27 days, lab ambient	Good (5A)
20 days, lab ambient plus 7 days accelerated, weathering	Poor (0A)

duced in the laboratory (where there were no issues of contamination), simply by weathering the primer for 1 week in an accelerated weathering devise prior to topcoating. When the primed panels were allowed to age on the lab bench, the topcoat adhered well to the primer, but a short exposure to heat, water and UV light caused a drastic loss of inter-coat adhesion.

Normally, one alkyd can be top-coated with another alkyd even after prolonged periods of outdoor ageing. Indeed, there were no warnings about the re-coat window on the coating manufacturer's product data sheets. It is possible that the sensitivity towards ageing demonstrated by this coating system might have something to do with the calcium sulfonate modification to the alkyd, although this would require additional research, which was not needed to solve the particular client's problem. However, the project does underscore a potential industry problem, in that sometimes manufacturers do not test a coatings system's adhesion properties under different environmental scenarios. Indeed, as the above results show, making up panels in the laboratory and allowing them to cure on the lab bench, which would be a typical procedure, would indicate that this system has very good inter-coat adhesion. Obviously, however, its outdoor weathering properties are much different.

10.17 DISCOLOURATION OF FURNITURE LACQUER

Large quantities of new furniture were purchased for the offices of a firm located in Canada. The furniture had been stained a dark mahogany and sealed with a vinyl sealer. The final coating applied to the furniture was a clear, post-catalysed lacquer. The data sheet for the lacquer called for catalyst to be added at a rate of 1.0 oz qt^{-1}.

Within a few months of delivery, problems were noticed with a splotchy discolouration, or dulling, of the furniture. In the worst areas, the finish had taken on a milky appearance. When asked if there was any pattern to the failure, the client replied that it was occurring on the tops of credenzas located adjacent to the windows, but not on desks or other items of furniture even a few feet removed from the windows. Two pieces of wood from a discoloured item were sent to the laboratory, and the client was also able to obtain pint quantities of the stain, sealer and lacquer from the company which had finished the furniture.

The first step in the investigation consisted of a visual and microscopic examination of the finished wood samples. The surface of the finished wood had a very non-uniform, splotchy appearance, with dull whitish or milky areas. The appearance was especially objectionable considering the dark mahogany nature of the furniture. In contrast, a door front from the same unit was in very good condition, with no discolouration or fading. When examined with a stereo zoom microscope, the milky discolouration was on or in the clear finish. Indeed, using a sharp scalpel, this discoloured finish could be carefully sliced away to reveal non-discoloured, stained wood underneath.

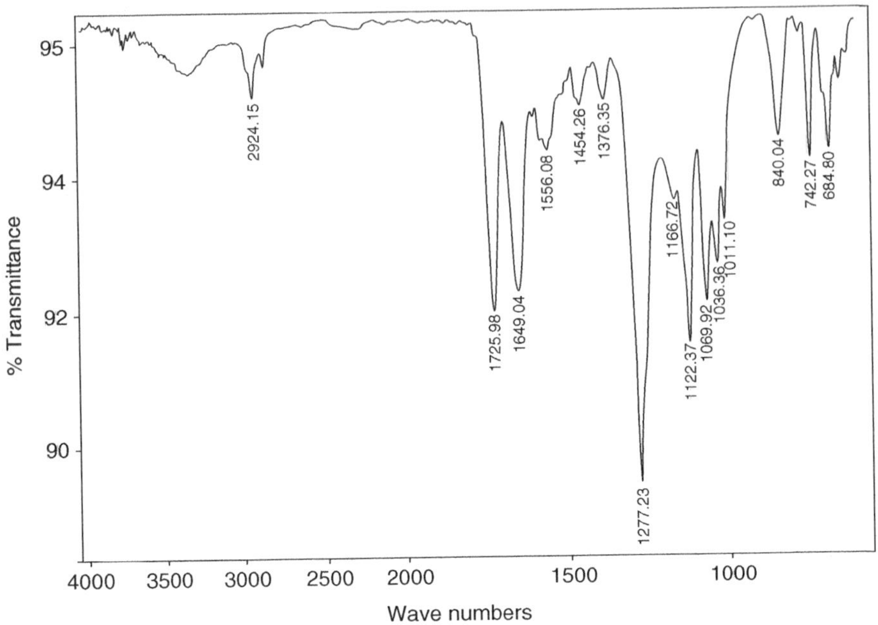

Figure 10.18 Infrared spectrum of non-discoloured furniture lacquer.

The next step in the investigation consisted of IR spectroscopy. Most of the spectra obtained were of surfaces, using an ATR attachment.

The first spectrum obtained was of the non-discoloured lacquer topcoat from the door front. This spectrum (Figure 10.18) is very interesting and has features of an alkyd (or possibly a phthalate plasticizer), nitrocellulose and some type of nitrogenous cross-linking resin, such as a urea–formaldehyde resin. Alkyd (or phthalate plasticizer) is evident by characteristic bands near 1725, 1275, 1120, 1070 and 740 cm^{-1}, while the strong bands near 1650 and 840 cm^{-1} are due to nitrocellulose. The presence of an amino resin, such as urea–formaldehyde, is indicated by the weak band near 1555 cm^{-1}. Its presence is consistent with the label on the bottle of catalyst received for this project, which listed para-toluene sulfonic acid (pTSA). This is a common catalyst for nitrogenous cross-linking resins such as urea–formaldehyde resins. Based on this information, the catalysed lacquer finish appears to be a nitrocellulose modified alkyd cross-linked with a urea–formaldehyde resin.

The next spectrum obtained (Figure 10.19) was from one of the milky areas of the credenza top. While there are similarities between the spectra from the good and bad areas, the spectrum of the discoloured area has several new bands, or at least bands which are very weak in the non-failing area. These include a broad band from 3200 to 2700 cm^{-1}, along with bands near 3170, 3060, 1450, 1150 and 1025 cm^{-1}.

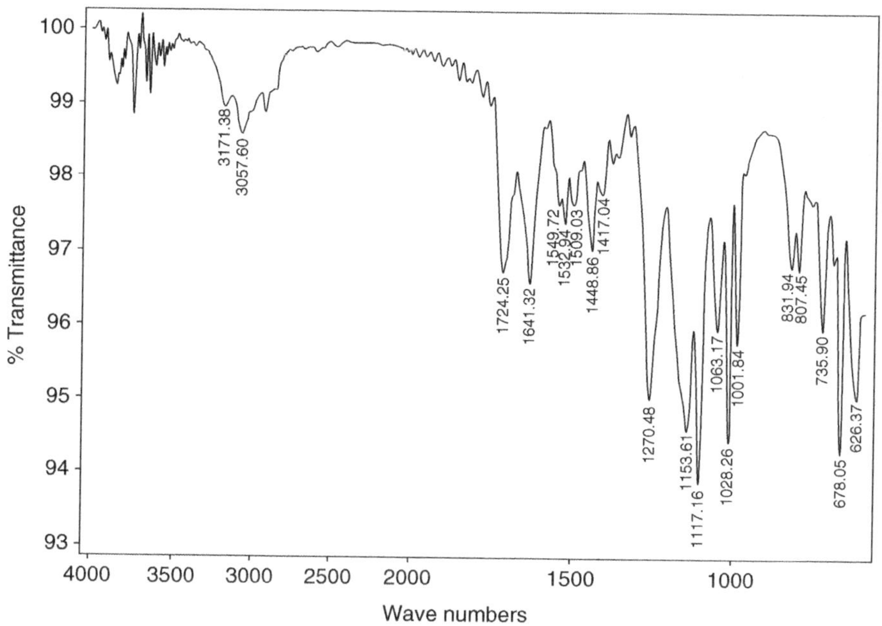

Figure 10.19 Infrared spectrum of milky, discoloured furniture lacquer.

The additional bands present in the IR spectrum of the failing, milky area would certainly seem to be a clue as to the nature of the discolouration, but their origin was unclear. A spectrum was obtained of the catalyst for the lacquer by allowing a few drops of it to evaporate and obtaining a potassium bromide pellet spectrum of the residue. This spectrum (Figure 10.20) was similar to that of pTSA, as expected. However, its spectrum could not account for the additional bands in the spectrum of the discoloured lacquer. Thus, the discolouration did not appear to be due simply to an excess of catalyst in the finish.

Since the discolouration was primarily on the tops of the furniture located directly adjacent to the windows, this suggested that the problem might be due to some type of coating degradation caused by sunlight or UV light. Therefore, a small piece was cut from the non-discoloured door front and exposed in an accelerated weathering device using a UV light source selected to more or less match the spectral fingerprint of sunlight through the window glass. After 240 hours of this accelerated exposure, the milky discolouration had not developed. Therefore, sunlight exposure did not appear to be responsible for the problem.

At this point, more discussions ensued between the laboratory and the client. During this discussion, it was found that a few other pieces of furniture, not located next to the windows, were also beginning to show milky discolouration. These pieces of furniture were desks. When asked if it were the tops of the desks which discoloured (similar to the credenzas), the client replied no, but that the

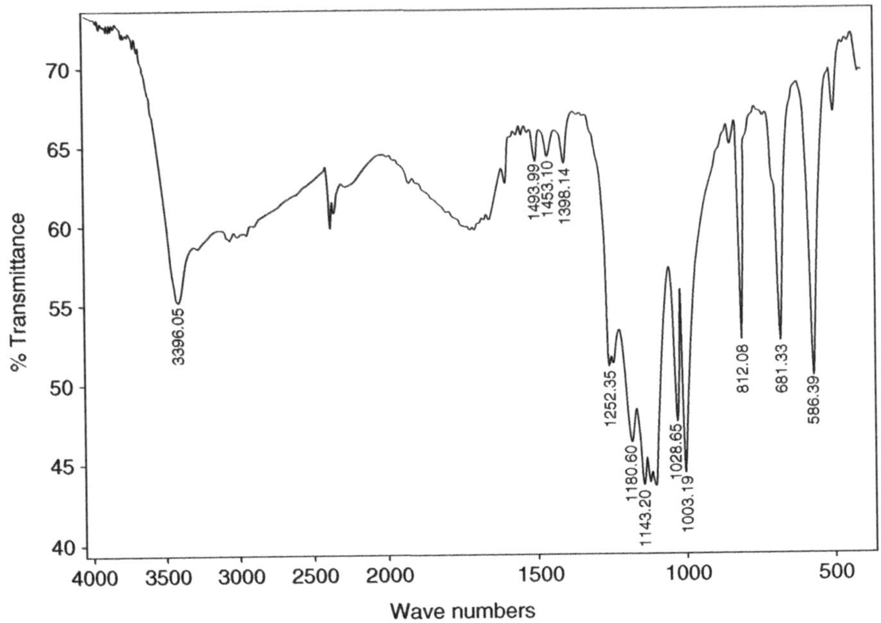

Figure 10.20 Infrared spectrum of dried catalyst component for furniture lacquer.

tops of the desks were protected with glass tops. This new piece of information suggested that the failure was not related to sunlight, but to glass. Further conversation revealed that the janitorial people often used glass cleaner containing ammonia to clean the windows as well as any glass furniture tops.

As discussed above, the spectrum of the pTSA catalyst was not able to account for the spectral differences between the discoloured and non-discoloured samples. However, the additional features could be explainable based not on the presence of a sulfonic acid, but rather on the presence of a sulfonamide. It seemed reasonable that reacting a sulfonic acid such as pTSA with a base such as ammonia could result in the formation of a sulfonamide (specifically, para-toluene sulfonamide). If such a reaction occurred, could it account for the differences in the IR spectra between failing and non-failing samples? More importantly, could it account for the milky discolouration?

A small portion of the pTSA catalyst was spread across a glass slide and allowed to air-dry. This resulted in a thin, clear, colourless material. This slide was then placed face down over a small beaker containing a few millilitres of household ammonia. The next morning, the clear, colourless film of catalyst was milky white. An IR spectrum of this discoloured material was obtained (Figure 10.21) and was found to differ dramatically from the original spectrum of the catalyst. Specifically, the discoloured, milky catalyst had new bands near 3170,

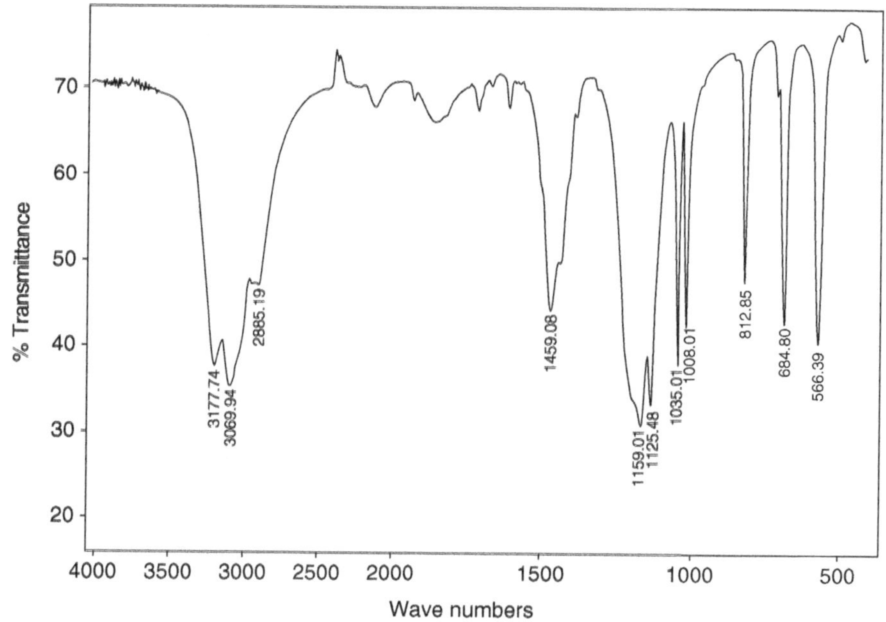

Figure 10.21 Infrared spectrum of catalyst component after exposure to ammonia fumes.

3060, 1450 and 1150 cm^{-1} – the same bands which were present in the spectrum of the discoloured lacquer from the top of the credenza.

The discolouration of the finished furniture was therefore due to a chemical reaction of the pTSA catalyst used in the catalysed lacquer topcoat, with ammonia used in the window cleaner. Apparently, enough over-spray of this material got on the furniture during the cleaning of the windows, or perhaps the tops were even wiped with it, to cause the production of the milky white sulfonamide. Indeed, subsequent to this investigation, the author encountered technical literature for a similar conversion varnish, produced by a different manufacturer, which specifically warned against the use of ammonia-based cleaners.

10.18 FAILURE OF TANK CAR LINING

The interiors of a number of tank cars intended to haul corn syrup were blast cleaned and then subsequently coated with a 100% solids amine-cured epoxy. The mix ratio of the epoxy was 1 : 1 by volume, and it had a recommended one coat thickness of 8–10 mil. The coating was specifically advertised as being an appropriate lining for tank cars and was said to be resistant to a variety of cargos, including corn syrup. The coating required a bake of 4 hours at 150 °F to achieve cure. Reportedly, the cars were baked for this length of time in large ovens at the coating applicator's shop.

Like several cargos shipped by rail, corn syrup, being very viscous, is loaded and unloaded hot, in order to reduce its viscosity. Tank cars used for this type of service have exterior heating coils running along the length of the car, where steam (either regular or super-heated) can be introduced. An individual coil runs the length of a car, makes a u-turn near the end of the car (referred to as the 'bolster' area) and then runs back in the opposite direction. The coil may make two or three u-turns, but it is usually located only in the lower hemisphere of the car and not in the bolster areas.

Because the corn syrup is loaded and unloaded hot, the lining must be resistant to elevated temperature. The amine-cured epoxy used on the cars in question had a listed dry temperature resistance of 240 °F continuous and 280 °F non-continuous. No data were available concerning the temperature ceiling for immersion in hot corn syrup.

Sometime after the cars were put in service, many of them began to show failures. The failures involved cracking and delamination of the coating from the steel substrate. Several shipments of corn syrup were ruined due to contamination from chips of coating.

The first step in the investigation of this failure involved an examination of two failing tank cars. The cars had been drained of cargo and steam cleaned prior to the inspection. In addition to visual observations and sample taking, thickness readings were obtained using a non-destructive magnetic DFT gauge. Coating adhesion was also evaluated by probing with a knife, often after first cutting an 'X' into the coating and then probing at the intersection of the cuts (similar to ASTM D6677, Test Method for Evaluating Adhesion by Knife).

Upon entering the first car, it was obvious that there had been considerable cracking and flaking of the blue epoxy lining. Not only were there patches of failure as large as a square foot or so in size (Figure 10.22), but there were large streaks of failing coating, up to a foot wide and running almost the length of the car (Figure 10.23). Standing at one end of the car and looking towards the opposite end, it was immediately obvious that all of the failure occurred in the lower hemisphere of the car. No failure, defined either by cracking or flaking, was observed on the upper hemisphere, or on either of the bolster areas. The position of the failure coincided with that portion of the car which would have been heated by the heating coils. Furthermore, it appeared that the worst areas of failure probably occurred immediately adjacent to where the coils themselves were positioned on the other side of the steel shell.

Adhesion testing was performed in the good areas on the ends of the cars (bolsters) and in the upper hemisphere. In these non-failing areas the adhesion of the coating to the steel was very good, and it was extremely difficult to remove even small chips. Adhesion was also very good even at the bottom hemisphere of the car, at the midway point between adjacent failing areas. However, the

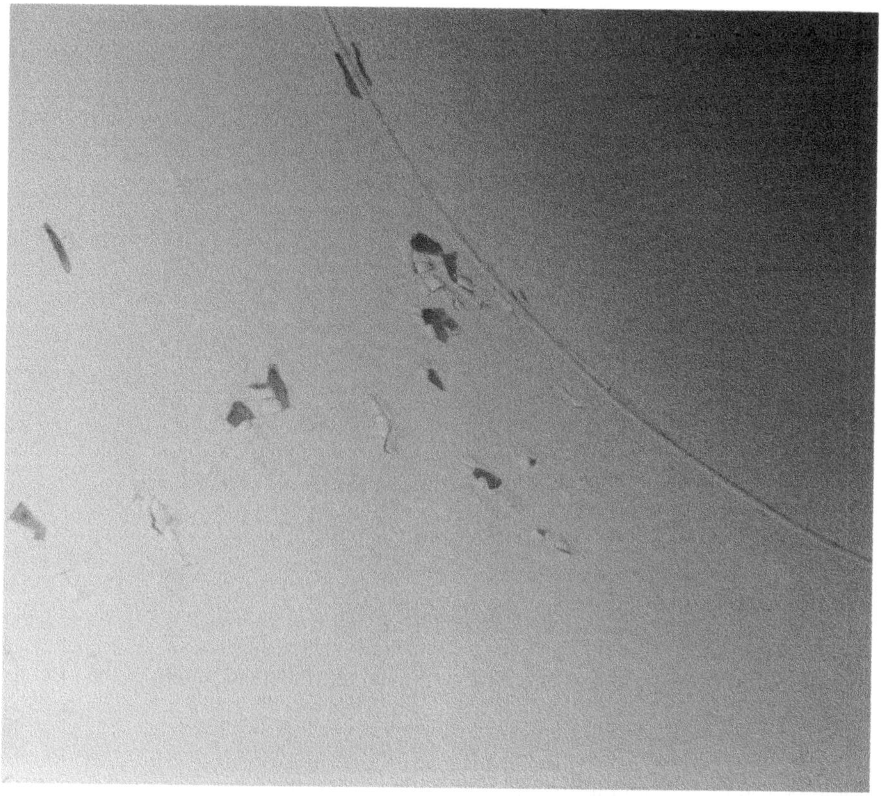

Figure 10.22 Cracking and flaking of a railroad tank car lining.

cracked, failing coating itself often had such poor adhesion that chips could be removed simply using one's fingers.

At failing areas, the exposed steel was rusting. However, this likely occurred after the coating had failed. Indeed, when small chips of coating were forcibly removed with a knife an inch or two away from a rusted, failing area, the underlying steel was clean and bright, and a blast profile was evident.

Several thickness measurements were made in this first car and were found to vary considerably. Generally, the coating thickness in both failing and non-failing locations was 10–18 mil, compared with the recommended thickness of 8–10 mil.

The interior of the second car was similar in appearance to that of the first car. The coating had failed very badly in the lower hemisphere, in the same general pattern as seen in the first car. Again, there was no failure on the two ends of the car, or in the upper hemisphere. Although some areas of coating as thick as 17 mil

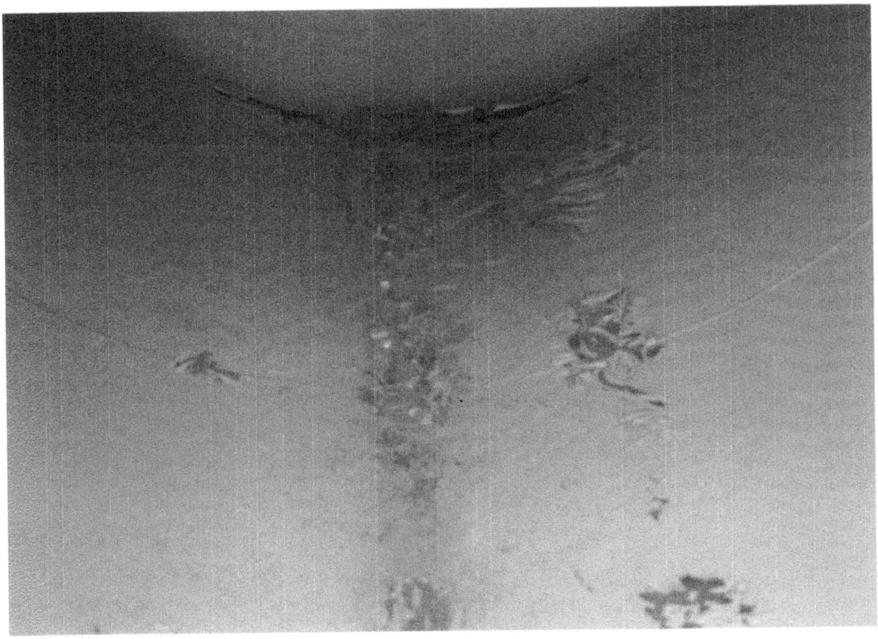

Figure 10.23 Massive failure of a tank car lining, reflecting the pattern of the external heating coils.

could be found, in both failing and non-failing areas, the typical coating thickness throughout the car was 8–12 mil, essentially in agreement with the recommended thickness of 8–10 mil.

As is typical in a failure analysis, the first step in the laboratory investigation consisted of visual and microscopic observations. The failing chips ranged up to about 1×4 in. in size. The front side was smooth and glossy, whereas the back side was rough due to the replication of a blast profile. The chips were hard and brittle. Some had a slight amount of rust staining on the back side, while others had no rust. There tended to be only a few small voids in the cross section. Some of the chips had some beige or whitish debris on the back. However, this debris was loose and could be brushed away, suggesting that it got behind the loose coating after it had failed. Thickness measurements were similar to those obtained during the field investigation.

The next step in the investigation consisted of IR spectroscopy. This was done primarily to confirm whether or not the specified coating had been applied (a liquid sample of this coating had been supplied to the laboratory) and to check for the presence of any type of organic contamination on the back of the failing chips. Spectra were obtained via ATR.

IR spectra were obtained of both a laboratory applied control sample of the coating in question (Figure 10.24) and of a failing paint chip from one of the tank

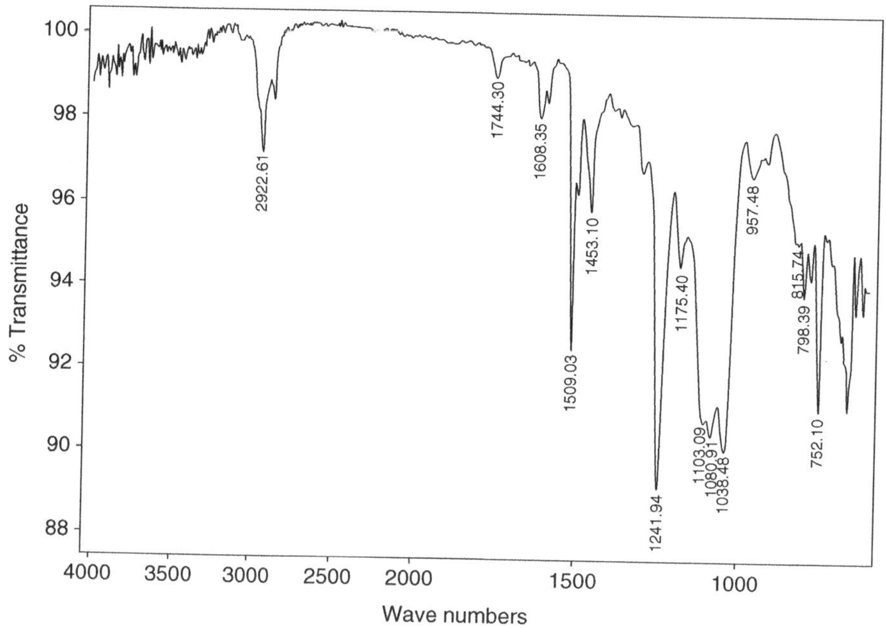

Figure 10.24 Infrared spectrum of epoxy coating used in a railroad tank car.

cars. The two spectra were very similar to one another and consistent with an epoxy (characteristic bands near 2900, 1610, 1510, 1240, 1175 and 815 cm^{-1}), pigmented with crystalline silica (bands near 1100, 800 and 780 cm^{-1}) and possibly with another silica/silicate extender pigment (bands near 1080 and 1040 cm^{-1}). The lack of a band near 1640 cm^{-1}, which is where many amide curing agents absorb, suggests that the curing agent is an amine, which would be consistent with the manufacturer's description of the coating. Furthermore, the analysis of the front and back sides of the failing chips showed only some differences in the relative amounts of certain pigments and some variations in the relative amounts of pigment and binder. There was no indication of any type or organic contamination, such as grease or oil.

Although in some cases IR spectroscopy can be used to determine the mix ratio of an epoxy, the technique seldom works for an amine-cured epoxy. This is because the bands due to the amine-curing agent are either too weak or tend to overlap too much with bands from other ingredients of the coating. Therefore, in order to determine whether or not the coating had been applied at the correct 1 : 1 mix ratio, samples were analysed for % nitrogen. This requires not only an analysis of chips from the cars, but also the coating applied in the laboratory at three or more different mix ratios, in order to 'calibrate' the technique. Since the amine-curing agent contains nitrogen and the epoxy base component does not

Table 10.9 Nitrogen analysis of laboratory controls, and chips from tank cars.

Sample	**% Part B (curing agent)**	**% Nitrogen**
Lab control mixed 2 : 1	33.3	1.30
Lab control mixed 1 : 1	50.0	1.85
Lab control mixed 1 : 2	66.7	2.73
Failing chip #1	46.3 (calculated)	1.80
Failing chip #2	52.3 (calculated)	2.06
Non-failing chip #1	46.5 (calculated)	1.82
Non-failing chip #2	51.2 (calculated)	2.01

(unless it contains a nitrogen-bearing pigment), the amount of nitrogen in the cured film should correspond to the mix ratio of the applied coating.

The results of the nitrogen analysis were as follows (Table 10.9):

As the above results show, there is a direct relationship between mix ratio and % nitrogen. A simple least squares analysis of the data from the laboratory controls using a hand-held calculator shows that the correlation coefficient for the calibration curve (% part B versus % nitrogen) is 0.991. A perfect straight line relationship would show a correlation coefficient of 1.000. Therefore, while not perfect, the analysis allows for a very reasonable determination of the amount of part B in the field samples. As shown above, these data clearly show that the coating was mixed very close to the specified 1 : 1 mix ratio. Therefore, the failure is not due to mis-mixing of the coating.

So far, the investigation has shown that the failure is not due to lack of blast cleaning (field and laboratory microscopic observations) or of painting over contamination (laboratory microscopic observations, plus IR spectroscopy). It is also not due to the use of the wrong coating (IR spectroscopy) or to mis-mixing of the coating (% nitrogen analysis). While some of the failing samples were considerably thicker than recommended, which could have contributed to the failure by imparting both more weight and additional stress, in the second car examined, most of the failing coating was at or very close to the recommended thickness. Therefore, coating thickness also does not appear to be the cause of the failure. One key variable which has not yet been addressed, however, is whether or not the coating was properly baked.

Baking would seem to be a very pertinent issue. Not only is it conceivable that this could have been done incorrectly, but the extensive cracking of the failing coating also shows that it is quite brittle (a fact confirmed in the laboratory upon trying to flex the coating). Over-baking of a thermoset coating can often result in an overly brittle film.

The degree of cure of the coating, a property directly related to both its mix ratio and its amount of bake, was investigated in the laboratory via DSC. Since it has already been shown that the coating was applied at the correct mix ratio,

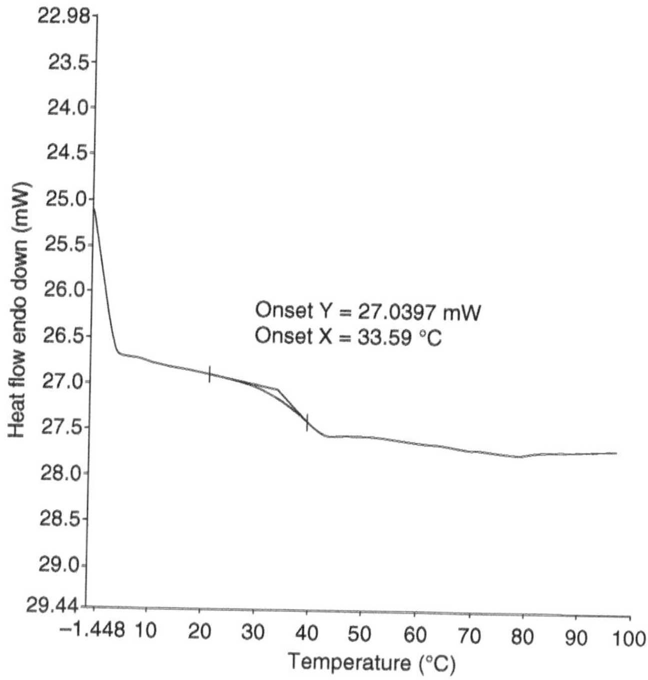

Figure 10.25 Differential scanning calorimetry analysis of cured laboratory control sample of epoxy coating specified for the lining of the railroad tank car.

any deficiencies in the degree of cure could only be attributed to its bake or perhaps to a defective batch of coating.

In performing this analysis, the liquid coating was mixed and applied in the laboratory and subsequently cured for various times/temperatures. These controls, along with failing and non-failing chips from the cars, were then analysed by DSC in order to determine their glass transition temperatures (T_g). As previously discussed, the T_g of a thermoset coating is related to its degree of cure. A typical DSC heating curve from one of the laboratory control samples is shown in Figure 10.25. The shift in baseline near the glass transition is very evident, and the glass transition at onset was found to be 34 °C.

The results of the DSC testing were as follows (Table 10.10):

As the above results show, when cured according to the manufacturer's literature, the coating has a T_g of 34 °C. When under-baked, the T_g is lower, and when over-baked, the T_g is higher, as expected. All of the samples tested from the tank cars had a glass transition temperature significantly higher than 34 °C.

Because the lining was exposed not only to elevated temperatures during its baking at the applicator's shop, but also to many hours of exposure to hot cargo, there is some difficulty in interpreting the above data. However, it certainly

Table 10.10 Results of DSC testing of laboratory controls, and samples removed from tank car.

Sample	Glass transition (°C)
Control baked for 3 h at 125 °F	30
Control baked for 4 h at 150 °F (specified bake)	34
Control baked for 3 h at 125 °F plus 4 h at 175 °F	45
Control baked for 4 h at 150 °F plus 16 h at 215 °F	62
Control baked for 4 h at 150 °F plus 24 h at 255 °F	79
Failing #1	60
Failing #2	57
Non-Failing #1	43
Non-Failing #2	45

appears as though the coating had been properly baked (cured) by the applicator, since failing and non-failing samples all had glass transitions above the 34 °C of the properly baked control. While it could be argued that the coating achieved this T_g not during the bake, but by being exposed to warm cargo, this seems very unlikely. Had the coating been under-cured at the time that it was put in service, it is likely that it would have failed everywhere, not just in a specific pattern at the bottom half of the car.

The DSC data, along with visual observations of the cars themselves, show that the failure is the result of elevated temperature exposure from the tank car's heating coils. This was certainly suggested by the pattern of failure, which corresponded to the position of the heating coils. Furthermore, the T_g of non-failing samples taken from areas where there were no heating coils ranged from 43 to 45 °C, whereas failing coating from areas adjacent to heating coils had glass transitions of 57 and 60 °C. This is proof that the failing coating adjacent to the heating coils was exposed to much higher temperatures during the hauling of cargo than was the non-failing coating, in the same car.

No information was available concerning the details of the heating operations during cargo loading and unloading, but the DSC data suggest that the coating in the failing areas was not exposed to temperatures much higher than that of regular steam (212 °F). This is because when the coating was baked overnight in the laboratory at approximately this temperature, its T_g was 62 °C, very close to that of the two failing samples. Baking it at a significantly higher temperature than that of regular steam resulted in a much higher T_g of 79 °C, which was not observed on any of the samples from the cars. Therefore, it does not appear that the failure was due to heating the car or cargo to unusually high temperatures. Apparently, the coating adjacent to the external heating coils continued to cure upon exposure to temperatures close to that of regular steam. The additional cure resulted in a very brittle coating with a much higher than normal glass transition temperature. The embrittled coating subsequently developed cracks and eventually lost adhesion to the underlying steel.

REFERENCES

1. G. M. Powell, *Vinyl Resins*, Federation of Societies for Coatings Technology, Blue Bell, Philadelphia, PA, USA, 1972, p. 25.
2. D. G. Weldon, A. J. Bochan, W. D. Corbett and R. O. Lackey, *J. Prot. Coat. Linings*, **6**, 21 (April 1989).
3. T. J. Veneri and J. A. Kramer, *J. Coat. Technol.*, **66**, 23 (1994).
4. A. Skaja, D. Fernando and S. Croll, *J. Coat Technol. Res.*, **3**, 41 (2006).
5. G. Winterbottom and F. Goodwin, *J. Prot. Coat. Linings*, **22** (1), 40 (January 2005).
6. American Concrete Institute, *State of the Art Report on Alkali-Aggregate Reactivity*, American Concrete Institute Paper 221.1R-98, American Concrete Institute, Farmington Hills, MI, USA.
7. R. Tombaugh and J. Kramer, *Coat. Pro*, 52 (March 2008).

11

The Repair of Coating Failures

It is a very difficult task to provide universal guidelines on how to repair coating failures. There are several reasons for this, not the least of which is the seemingly endless variety of coating failures and the multitude of factors that can cause them. Furthermore, a repair recommendation to fix a blistered vinyl coating may be totally inappropriate for fixing a blistered epoxy coating. In addition to technical constraints, there may be economic ones as well. If the coating on a 20-storey building in an urban metropolis is failing because of defective primer, the 'technically correct' repair procedure may be to blast clean all of the coating off and start over. However, this is not an economically sound solution. Therefore, the most that can be put forth in this chapter are some general guidelines and considerations.

One of the first things to consider in developing a repair plan, after first establishing the reason for the failure, is the extent of the failure. If the failure is small and localized, spot repair or touch-ups may be adequate. This can usually be accomplished by using simple equipment such as hand or power tools, as opposed to abrasive blast cleaning. If the areas are small, touch-up may often be carried out by brush or roller (provided aesthetics is not a primary concern). Spot repair by such methods, at least on large structures, has the added advantage of minimal environmental impact. This can be a major concern if the coating being repaired has hazardous metals, such as lead or chromium, contained in it.

When carrying out spot repairs, it is common practice to remove all failing coating until an area of sound coating is reached, and then to feather the edges of the sound coating. Failure to do this can result in additional failure in the worst case, and in the best case is simply unattractive. It is also important to choose a repair coating system that is compatible with the existing system in order to avoid failures in the overlap areas.

If the failure area is 10% or more of the surface area, it usually makes sense to carry out a complete repair rather than a spot repair. It is very important, when

Failure Analysis of Paints and Coatings, Revised Edition Dwight G. Weldon

estimating the % failure, to also include those areas that have not yet spontaneously failed but show clear signs of impending failure. If the failure mode is one of peeling, for instance, it may be that the obvious peeling comprises less than 1% of the surface areas. However, if adhesion testing is done at randomly selected 'good' areas, one may find large areas of very poorly adherent coating that simply have not had a chance to fall off yet.

Depending on the type of failure, it is not always necessary to remove the entire coating system, even if the extent of coating failure is massive. If the topcoat is peeling from a primer because the former is defective (perhaps it is too thin, too thick, 'mis-mixed' or simply defective paint), it may make sense to remove only the topcoat and leave the primer intact. Not only does this save money on repainting (it is cheaper to apply one coat than two), but it also spares exposing the substrate to the environment. Such a scenario might involve high-pressure water blasting to remove the loose topcoat or the use of non-aggressive (soft) abrasive media.

If the failure is cosmetic (such as discolouration) rather than structural (such as peeling or blistering), the repair may simply be to cover it up. After cleaning to remove dirt or other contamination, a fresh coat of topcoat may be applied, provided that it is compatible with the existing coating.

The astute reader has probably noticed how many warnings have been given concerning 'compatibility'. This is a key issue when repainting old or failed coatings. In some cases, lack of compatibility is obvious. One would not, for instance, topcoat an acrylic latex with a solvent-based two-component epoxy. The combined effects of solvent and curing stress would be likely to lift the solvent-sensitive acrylic.

Sometimes, compatibility issues are not so obvious. Could a coal tar epoxy be repaired by coating it with the same coal tar epoxy? Surely the two coats would be compatible. Unfortunately, aged coal tar epoxies are difficult to re-coat, even with themselves. This is due both to their very hard nature and to surface deterioration as a result of sunlight exposure.

At a minimum, when contemplating a re-coating, one should obtain the opinion of the coating manufacturer. It is also good practice to apply a test patch prior to wholesale re-coating of a large structure. This test patch should be applied via the same technique that will be used on the whole structure. It should also be given adequate time to fully cure before it is assessed for adhesion (rarely will a defective test patch simply 'fall off' – it needs to be stressed for proper assessment).

A common objection to test patches is that they take too long to evaluate. At a minimum, the newly applied coating should be allowed to cure for a week prior to evaluation, and ideally for several weeks. The properties of a coating, especially a two-component coating or one that cures by air oxidation, can change substantially over time. A coating that displays good initial adhesion may become poorly adherent as it cures and embrittles. The longer one can wait to examine a test patch, the better.

One of the most difficult types of failures to repair is a coating that, for whatever reason (usually improper catalysis or mixing), has remained soft and gummy. This author was involved in one such failure on several thousand square feet of concrete floor, which was the result of the contractor trying to extend a short pot life by cutting the catalyst level in half.

Soft, gummy coatings are extremely difficult to remove by blast cleaning. Either laborious hand tool methods are required, or when odour is not a concern, chemical strippers can be used. If the substrate is porous (such as a concrete floor), chemical stripping may result in uncured coating residues penetrating into the substrate. A final scouring, such as by blast cleaning, may be required in order to ensure cleanliness.

When the extent of the failure justifies total removal, the removal method itself must be given careful consideration. Things to be considered include cost, environmental impact, damage to the substrate and damage to nearby machinery. Using a paint stripper and sanding to remove a defective furniture lacquer makes sense, while adopting the same procedure on a bridge would be an economic, and perhaps an environmental, disaster.

An example of a repair that could damage the substrate would be the removal of defective coating from coil-coated galvanized roof panels. Normal abrasive blast cleaning would be efficient, but it could also remove substantial amounts of the protective galvanizing, as well as potentially deform the relatively thin-gauge steel. The use of high-pressure water blasting, or a soft abrasive such as crushed walnut shells, would be a better choice.

It is usually important to know why something has failed before trying to fix it, so as not to make the same mistake twice. A water tank that has blistered because the epoxy primer contained water-soluble solvents will blister again if the same type of coating is applied under the same conditions.

The investigation of coating failures can be a fascinating pursuit. It requires knowledge of application methods, service environments, coatings chemistry, analytical methods and logic. Unfortunately, it sometimes requires things that the investigator may have little control over, such as accurate, detailed background information and good samples. Hopefully, this book has provided the reader not only with some of the tools needed to decipher coating failures, but also a sense of the enjoyment that such a challenge can bring.

Index

Page numbers in *italics* refer to figures and tables

abietic acid 66, *67*
abrasive blasting 22 (*see also* blast cleaning)
abrasives 15
accelerated testing 266, 268–9
acetic acid *234*
acid etching 22, 130, 294, 296
acidity 66
acrylamide *91*
acrylic acid *91*
acrylic epoxies 84–5
acrylic-modified vinyl latexes 111–12
acrylic–polyurethanes 103, 105
acrylics 90–6
  delamination 231
  hydroxyl-functional 103
  infrared spectra 178, 180
  lacquers 91–2
  latexes 92–5
  powder coatings 123
  silicone-modified 116
  silicon-modified 119–20
  solution 91–2
  thermoset 95–6
  water-soluble 95
additives 5, 51–7
adhesion 9–10 (*see also* delamination)
  failure investigation 183, 187
  inter-coat *see* inter-coat adhesion failure
  related to cure conditions and thickness 308
  testing 136–8, 261–3, 279
adipic acid *98*
ageing, physical 30–1, 257–8 (*see also* weathering)
air compressors 126
air entrainment 58
air sprays 126
airless sprays 126–7
alcohol hydrolysis *see* saponification
alcohols *63*, 64
aliphatic amines 76–7, 79
aliphatic hydrocarbons 59
aliphatic isocyantes 103
alkali-aggregate reactivity (AAR) 321–2
alkaline conditions *see* salts
alkaline hydrolysis *see* saponification
alkali-silica reactivity (ASR) 321–2
alkyds 2, 70–3
  calcium sulfonate modified 331–3
  chemical attack 32
  cracking 146, *147*
  dry-fall 296–300
  drying 71
  failure mechanisms 72–3
  infrared spectra *174*, *179*
  lacquer discolouration 333–7
  modification 73
  production 70–1
  saponification 72
    detection 191, 193, *194*
    failure example 296–300
  silicone-modified 115

Failure Analysis of Paints and Coatings, Revised Edition Dwight G. Weldon

urealkyds 106–7
uses 72
aluminium
corrosion protection 45
discolouration 285–8
surface preparation 18
aluminium flake 32
amides 78
amido-amines 78
amine adducts 77
amine blush 77, 81–2, 183–5
amine-functional amides 77–8
amines 76–9, 256
amino resins 88–90
ammonia-catalyst reaction 336–7
anodizing 18
anti-foaming agents 54
anti-settling agents 51–2
anti-skinning agents 55–6
application-related problems 125–34
applicator errors 132–4
epoxies 83–4
mixing errors *see* mixing errors
spraying 126–7, 145
testing 269–73
thinning errors *see* thinning errors
unauthorized solvents 219–20
aromatic hydrocarbon solvents 59, 62
aromatic isocyanates 103
asphaltic coatings 112–13
ASTM standards 259, 261
gas chromatography *218*
C868 267
D2370 264
D2697 7
D2794 264
D3132 59
D3359 137, 262
D4145 264
D4258 22
D4259 22
D4541 137
D4752 265
D522 264
D5402 265
D7127 15–16
D7396 19–20
D822 266
Dl308 267
G23 266
G26 266
G53 266

attenuated total reflectance (ATR) 171–3, *174–5*, 277, *278*
autoxidation 69–70, 71–2
azelaic acid *98*

baking errors 133, 273
barium lithol red 47
barium metaborate 46, 56
barium sulfate pigment 43
barrier coatings 44–5
baryte pigments 43
batching errors investigation 216, 227–9
benzimidazolone yellows 49
benzoguanimine–formaldehyde (BF) resins 89–90, 95, 96
binders 2–3, 32
biocides 56
biological growths 56, *142*
bismuth vanadate 40
bituminous coatings 112–14
biurets 102, *103*
blast cleaning 15–16, 19, 129
blister liquid, analysis 280, 281 (*see also* liquid samples)
gas chromatography 221
infrared spectroscopy 315–18
ion chromatography 237–8
blistering, cathodic 36
blistering, osmotic 34–5, 220–1, 281
blistering investigation 139, 145, *148*, 279–82
blue pigments 41, 49
blushing *see* amine blush
BON red 47, *48*
bridge coatings 275–9, 304–9, 331–3
brittle failure 26, 27–8
and cross-link density 83
dry-fall alkyd 300
epoxies 83
low temperature 311–13
phenolics 87
physical ageing 31
PVC coating 292–3
brittleness testing 151–2, 263–4
brush application 125
butyl acetate *63*, 64

cadmium orange 40
cadmium yellows 40
calcium BON red 47
calcium borosilicate 45

calcium carbonate
  extender pigment 42–3
  infrared spectra 180, *181*
calcium sulfonate modified alkyds 331–3
calcium zinc molybdate 46
caprolactam 229
carbamic acid 102
cast films 176–7
castor oil *69*
catalysts
  dryers 54, 70
  natural oils 70
cathodic blistering 36
cathodic protection 46
chairs, metal, coating failure 322–7
chalk *see* calcium carbonate
chalking 80, 195
chemical attack 31–2, 191–5 (*see also* corrosion)
chemical bonding 9–10
chemical resistance testing 266–9
chemical stripping 349
chemically resistant coatings 32
  pigments 4, 43–6
china clay *see* kaolin
chlorinated hydrocarbon solvents 62
chlorosilanes 114
chromate pretreatment 19
chrome green 41
chrome orange 40
chrome yellow 40
chromium oxide pigment 41
chromophores 47
cladding *see* sidings
cloud chambers 128
CMC (critical micelle concentration) 92
coal tar coatings 113–14
coal tar epoxies 85, 113–14, 191, 300–4
coalescing aids 93, 94
coats, number of 146
coconut oil *69*, 178, *181*
cohesive splitting 23, 82, 117, 304–9
cohesive strength 23
coil coated stock 128
  aluminium, discolouration 285–8
  baking 273
  flexibility testing 264
  galvanized, paint removal 349
  stress–strain behaviour 27–8
cold-wall effect 268
colophony *see* rosin
colour pigments
  inorganic 39–41
  organic 47–50
combustion analysis 259
compatibility issues 348
concrete
  abrasive blasting 22
  acid etching 22, 130
  coating failures 130, 186, 293–6, 314–22
  surface preparation 21–3, 130, 294, 296
condensation 268, 271, 272
conductivity detectors 235–6
contact angle 11
contamination 269–73 (*see also* surface preparation)
  biological growths 56, *142*
  coating failures 129
  effect on wetting 13–14
  investigation 272
    infrared spectroscopy 182–7
    microscopy *143*, 144
contraction
  freeze–thaw testing 269
  stress–strain behaviour 27–8
control samples 186–7
conversion coatings 17–18
copal 66
copper phthalocyanine blue 49
copper phthalocyanine greens 50
corrosion
  cathodic blistering 36
  coal tar epoxy failure 301
  mechanisms 43–4, 144–5
  protection 17–18, 44–6
  resistance 227
    epoxy–siloxane hybrids 119
    zinc-rich primers 116–17
  salt contaminants 237
corrosion-resistant pigments 4, 43–6
co-solvents 93, 94
covalent bonds 9, 31–2
CPVC (critical pigment-volume concentration) 6, 23, 24
cracking (*see also* brittle failure)
  alkyds 146, *147*
  epoxies 282–5
'cratering' 57
'crawling' 57
critical micelle concentration (CMC) 92

critical pigment-volume concentration (*CPVC*) 6, 23, 24
cross-link density
  and brittleness 83
  and cohesive strength 23
  and permeability 24
  and stress–strain behaviour 29
cross-linking agents
  for amino resins 89–90
  for epoxy resins 76–9
  for polyesters 99, 100
  for polyurethanes 102–3
  thermoset acrylics 95–6
cross-linking mechanism 69–70 (*see also* degree of cure investigation)
curing process
  alkyds 2, 71–2
  amino resins 88–90
  coating failure 139
  cross-linking mechanism 69–70
  epoxies 75–80, 271
  investigation *see* degree of cure investigation
  natural oils 69
  phenolics 87
  solvent reaction 58
  temperature effects 271–2
curing time 81, 271
cycloaliphatic amines 79

dammar 66
de-foaming agents 54
degree of cure investigation
  differential scanning calorimetry (DSC) 252–3
  infrared spectroscopy 195–202
  microscopy 151
  scanning electron microscopy 245–6
  solvent resistance testing 265
delamination 133, 138 (*see also* adhesion)
  acrylics 231
  cohesive splitting 23, 82, 117, 304–9
  epoxies 282–5
  inter-coat *see* inter-coat adhesion failure
  investigation 145–6
  plastisols 288–93
  stress relief 28
density 7
DETA (diethylene triamine) 76
DGEBA (diglycidyl ether of bisphenol A) 74
diamine cyclohexane 79
diarylide yellows 48–9
dicyandiamide 79, *80*, 122
diethylene triamine (DETA) 76
differential scanning calorimetry (DSC) 247–58
  applications 252–8, 286–7, 342–4
  calibration 250–1
  degree of cure investigation 252–5
  environmental tests 256
  physical ageing investigation 257–8
  sample preparation 251–2
  theory 247–50
differential thermal analysis (DTA) 248–9
diglycidyl ether of bisphenol A (DGEBA) 74
'diluents' 59
diphenyldichlorosilane 114
diphenylmethane diisocyante (MDI) 103, *104*
disbonding *see* delamination
discolouration
  aluminium siding 285–8
  furniture lacquer 333–7
  plastisols 288–93
dispersive infrared spectrometer 166–7, *168*
dolomite extender pigment 43
driers 54, 70
dry-fall alkyds 296–300
drying alkyds 71
drying oils 69–70
drying process 58
  alkyds 71–2
  natural oils 69
dry-spray 145
DSC *see* differential scanning calorimetry (DSC)
DTA (differential thermal analysis) 248–9
ductility 25–8

EDS (energy-dispersive spectrometer) 243
efflorescence 21
'electrochemical barrier' 45
electroendosmotic blistering 35–6
electromagnetic spectrum 153, *154*
electrostatic fluidized beds 128
electrostatic sprays 127, 128–9
elemi 66
elongation *see* expansion
embrittlement *see* brittle failure
emulsifying agents 53–4

emulsion polymerization 92–3
enamel 2
energy-dispersive spectrometer (EDS) 243
epichlorohydrin *86*
epoxies 75–84
  acrylic 84–5
  adhesion related to mix ratio and cure *279*
  amine blush 77, 81–2, 183–5
  amine-cured 4, 76–80, 84
    failure examples 282–5, 300–4, 337–44
    glass transition temperature 256
    infrared spectra 185–6
  blistering 279–82
  brittle failure 83
  chalking 195
  coal tar 85, 113–14, 191, 300–4
  on concrete 293–6, 314–22
  cracking 282–5
  curing agents 76–9
  curing time 271
  cycloaliphatic amine 256
  degree of cure investigation 199, *201*, 202
  delamination investigation 245–6, 282–5
  disbonding from 275–9
  failure mechanisms 80–4
  gel permeation chromatogram 229, *230*
  impact resistance 256
  infrared spectra *175*, *179*, 341
  inter-coat adhesion failure 327–30
  mix ratio investigation 188, *189–90*, 270
  modified 84–6
  phenolics 85–6
  photodegradation 33–4, 80, 279
  physical ageing investigation 257
  physical properties 256
  polyamide 256
  powder coatings 79–80, 122–3, 253–5, 256–7
  shrinkage stress 82
  structure *74*
  wall coating 282–5
epoxy esters 74–5
epoxy phenolics 85–6
epoxy–siloxane hybrids 119
equipment for site visits 135
equivalent weight, binders 3–4
ester gum 66–7
esters 64
ethyl acrylate *91*
ethyl silicate 116
evaporation rate, solvents 5, 57–8, 59, *63*
expansion
  freeze–thaw testing 269
  investigation 256
  stress–strain behaviour 27–8
extender pigments 4, 41–3

failure mechanisms 24–36
  acrylic latex coatings 94
  alkyds 72–3
  application problems 129–34
  brittle failure *see* brittle failure
  chemical attack 31–2
  delamination *see* delamination
  epoxies 80–4
  internal stress 29–31
  mechanical stress 25–9
  phenolics 87–8
  polyurethanes 106
  weathering stress 33–4
failure patterns 136
fatty acids 68, *69*, 71
fibreglass 101
field methods 135–9
flame-ionization detector (FID) 211–12
flexibility 25–6 (*see also* brittle failure)
  testing 151–2, 263–4
'floating' 57
floor coatings 293–6, 314–22
flow coating 127–8
flow-modifying agents 56–7
fluidized beds 128
formaldehyde *85*
formic acid *234*
Fourier-transform infrared (FTIR) spectrometers 167–9
freeze–thaw testing 269
fumaric acid 100
fungi 56, *142*
furniture lacquer *see* lacquers

galvanized steel
  blast cleaning 19
  coating failures 288–93, 296–300, 309
  paint removal 349
  pretreatments 19
  storage stain 20

surface preparation 18–21
weathering 20
gas chromatography 202–21
batching errors investigation 216
GC-mass spectrometry 213–16
headspace analysis 219, 284–5
instrumentation 208–16
pyrolysis-GC 216–17
solvents investigation 217, 219–21
theory 202–8
gel permeation chromatography (GPC) 221–32
additives investigation 230
application example 291–2
batching errors investigation 227–9
control samples 231
failure analysis 231–2
GPC–mass spectrometry 231
instrumentation 226–7
pigmented samples 230
theory 222–6
water-based samples 230
generic identification 177–82
gilsonite 112
glass transition temperature 26–7, 30–1
acrylic latexes 93
differential scanning calorimetry (DSC) 252, 253–5, 256
epoxies 256
polyester–melamine coatings 287–8
polyurethanes 105
solution acrylics 91–2
glycerol *67*, 70
glycol acetates *63*, 64
glycol ethers *63*, 64, 221
glycoluril–formaldehyde resins 90
GPC *see* gel permeation chromatography (GPC)
grease detection 182–3
green pigments 41, 50

HALS (hindered amine light stabilizers) 55
Hansa Yellow 6: 48
hardness measurement 151
HDI (1,6-hexamethylene diisocyanate) 103, *104*
headspace analysis 219, 284–5
HEMA (2-hydroxyethyl methacrylate) 103
heterocyclic yellows 49
hexamethoxymethylmelamine (HMMM) 88–9
1,6-hexamethylene diisocyanate (HDI) 103, *104*
hexamethyol melamine 88–9
hexane 299
1,6-hexanediol *97*
high-pressure water jetting 16
hindered amine light stabilizers (HALS) 55
HMMM (hexamethoxymethylmelamine) 88–9
hydrocarbon solvents 59, 62
hydrolysis, alkaline *see* saponification
hydroxyethyl acrylate *91*
hydroxyethyl methacrylate 95
2-hydroxyethyl methacrylate (HEMA) 103
hydroxyl-functionality 103, 105
acrylics 95, 103, 105
polyesters 105

ICRI (International Concrete Repair Institute) 22–3
immersion service 35–6, 58, 237, 252
immersion testing 267–8
impact resistance
epoxies 256
testing 264–5
induction time 271
infrared spectroscopy 152–202, 258–9
coating degradation investigation 191–5
contaminant detection 182–7
correlation charts 158, *159–64*
degree of cure investigation 195–202
generic identification 177–82
instrumentation 166–9
mix ratio investigation 187–91
sample handling 169–77
theory 152–66
inhibitive pigment protection 45
inorganic pigments 4, 39–41 (*see also* extender pigments)
inter-coat adhesion failure 183
alkyd 331–3
coal tar epoxy 301–4
epoxy 327–30
urethane on epoxy 275–9
internal stress 29–31, 82
International Concrete Repair Institute (ICRI) 22–3
ion chromatography 232–8
suppression-based 236–7
ionic bonds 9

IPDI (isophorone diisocyanate) 103, *104*
iron blue *see* prussian blue
iron corrosion *see* corrosion
iron oxide 4
iron oxide pigments 40, 41
iron phosphate conversion coatings 17
isocyanates 102–3, *104*, 106, 121, 227
  degree of cure investigation 196–9, 255
  mix ratio investigation 191, 219–20, 228–9
isocyanurates 103
isophorone diamine 79
isophorone diisocyanate (IPDI) 103, *104*
isophthalic acid 70, *71*, *98*, 100

kaolin extender pigment 42
KBr (potassium bromide) pellet technique 170–1, *174*, 193, 195
ketimines 78
ketones *63*, 64 (*see also* methyl ethyl ketone (MEK))
knife probing 137, 263

lacquers 66, 67
  acrylic 91–2
  discolouration 333–7
  infrared spectrum 334–5
  polyurethane 108
laitance 21
layers, number of 146
lead carbonate 40
lead chromate pigment 40
lead silicochromate 45
lead sulfate 40
light poles, coating failure 309–13
linseed oil 69, *157*
linseed oil alkyd 2
liquid samples 173, 176–7
  gas chromatography 217, 219, 220–1
  gel permeation chromatography (GPC) 227–30

magnesium silicate pigment 42
maleic acid *234*
maleic anhydride *67*, 100
maleic esters 67
malic acid *234*
manganese BON red 47
Mannich bases 79
mass spectrometry 213–16
MDI (diphenylmethane diisocyante) 103, *104*
mechanical bonding 10
mechanical stress 25–9
MEK *see* methyl ethyl ketone (MEK)
melamine–formaldehyde (MF) resins 88–9, 95–6
melamine–polyester coatings 285–8, 322–7
membrane-curing compounds 130, 294
metal chairs, coating failure 322–7
methyl ethyl ketone (MEK) *63*, 64, 78–9
  infrared spectrum *156*
  solvent resistance testing 265
methyl isobutyl ketone (MIBK) 213, *215*
methyl methacrylate *91*, 93–4
methyl silicones 195
MF (melamine–formaldehyde) resins 88–9, 95–6
MIBK *see* methyl isobutyl ketone (MIBK)
mica extender pigment 42
microscopy 141–52
mildew *142*
mildewcides 56
mill scale *149*
mis-mixing *see* mixing errors
mix ratio investigation
  infrared spectroscopy 187–91, 258–9, 281
  physical methods 270–1
  related to adhesion and cure *279*
mixing errors 131
  adhesion testing 262–3
  epoxies 84
  example 280–2
modulus 25
moisture-cured polyurethanes 107–8
molecular weight 3, 23
molybdate orange 40
monoarylide yellows 48
monoglycerides 70–1
mulls 171

National Association of Corrosion Engineers (NACE)
  RP0394-94 253
  TM-01-74 267–8
  Wet Abrasive Blast Cleaning 16
natural oils 68–70
natural resins 66–8
'neat' spectrum 176
neopentyl glycol *97*
nitrogen, combustion analysis 259

non-drying oils 69, 71–2
novolac phenolics 87
number of coats 146

oil detection 182–3
oil length 72
oleic acid 68
oleoresinous varnishes 67–8
'orange peel' 56–7
orange pigments 40
organic pigments 4–5, 46–50
organosols 110
osmotic blistering 34–5, 220–1, 281
over-baking 133, 273
over-spray 145
oxalic acid *234*
oxygenated solvents 62, 64

paint removal 349
para-toluene sulfonic acid (pTSA) 334, 336–7
passivation barrriers 45
patterns, failure 136
P/B (pigment-to-binder ratio) 6, 23
peeling *see* delamination
pentaerythritol 70, *71*, *97*
per cent binder 8
per cent pigment 8
permeability 23–4
permeation 34–5, 36
phenolic alkyds 73
phenolic resins 73
phenolics 86–8
  epoxy 85–6
  failure mechanisms 87–8
phenols *73*
phenoxy resins 75
phenyl mercury biocides 56
phosphate conversion coatings 17
photodegradation 33–4
  epoxies 80, 279
  testing 266
  ultraviolet stabilizers 55
photomicrography *see* microscopy
phthalate plasticizers 291–2
phthalic anhydride 70, 71, *98*
phthalocyanine greens 50
physical ageing 30–1, 257–8 (*see also* weathering)
physical constants 7–8
physical methods 261–73
  adhesion testing 136–8, 261–3, 279
  application-related problems 269–73
  chemical resistance testing 266–9
  flexibility testing 151–2, 263–4
  freeze–thaw testing 269
  impact resistance testing 264–5
  solvent resistance testing 265, 283
  weathering resistance testing 265–6
pickling 16–17
pigment volume concentration (*PVC*) 6
Pigment Yellow 1: 48
Pigment Yellow 151: 49
pigments 4–5, 39–50
  barrier coatings 44
  chemical attack 32
  infrared spectra 180
  inorganic 4, 39–41
  organic 4–5, 46–50
  zinc dust 116–17
pigment-to-binder ratio (*P/B*) 6, 23
pine oil 213, *214*
plaster failure *150*
plasticizers 54, 195, 291–2
plastisols 110, 288–93
plural sprays 127
polar bonding 10
polar molecules 11
polyamide epoxies 256
polyamides
  cross-linking agents 77–8
  mix ratio investigation 188, *189–90*
polyaspartic polyureas 121–2
polyester–melamine coatings 285–8, 322–7
polyester–polyurethanes 105
polyesters 96–101
  epoxy hybrids 123
  failure example 309–13
  hydroxyl-functional 105
  infrared spectra 178, *313*
  powder coatings 123
  saponification detection 193
  saturated 97–9
  silicone-modified 115–16
  unsaturated 100–1
polyether–polyurethanes 105
polyethers, hydroxyl-functional 105
polyisocyanate/polyol coatings 103–9
polyols
  alkyds 70
  polyesters *97*
  polyisocyanate 103
  polyurethanes 108

polysiloxanes 118–20
polyureas 120–2
polyurethanes 101–9
  degree of cure investigation 196–9, 255
  on epoxy 275–9
  lacquers and dispersions 108
  mix ratio investigation 191, *192–3*
  moisture-cured 107–8
  photodegradation 33–4
  physical ageing investigation 257–8
  polyester 105
  polyurea hybrids 120
  solvents 58
    investigation 219, 220
  two component polyisocyanate/polyols 103–6
  two-component water-borne 108–9
  urealkyds 106–7
  voids 143
poly(vinyl acetal) 111
poly(vinyl butyral) 111, 246
poly(vinyl chloride) (PVC) 110, 288–93
poly(vinylidene fluoride) (PVDF) 110–11
porosity 58, 143–4
  investigation 303
  zinc-rich primer *150*
pot life 83, 127, 271
potassium bromide (KBr) pellet technique 170–1, *174*, 193, 195
potassium ferric ferrocyanide 41
powder coatings 122–4
  coating application 128–9
  differential scanning calorimetry (DSC) 253–5
  epoxy 79–80, 122–3, 253–5, 256–7
predictive tools 138
pretreatments
  aluminium 18
  conversion coatings 17–18
  galvanized steel 19
  'wash primers' 111
primary chemical bonding 9–10
primers
  'shop' 72
  shrinkage stress 30
  stress relief 28–9
  zinc-rich *see* zinc-rich primers
propylene glycol 100
prussian blue 41
pTSA (para-toluene sulfonic acid) 334, 336–7
PVC (pigment volume concentration) 6
PVC (polyvinyl chloride) 110, 288–93
pyrolysis-GC 216–17

re-coat window 131
  amine-cured coal tar epoxy 302
  calcium sulfonate modified alkyds 332–3
  epoxies 82
  polyamide cured epoxy 279
  zinc-rich primers 307
re-coating 348
red lead inhibitive pigment 45
red pigments 41, 47–8
removal methods 349
repair of coatings 347–9
residual heat of reaction 255
resole phenolics 86–7
reverse roll coating 128
RI detectors 226–7
roll coating 128
roller application 125–6
roofing panels 288–93
rosin 66–7
roughness 15–16, 132
rust *see* corrosion

safflower oil *69*
salts
  contaminants 35, 237
  electroendosmotic blistering 36
  ion chromatography 237–8
  osmotic blistering 35, 237–8
  SEM-EDS 247
samples 138
  acetone extract 219–20
  blister liquid 139
  cross-sections 147, 151
  differential scanning calorimetry (DSC) 251–2
  infrared spectroscopy 169–77
  liquid *see* liquid samples
  scanning electron microscopy 244
saponification
  acrylic lacquers 92
  alkyds 72, 191, 193, *194*
    failure example 296–300
  detection 191–5
  epoxies 316–18
  polyesters 98–9, 193
scanning electron microscopy (SEM) 238–47
  applications 244–7

elemental analysis 241–3
imaging theory 239–41
sample preparation 244
SEM-EDS 243, 244–7
SEC (size exclusion chromatography) *see* gel permeation chromatography (GPC)
secondary chemical bonding 10
shellac 67
shop primers 72
shrinkage 29–30
shrinkage stress 30, 82, 132
sidings 27, 28, 285–93
silanes 118
silicas
extender pigments 42
infrared spectra 180, *182*
silicates
coatings 116–18
extender pigments 42
silicone-modified alkyds 115
silicones 114–16
alkyds 73
degradation 195, *196*
nomenclature 118
resistance to photo-oxidation 33
siloxanes 118–20
site visits 135–6
size exclusion chromatography (SEC) *see* gel permeation chromatography (GPC)
Society for Protective Coatings (SSPC)
SSPC-SP1 14
SSPC-SP2-10 15
SSPC-SP12 16
surface preparation standards 14–15
Wet Abrasive Blast Cleaning 16
softness *see* brittle failure
solubility parameters 5, 58–9, *60–2*
solution acrylics 91–2
solvent extraction 219–20, 299, 315, *316*
solvent resistance
related to cure conditions *307*
testing 265, 283
solvent rub tests 139, 151, 265
solvents 5, 57–64
alcohols *63*, 64
esters 64
evaporation rate 5, 57–8, 59, *63*
gas chromatography 217, 219–21
glycol ethers/acetates *63*, 64
hydrocarbons 59–62
hydrogen bonding 59
inter-layer leaching 220–1
retained 35, 219–20, 252, 284–5
selection 58
solubility parameters 5, 58–9, *60–2*
surface tension *12*
soybean oil *69*, *158*
specimens *see* samples
splitting, cohesive 23, 82, 117, 304–9
spot repair 347
spray application 126–7, 145
spray distance 271
SSPC *see* Society for Protective Coatings (SSPC)
steel (*see also* galvanized steel)
acrylic latexes 94
blast cleaning 15–16, 129
coating failures 129–30
conversion coatings 17–18
corrosion 43–4, 144–5
protection 17–18, 44–6
pickling 16–17
surface preparation 14–18, 129–30
water jetting 16
wet blasting 16
stiffness 25–6
stratification 173
stress–strain behaviour 25–9
stripping 349
strontium chromate pigment 45
styrene *73*
in acrylics 91, 94
alkyds 73
in polyesters 100
substrate 10
failure *150*, 282–5
surface tension 11–12
succinic acid *234*
sulfonamide 336
surface cracking *see* cracking
surface driers 54, 70
surface energy 11
surface preparation 14–23, 129, 133
aluminium 18
concrete 21–3, 130, 294, 296
failure investigation 272–3
galvanized steel 18–21
steel 14–18
surface roughness 15–16, 132
surface tension 11–12, *13*
surfactants 53–4
swelling behaviour 29

talc (magnesium silicate) 42
tall oil *69*
tank linings 279–82, 327–30, 337–44
tape adhesion testing 262–3
T-bend test 264
TCD (thermal conductivity detector) 210–11
TDI (2,4-toluene diisocyanate) 103, *104*
temperature
  brittle failure 313
  chemical resistance testing 267, 268
  coating application 271–2
  curing reactions 271–2
  differential 268
  freeze–thaw testing 269
  impact tests 265
  implicated in failure 344
  substrate 131
temperature resistance
  amine-cured epoxy 344
  silicones 115, 195, *196*
tensile strength 25–7
tensile testing 26, 263
test patches 348
test samples *see* samples
tetraethylorthosilicate 116
tetramethylolglycoluril (TMGU) 90
TGIC (triglycidylisocyanurate) 123
thermal conductivity detector (TCD) 210–11
thermoplastic binders 2
thermoplastic coatings 29
thermoset coatings 29–30
thermosetting binders 2, 24
thickness, coating 24
  application problems 131–2
  curing time 81
  failure investigation 273
  implicated in failures 282
  measurement 138–9, 146–7, *148*, 151
thinner 272
thinning errors 108, 132–3
  examples 272, 285, 303–4
thixotropic agents 53
through driers 54, 70
timber coatings 130
titanium dioxide pigment 4, 39–40
TMGU (tetramethylolglycoluril) 90
toluene *157*
2,4-toluene diisocyanate (TDI) 103, *104*
toluidine red 47, *48*
tributyl tin biocide 56
triglycerides 68–9
triglycidylisocyanurate (TGIC) 123
trimethylchlorosilane 114
trimethylol propane *97*
'true solvents' 59
tung oil *69*
two-component coatings (*see also* mix ratio)
  epoxies 3, 75
  urethanes 58, 105–6

ultra-high-pressure water jetting 16
ultramarine blue 41
ultraviolet light
  absorbers 55
  detectors 226
  lignin degradation 130–1
  photodegradation *see* photodegradation
  stabilizers 55
  weathering resistance 266
under-baking 133, 273
'universal' metal primers 73
universal testing machines 264
urea, disubstituted 102
urea–formaldehyde crosslinking resins 90
urealkyds 106–7
urethanes *see* polyurethanes

varnishes 67–8
vegetable oils *see* natural oils
vinyl acrylic substrate 282–5
vinyl butyral 111, 246
vinyl esters 101
vinyl toluene 73, *73*
vinyls 109–12
  fluorides 110–11
  infrared spectra *179*, 180
  latexes 111–12
  plastisols and organosols 110
  poly(vinyl butyral) 111, 246
  poly(vinyl chloride) (PVC) 110, 288–93
  solution 109
viscoelasticity 25, 29
viscosity
  binders 2–3
  modifiers 52–3
  solvent properties 57
  and wetting 12–13, 14
voids 58, 143, *144* (*see also* porosity)
volume solids 7–8

walls, coating failure 282–5
water immersion 35–6, 58, 237, 252
water jetting 16
water permeation 23–4, 34, 36
water-based samples
  gel permeation chromatography (GPC) 230
  infrared spectroscopy 176–7
water-soluble acrylics 95
water-soluble solvents 58
weathering 33–4 (*see also* photodegradation)
  failure example 292
  testing 265–6
weight solids 7
wet abrasive blasting 16
wet blasting 16
wetting 10–14
white lead 40
white pigments 39–40
whiting *see* calcium carbonate
wood coatings 130
wrinkling 54, 56–7, 72–3

X-ray spectroscopy 241–3, 244–7

yellow pigments 40, 48–9
yield point 27

zinc
  cathodic protection 46
  oxidation 20
zinc chromate 45
zinc dust pigment 116–17
zinc hydroxy phosphite 46
zinc molybdate 46
zinc molybdenum phosphate 46
zinc oxide
  biocidal properties 56
  pigment 40
zinc phosphate
  conversion coatings 17–18, 309
  inhibitive pigment 46
  pretreatments 19
zinc potassium chromate 45
zinc tetraoxychromate 45
zinc–epoxy–urethane systems 117
zinc-rich primers 116–18
  curing failure 304–9
  degree of cure investigation
    infrared spectroscopy 199, *200*
    scanning electron microscopy 245–6
    solvent resistance testing 265
  failing *149*
  porosity *150*
  solvent resistance *307*

Printed and bound in the UK by
CPI Antony Rowe, Eastbourne

Printed and bound by CPI Group (UK) Ltd, Croydon, CR0 4YY
16/09/2026
14963379-0007